MECHANICAL TWINNING OF CRYSTALS

MEKHANICHESKOE DVOINIKOVANIE KRISTALLOV

МЕХАНИЧЕСКОЕ ДВОЙНИКОВАНИЕ КРИСТАЛЛОВ

MECHANICAL TWINNING

of

CRYSTALS

by

M.V. Klassen-Neklyudova

Authorized translation from the Russian
by J.E.S. Bradley, B.Sc., Ph.D

CONSULTANTS BUREAU
NEW YORK
1964

The original Russian text, published by the Press of the
Academy of Sciences of the USSR for the Institute of
Crystallography in Moscow in 1960, has been revised and
expanded by the author for the English edition.

Марина Викторовна Классен-Неклюдова

Механическое двойникование кристаллов

ISBN-13: 978-1-4684-1541-4   e-ISBN-13: 978-1-4684-1539-1
DOI: 10. 1007/978-1-4684-1539-1

FOREWORD

This monograph is not confined to mechanical twinning in the narrow sense (lattice reorientation in response to mechanical stress); it deals also with many effects related to mechanical twinning, such as formation of reoriented regions in response to high temperatures (martensite transformations, recrystallization twins), electric fields (ferroelectric domains), and magnetic fields (magnetic domains).

Mechanical reorientation is discussed for classical twinning and also for an inhomogeneous distribution of residual stresses (irrational twinning, kinking, and so on).

Mechanical twinning in the narrow sense (regular, symmetrical lattice reorientation in response to mechanical stress) was for many years a specialist topic for mineralogists, petrographers, and crystallographers. Mineralogists and crystallographers carried out the study of the basic geometrical relationships in twinning; the principal names here are Mügge, Niggli, Johnsen, Reusch, Baumhauer, Churchman, Wallerant, Evans, and Friedel.

The laws of mechanical twinning are now widely used in mineral identification and in elucidating the conditions of formation of rocks from the minerals they contain. The distribution of the twin bands in rock-forming minerals enables one to establish the later processes that have occurred in the rock. Mechanical twinning is discussed by geologists and petrologists in the analysis of flow effects.

The importance of mechanical twinning in the plastic deformation and rupture of crystalline solids was stressed by Academician V. I. Vernadskii in 1897 and by Kirpicheva in a paper entitled "Fatigue in Metals" in 1914.

Physicists and metallurgists became interested in mechanical twinning from 1920 onwards; papers appeared in Britain and Germany on the geometry of twinning in metal monocrystals (Harker, Gough and Cox, Wasserman and Schmid, Matthewson and Phillips, and so on). This work was started in the USSR at this time by Academician I. V. Obreimov and was extended by his students (Brilliantov, Garber, Startsev), who made a systematic study of twinning in crystals of metals and nonmetals to lay the basis of the physical theory of plasticity for crystalline solids. Academician N. N. Davidenkov and his students (Yakutovich, Yakovlava, Shevandin) studied twinning in metals to elucidate rupture in crystals; for a time, the Soviet school was the only one working in this field. The last 10-15 years have seen a rapid increase in the interest in twinning among physicists and metallurgists; Cahn's paper "Soviet Work on Mechanical Twinning" was published in Italy in 1953, and his "Twinned Crystals" review appeared in Britain in 1954.

About 200 papers on this topic for various crystals have been published in physics journals abroad during the past 15 years; there is also Hall's "Twinning and Diffusionless Transformations in Metals" of 1954.

The increase in interest was particularly great during the Second World War, on account of the large increase in the demand for quartz crystals. Wooster in Britain, Pérez in France, and Frondel and Armstrong in the USA examined the mechanical twinning of quartz, which was discovered in 1932 in the USSR by Shubnikov and Tsinzerling. Many years' working by Tsinzerling provided means of detwinning discarded crystals to yield first-grade material.

Studies of the actual structures of ferroelectrics and ferromagnetics have also increased the requirements for knowledge of twinning in various fields.

In 1947-8 it was observed in the laboratory (of the mechanical properties of crystals) under my direction at this institute that slices of Rochelle salt show a twinned structure in polarized light; it was shown at that time that the components of a polysynthetic twin are regions of spontaneous polarization (ferroelectric domains). In 1948 Matthias and Hippel discovered the twinned (domain) structure of barium titanate. (It had been known since the time of Bitter (1930, 1932) and Akulov (1932) that the magnetic domains in a ferromagnetic are related to the twin components.)

For many years Academician Shubnikov directed work in this field at this institute, where the mechanical twinning of quartz was discovered and studied in detail; this work has been closely linked with studies on ferroelectric domains, irrational twins and deformed crystals (Obreimov), and related topics for graduate studies (Chernysheva, Zemtsov, Urusovskaya). This extensive research has provided me with much information on lattice reorientation, and this is surveyed here as being of general interest to crystallographers, physicists, metallurgists, mineralogists, and petrographers.

This monograph takes the form of a review of the experimental and theoretical work on twinning. Assistance in the compilation has been freely given by my colleagues in this laboratory (V. L. Indenbom, A. A. Urusovskaya, G. E. Tomilovskii) and by E. G. Ponyatovskii of the Laboratory of High Pressures. The chapters on the "Theory of Twinning" were written by V. L. Indenbom; A. A. Urusovskaya wrote the sections "Recrystallization Twins" and "Prediction of Twin Elements for Metal Crystals"; and E. G. Ponyatovskii wrote the section "Martensite Phase Transitions."

The three parts of the book are as follows: I. Experimental Data on Mechanical Twinning; II. Effects Related to Mechanical Twinning; and III. Theory of Twinning. An appendix contains a table of the mechanical twinning elements for metals and minerals, which was compiled by G. E. Tomilovskii.

The first part surveys the experimental evidence on twinning with and without change of shape (geometry, crystallography, relation to atomic structure, origin and development of mechanical twins).

The second deals with related effects: lattice reconstruction in martensite-type (diffusionless) transformations, formation of twins by recrystallization, and formation of reoriented regions (irrational twins, kinks, deformation bands, and so on) in inhomogeneous deformation.

The third deals with macroscopic and microscopic theories of twinning, including dislocation concepts.

Oriented regions occur (as well as twinning) in diffusionless transformations; there is a strict relation between the lattice orientations of the old and new phases. The process resembles mechanical twinning in the dynamics of the conversion.

The reverse process (restoration of the former phase) also occurs; this resembles detwinning.

Analogy with mechanical twinning suggests that two types of transition should be observed, namely ones that are associated with change of form and ones that are not. In fact, Bagaryatskii, Tagunova, and Nosova have observed in titanium-base alloys the formation of a new phase that causes no change in the surface relief of the specimen (Problemy metallovedeniya i fiziki metallov, 1958, No. 5, p. 210-235).

Work is at present in hand on a general theory of martensite transitions and mechanical twinning.

In this monograph I have drawn extensively on E. V. Tsinzerling's dissertation "Morphologic Study of Artificial Twinning in Quartz in Response to Various Factors." I have also made use of V. P. Konstantinova's work (in this institute) on the domain structure of triglycine sulfate in relation to the general laws of twinning. I should like to record my indebtedness to them.

When the manuscript had been completed, V. L. Indenbom performed an interesting study on a general method of relating the properties and structure of a crystal. Details of this are to be found in his paper "Phase Transitions Without Change in the Number of Atoms in the Unit Cell" (Kristallografiya, 1959, 4, No. 6).

A few points from this may be noted here. Indenbom has shown that the theory of representation can be used to describe (in very general form) the physical properties of a crystal from the space group. His study of changes in physical properties (ones related to various degrees of freedom) has enabled him to give a classification of the twins formed in polymorphic transitions in which the number of atoms in the unit cell does not alter. The following types of twins can be formed.

1. Common twins, whose components show differences in spontaneous deformation and which have differently oriented optical indicatrices. Such twins can be described by reference to twin planes (ones in which the ellipsoids of spontaneous deformation intersect) and twin directions (ones lying in twin planes and perpendicular to the line of intersection of conjugate twin planes).

2. Twins whose components have the same spontaneous deformation and identical optical indicatrices. Two cases are possible here.

In the first, the components differ in their elastic constants. A characteristic here is that the sense of motion of the twin boundaries does not alter when the sign of the stress is reversed. The only known example is that of the Dauphiné twins formed in the $\alpha \rightarrow \beta$ transition of quartz. Indenbom has enumerated all possible crystal structures in which a phase transition should give rise to such twins.

In the second, all the elastic constants are the same, as well as the indicatrices and spontaneous deformation. Such twins are easily overlooked; a careful study of the structure and properties is needed to detect them. Examples of this type are all inversion twins and some kinds of rotation and reflection twins. Indenbom has listed all possible symmetry classes in which phase transitions that do not alter the number of atoms in the unit cell can give rise to such reflection and rotation twins. Triglycine sulfate and the 180° domains in barium titanate provide examples of this. The deductions are in good agreement with the experimental evidence presented here.

Indenbom has also used the theory of groups to examine second-order phase transitions for the cases of production of a ferroelectric phase and occurrence of ferromagnetic transitions.

One of the difficulties of recent years is the rapidly increasing number of publications. Many of us now find that one can either concentrate on bench work or concern oneself solely with published papers. The Institute of Scientific Information has done something to overcome this difficulty, but the principal requirement of research workers and of instructors is for books that enable them to judge the current state of the art; these are ones that are simultaneously indispensable works of reference and also sources of inspiration for further work. The exposition should be such that the reader can follow the line of development, which often throws light on the working hypotheses and enables one to overcome difficulties.

I have tried to arrange the material to provide information on each of the major aspects of twinning while not omitting the links between related sections.

A topic not considered here is that of growth twins, for this is adequately dealt with in "Leçons de cristallographie" (Friedel, 1926) and in "Advances in Physics" (Cahn, 1954). A more detailed treatment of magnetic domains is to be found in "Ferromagnetism" (Vonsovskii and Shur, 1948) and in Shur's papers of 1958-9.

Attention must be drawn to various publications appearing after this book was sent to press; in these, selective etching has been applied to triglycine sulfate, calcite, bismuth, and antimony, which has provided solutions to some problems discussed here. A brief summary of this work is given at the end of the book.

Some information on mechanical twinning is to be found in "Dislocations and Mechanical Properties of Crystals" (Fisher et al., 1956) and in "Imperfections in Crystals" (van Bueren, 1960; Amsterdam).

Some additions and corrections have been made for the English edition. Section 13 of § 2 of chapter 1, "Effects of Size of Specimen on Twinning: Twinning of Crystals with Few Dislocations," is new, though it includes part of the material formerly in section 12. New data on the electron microscopy of rupture along twin boundaries (L. G. Orlov) has been added in § 6 of chapter 3. Chapter 4 of part II, "Martensite Phase Transitions," has been rewritten completely (by E. G. Ponyatovskii). In § 15 of chapter 6, section 6, "Kinks and Rupture," has been added. A second and third appendix deal with work published in 1960-3 on two topics: "Selective Etching Applied to the Dislocation Mechanism of Twinning" in Appendix 2 (compiled by V. I. Startsev); and "Selective Etching Applied to Twinning Without Change of Form" in Appendix 3.

The literature citations contain references introduced during the revision of the book for translation and also major papers on mechanical twinning appearing after 1960.

In conclusion, I wish to express my thanks to V. L. Indenbom for valuable advice during the work on the manuscript, to G. E. Tomilovskii and M. A. Chernysheva for assistance in preparing the material for publication, and to K. V. Flint for assistance in compiling the bibliography.

CONTENTS

PART II

EFFECTS RELATED TO MECHANICAL TWINNING

PART III

THEORY OF TWINNING

*   *   *

INTRODUCTION

The deformation of a crystal is a process that has many stages. At first, it is reversible and elastic; the atoms or ions return to their initial positions when the external force is removed, the displacement being proportional to the stress. Apart from this proportional change (Hooke's law), real crystals show nonlinear but reversible changes at fairly low stresses. This imperfect elasticity causes delay in the strain corresponding to a changed stress (or conversely): it is associated with migration in the lattice, and the displacements differ essentially from the elastic ones in an ideal lattice, being combinations of various elementary diffusion processes *.

In current ideas, delay effects are also associated with reversible displacement of the dislocations found in real crystals (§ 2.18).

Other reversible inelastic processes are the formation of elastic twin layers (§ 2.3) and the reversible displacement of the boundaries of polysynthetic twins (these are formed during polymorphic transitions in metals) or the boundaries of domains (twins) in ferroelectric crystals (§ 2.19).

The deformation becomes irreversible if the stress exceeds a certain limit (the elastic limit); the lattice then acquires residual changes. Larger stresses cause plastic flow (the plastic limit). The reversible elastic deformation is uniformly distributed throughout the crystal; the reversible inelastic deformation may be localized, as in elastic twin layers and elastic displacement of domain boundaries. Such localization is also characteristic of irreversible deformation.

Plastic deformation in an aggregate can occur by plastic flow of grains (crystallites) or by flow in layers between grains. The plasticity observed at high temperatures and low deformation rates is always the result of both processes. Deformation of grains is the usual process at normal or low temperatures, and also at high rates of loading.

Flow at boundaries is mainly † related to displacement of point defects, motion of individual atoms, and so on; it is always of diffusion type. Grain deformation involves rearrangement of the atoms resulting from the motion of line defects (e.g., dislocations). It is now considered that the plastic deformation of a monocrystal or grain occurs as follows:

1) by gradual slip of some parts of the grain (crystal) relative to other parts (shearing approximating to classical translational gliding);

2) by twinning (symmetrical reorientation of parts or of the crystal as a whole);

3) by diffusion of the lattice atoms or ions (this occurs only at very low deformation rates and near the melting point); and

4) reorientation of parts within the grain, the lattice being turned through various angles (deformation bands, kinks, substructures, twins on irrational faces, slip combined with rotation).

---

*Such as: diffusion caused by temperature differences, microscopic and macroscopic diffusion associated with uneven distribution of impurities, diffusion resulting from an anisotropic distribution produced by stresses, diffusion associated with ordering, magnetic diffusion, and so on (see; Elasticity and Inelasticity of Metals, 1952).
† There are indications of additional special effects, which are seen as stepwise deformation at grain boundaries.

We shall see (§ 3.3) that mechanically produced twins are not always associated with change in the shape of the specimen. The mechanical twinning of calcite (§ 2.2) and the twinning produced by the deformation of a metal (§ 1.13) cause changes in the size of the specimen. I call this type "twinning with change of form." Apart from macroscopic residual deformation, it is also possible for twinning (lattice reorientation) to occur without change in the external form of the specimen (chapter 2). The latter type was discovered by Shubnikov and Tsinzerling (1932).

Shubnikov and Tsinzerling named this effect as a special form of residual deformation. My term for it is twinning without change of form.

This effect is associated with changes in the elastic constants, on account of the lattice reorientation. A third possible form of twinning involves no change in shape or in the elastic constants, inversion twins (§ 4.8 and 4.9) being a special case of this.

There may be major changes in the process of deformation near the melting point; "plastic deformation increasingly degenerates into self-diffusion as the temperature increases" (Frenkel', 1950).

The first part of the book deals with the experimental evidence on mechanical twinning; chapter 1 deals with twinning with change of form, and chapter 2 with twinning without change of form. Chapter 3 is concerned with effects in polycrystals.

The second part deals with some effects closely related to mechanical twinning (diffusionless polymorphic transitions, recrystallization twins, lattice reorientation in inhomogeneous deformation).

The third part deals with theoretical work on mechanical twinning. The dislocation theory of lattice reorientation and of other simple effects in plastic deformation is to be found in a paper by V. L. Indenbom.[*]

---

[*]In: Plasticity of Crystals (editor M. V. Klassen-Neklyudova), Consultants Bureau, New York, 1962.

Part I

EXPERIMENTAL DATA ON MECHANICAL TWINNING

CHAPTER 1

TWINNING WITH CHANGE OF FORM

§ 1. Geometry, Crystallography, and Relation to Atomic Structure

1. Idealized Schemes for Translational Slip and Twinning with Change of Form

Plastic alteration of the shape and dimensions of a monocrystal is seen macroscopically as the relative displacement of thin layers, which resembles the displacement of cards in a pack. This division into macroscopic layers occurs in tension (Fig. 1) and also in response to compression, bending, torsion, and shear. This type of slip always occurs within the grains in a plastically deformed technical metal, rock, or other polycrystalline material; it is seen in polished sections as characteristic streaks with perhaps several directions (Fig. 2).

The shape of a deformed crystal gives the impression that the individual layers retain their orientation.

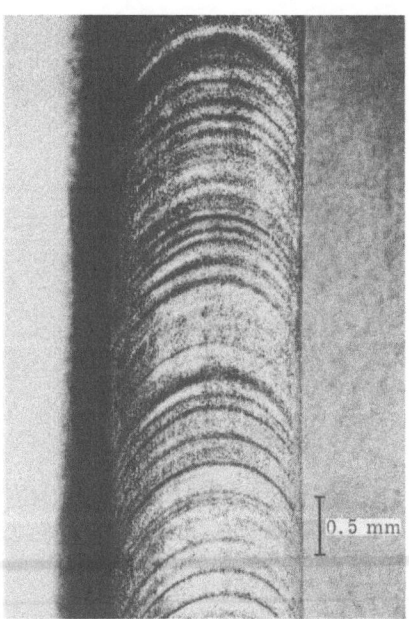

Fig. 1. Stretched monocrystal of zinc; deformation by slip, with signs of slip revealed by etching.

Figure 3 shows an idealized scheme for uniform translational slip; it is assumed that the lattice deformation corresponds closely to the external appearance and that the layers slide freely. Then plastic deformation by slip is equivalent to parallel displacement along a fixed crystallographic plane (the slip plane) by a multiple of the lattice parameter. Structurally speaking, this is an operation of translation; the displacement of any layer is dependent on the total deformation, but it is always a multiple of the lattice parameter. The displacement can occur in either sense; after displacement, the layers still have fully correlated lattices, as Fig. 3 shows; the lattice is not destroyed, but the surface of the specimen shows ridges.

We shall see later (§ 14.1 and § 16) that the process is actually rather more complicated; some lattice reorientation occurs. All the same, deformation in pure shear or in highly homogeneous extension can approximate to translational slip in the sense that no loss of orientation occurs. In any case, the x-ray patterns show no broadening in the spots (§ 15.5).

A large number of lines of slip in a small area may produce a macroscopic band on the surface. The subsequent development of the structure of these slip traces is not entirely clear.

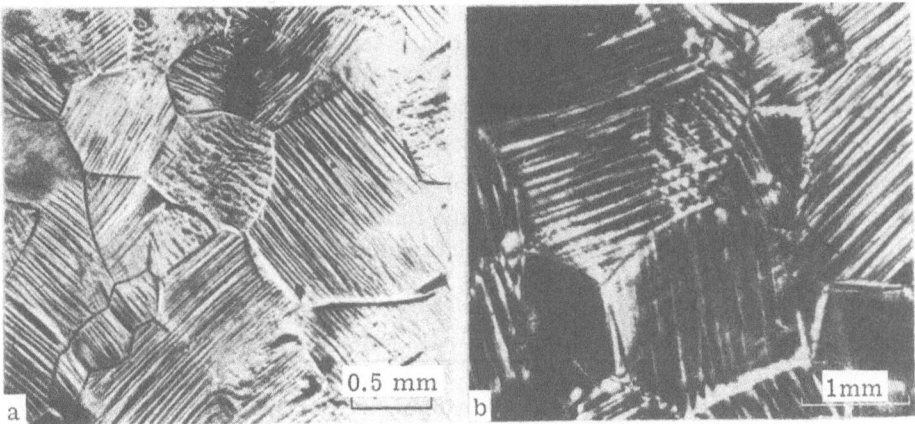

Fig. 2. Deformed polycrystals: a) etched section of 99.98% aluminum, with grains of various orientations, the lines in the grains being traces of slip (5% extension; Wood and Scrutton); b) polished section of silver chloride in transmission (polarized light). The slip planes are seen on account of localization of residual stresses; these cause birefringence (Nye).

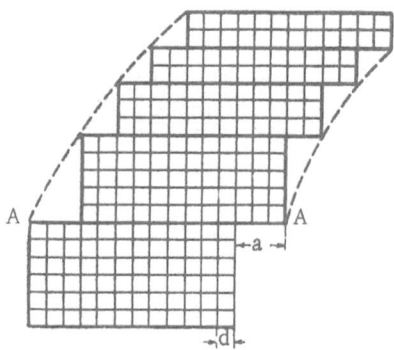

Fig. 3. Monocrystal deformed by translational slip; the lattice is not destroyed, but the slip gives rise to steps on the surface of the crystal (Ioffe's scheme). AA is a slip line; $a = nd$, in which a is the crystallographic displacement, d is the translation vector, and n is an integer.

A monocrystal (or a grain) can deform by twinning if plastic deformation by slip is hindered (e.g., because the possible planes are unfavorably oriented with respect to the force, or on account of structural features of the crystal, or because the loading rate is very high or the temperature is low). Twinning causes individual parts to take up a new orientation having a symmetry relation to the rest of the crystal; the lattice is reoriented in a regular fashion. The components are seen usually as plane-parallel or wedge-shaped bands (§ 1.16), whose reflectivity differs from that of the rest of the crystal; they sometimes take the form of biconvex lenses. Occasionally, the mechanical twins may be irregular in outline (§ 4.4 and 8).

The lattice within the twinned part is a mirror image of that in the other part; the conversion to the new position sometimes occurs very rapidly, as may be judged from the sound produced, e.g., by tin and zinc (§ 2.9).

Only the initial and final orientations are observed in twinning. Neglecting the actual movements of the atoms, we can represent the reconstruction (for the commonest case, namely twinning with change of form) as in Fig. 4; here we have a hypothetical crystal that has given rise to a twin. The operation can be represented as one of sequential displacement of the layers from right to left parallel to the a-a plane, which is the mirror plane and also the contact plane for the components 1 and 2.

It is assumed that the atoms are displaced parallel to the mirror plane; here the displacement is only a fraction of the lattice translation vector, so the process is one of shear (lattice reorientation). The result of twinning may thus be represented as an operation of simple shear (§ 1.2), in which all atoms in the displaced region are moved through distances proportional to the distances from the mirror plane. This plane is called the twin plane, and the direction of displacement is called the twin direction.

This direction differs from that of slip in being polar; the shear in twinning can occur to one side only, but a mechanical twin can be eliminated by a force applied in the reverse direction (§ 2.2 and § 4.3).

Here all atomic planes are displaced one relative to another by a constant, very small, amount; no visible signs of slip appear on the surface, and there are generally none at all. A twinned region is seen in a deformed crystal as a continuous and homogeneous region having an orientation differing from that of the rest of the crystal.

The scheme of Fig. 4 gives a correct macroscopic description of twinning with change of form, but it does not always represent the atomic mechanism correctly; for example, the CO_3 groups in calcite must be rotated, since displacement of the Ca atoms alone is insufficient. The paths taken by the atoms are not entirely clear (§ 1.8 and 9).

2. Twinning Equivalent to Homogeneous Deformation: Twin Ellipsoid and Elements of Twinning

The simplest types of deformation are described by reference to homogeneous deformation, in which the displacement is a linear function of the coordinates. Examples of this are deformation in extension and simple shear, while deformation in bending is an example of the inhomogeneous type. The definition implies that an imaginary cube within the body becomes a parallelepiped, while a sphere becomes a triaxial ellipsoid. Parallel edges and faces in a crystal must remain parallel after deformation. The volume is unaltered in homogeneous shear.

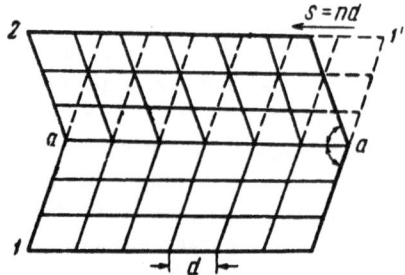
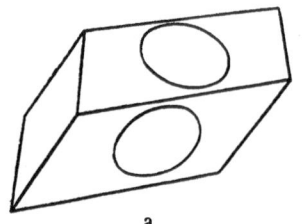
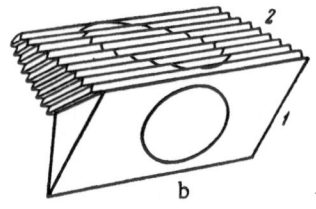

Fig. 4. Monocrystal deformed by twinning: 1-1' is the initial crystal, 1-2 being the twin; a-a is the slip plane and also the symmetry plane of the twin; d is the translation vector; s is the shear (a multiple of the lattice parameter); and n is an integer; 1-a is one component of the twin, and 2-a is the other component. The arrow indicates the direction of displacement.

Fig. 5. Spatial model of twinned calcite, with the twinning represented as homogeneous deformation in simple shear: a) cleaved crystal, undeformed; b) mechanically twinned crystal, in which the twin component 2 is produced by displacement of the atomic layers one with respect to another parallel to a certain plane in the crystal. The steps are of equal thickness.

The mutual disposition of mechanically produced twins may (Tertsch, 1949; § 3.1) be formally described as a result of homogeneous deformation in simple shear; the concept is usually illustrated by means of a macroscopic model for calcite (Fig. 5). This is analogous to Fig. 4, and it shows that the twinned crystal may be represented as the displacement of some layers relative to others parallel to a certain plane. For example, circles drawn on the surface before the crystal is deformed become ellipses on the faces parallel to the direction of shear but remain circles on the cleavage plane.

Homogeneous deformation is completely defined if we know the ellipsoid resulting from a sphere in the initial crystal. Let the radius R of this sphere be unity, and let the crystal be deformed along a slip (twin) plane K_1, which runs horizontally through the center 0 normal to the plane of the drawing (Fig. 6). As origin we take the center of the sphere, which is also that of the ellipsoid; as y axis we take the shear direction η_1, and as z axis the normal to the K_1 plane. The points in plane K_1 are not displaced, so the section of the sphere by this plane is unaffected by the deformation; it remains one of the circular sections of the ellipsoid. The second circular section K_2 lies symmetrically with respect to the principal axes a and c; it passes through the point A' where the ellipse and circle meet in the plane of the drawing. The b axis is normal to the plane of the drawing; its length is the radius of the sphere, and it coincides with the line of intersection of the two circular sections. The plane coincident with that of the drawing, which is defined by the major axis a and the minor axis c, is called the shear plane S. The trace of K_2 on S is usually denoted by η_2, which is called the axis of the principal zone. The shear is specified by the intercept s, which denotes the displacement of a point F at a distance R (unity) from the twin plane to the position F'. This s is called the crystallographic shear. The equation of the ellipse and the condition s:1 = AA':OB imply that

$$s = \frac{2}{\tan 2\varphi} = a - c, \tag{1}$$

in which φ is the angle between a and K_1.

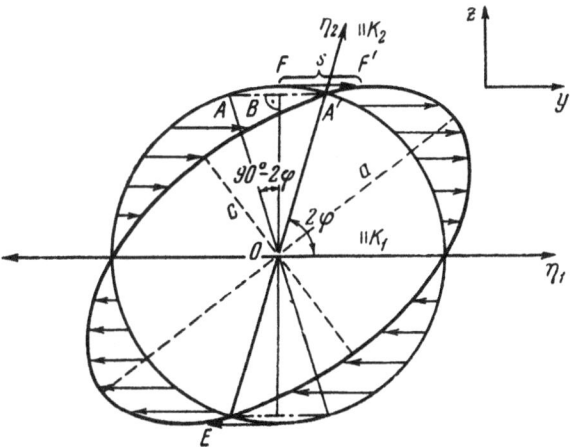

Fig. 6. Sections of the sphere and of the corresponding ellipsoid by shear plane S; the trace of the twin plane K_1 coincides with the shear direction η_1, while the trace of K_2 (the second circular section) coincides with η_2 (axis of the principal zone); s is the crystallographic shear, a and c are the axes of the ellipsoid, and φ is the angle betweeen a and the twin plane.

The twinning elements are then as follows: K_1, the twin (slip) plane; K_2, the second circular section; η_1, the direction of shear (twinning); and η_2, the axis of the principal zone (or simply the principal zone).

A more detailed description of the twinning usually requires a specification of S (the shear plane) and s (the crystallographic shear).

If now the center of the sphere lies at the boundary of a twin, then the hemisphere below the K_1 plane remains unaltered, while the upper hemisphere becomes half an ellipsoid. The twinning ellipsoid is often called the slip ellipsoid.

The formal crystallography of twinning was first fully expounded by Johnsen (1914a); Cahn (1953) has given a somewhat modified exposition.

3. Conjugate Twins, and Twins of the First and Second Kinds

The elements K_1, K_2, η_1, and η_2 define the deformation ellipsoid (§ 1.2); they were deduced by treating the crystal as a continuous medium. The actual structure is discontinuous, so these elements must be correlated with the lattice elements. If K_1 coincides with a lattice plane having small integers for its indices, and if η_2 corresponds to a direction also having small integral indices (that is, if K_1 and η_2 are rational*), then the twin is called one of the first kind; K_2 and η_1 can be rational or irrational in this case.

Twins of the second kind have K_2 parallel to a rational plane and η_1 parallel to a rational direction; K_1 and η_2 are irrational.

Crystals of high symmetry (cubic ones, and some hexagonal ones) often have K_1, K_2, η_1, and η_2 all parallel to rational elements; such twins may be considered as of the first or second kind. Then, if there is a twinning system having $K_1 = (hkl)$, $K_2 = (h'k'l')$, $\eta_1 = [UVW]$ and $\eta_2 = [U'V'W']$, there is often a second system whose elements are $K_1 = (h'k'l')$, $K_2 = (hkl)$, $\eta_1 = [U'V'W']$ and $\eta_2 = [UVW]$. Twins to which this applies are called conjugate; twinning in either of the conjugate systems results in the same s. Conjugate twins occur mainly in metals; they are rare in minerals.

Another classification is used in crystallography and mineralogy; here the lattice symmetry of the twin is used instead of the twinning elements. The twins are called reflection twins if the operation required to bring the lattices into coincidence is one of reflection at a plane (the twin plane); they are called rotation (or axial) twins (§ 1.4) if the operation is one of rotation through an angle characteristic of the particular structure about a crystallographic axis.

The axes and planes of symmetry in a twin usually coincide with twinning elements; for example, the mirror plane coincides with the twin plane K_1, so a reflection twin may be considered as a twin of the first kind.

An axial twin is the same as a twin of the second kind if η_1 coincides with a symmetry axis of the twin; then η_2 is a twofold (diad) rotation axis.

Twins in crystals of high symmetry may be described simultaneously as axial twins and as reflection twins, as in the case of calcite (§ 1.4).

Examples of axial twins often quoted in the mineralogical literature are albite twins in plagioclase and Dauphiné twins in quartz; but, in fact, the latter cannot be described as twins of the second kind, for mechanical twinning occurs in quartz without change of shape, so the concept of a deformation ellipsoid (and hence of K_1, K_2, η_1, and η_2) becomes physically meaningless. Dauphiné twins may be considered only as axial twins, and only the rotation axis need be stated.

4. Main Geometrical Laws of Mechanical Twinning

In nature we very seldom find single crystals; most mineral crystals are aggregates. Special conditions are needed to produce single crystals. Regular and irregular intergrowths occur under natural conditions and in the laboratory; several individual crystals may be present in the intergrowth. Regular intergrowths are called twins, threelings, or fourlings, in accordance with the number of components.

The components in an intergrowth may be transformed one to another by inversion, reflection, or rotation; the three kinds of twins are named correspondingly.

Shubnikov (1940) distinguishes five modes of formation of twins:

1) by accidental contact while the crystals are small;
2) by parallel deposition on a twin nucleus resulting from the coming together of two or more molecules in a symmetrical array;

---

* Irrational elements are planes and directions whose indices are irrational numbers or large integers. These irrational values are obtained when the plane or line does not pass through the nodes of the Bravais lattice. "Moreover, the higher the indices, the more difficult it is to say whether the ratios are rational or irrational" (Niggli, 1924, p. 177).

3) by deposition in a twin position on an existing large crystal;

4) by conversion of one modification to another;

5) in response to mechanical influences. *

Reflection and rotation twins occur in monoclinic and triclinic crystals, but there is no distinction between these types for cubic crystals and for some hexagonal ones, because all twinning elements are always rational (§ 1.3).

A mechanical twin in a crystal of one of the higher classes can be considered geometrically as a reflection twin or as a rotation twin; for example, calcite twins (Fig. 7) can be derived geometrically by reflection at the twin plane K_1, by rotation about axis a normal to the twin plane, by rotation about axis b lying in the twin plane and coinciding in direction with η_1 (the twinning direction), or by reflection in a plane perpendicular to η_1. The crystallographic axis that here acts as rotation axis thereby becomes the L^2 axis of the twin, so twinning adds L^2 as a macroscopic symmetry element. The twin (reflection) plane in this case is also the plane of junction.

Mechanical twins in crystals of low symmetry are usually rotation twins (§ 3.3), as in the case of quartz (§ 3.3) for Dauphiné twins. It is physically meaningless to describe these twins by the geometrical operation of reflection (§ 3.3).

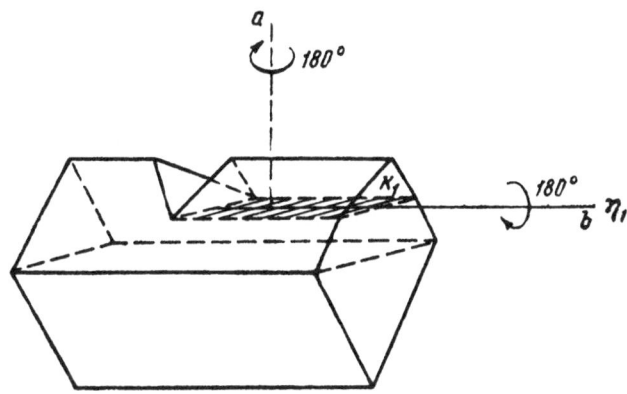

Fig. 7. Mechanical twinning in calcite; the twins can be considered as axial ones obtained by rotation through 180° about a = $[10\bar{1}2]$ or about b = $[01\bar{1}1]$, and also as ones formed by reflection in plane K_1.

Quartz also shows Brazilian and Japanese twins; the Brazilian type and one of the Japanese types are only reflection twins and cannot be produced mechanically; they are formed by the intergrowth of two crystals (see § 3.1 for details).

Mechanical twins are usually considered as reflection ones, but only for structures of high symmetry can mechanically produced rotation twins be described as reflection ones also.

Mechanical twins cannot be inversion twins (§ 19); the latter are formed by growth, in polymorphic transitions, and (in ferroelectric crystals) by electric fields.

---

* Twins formed by rapid cooling or heating, or by electrical breakdown, are also mechanical.

5. Some Additional Aspects of Twin Geometry

Some important rules can be formulated on the basis of the above survey of mechanical twinning. A crystal retains its lattice symmetry and structure during twinning (the shape and size of the unit cell are not affected), but the macrosymmetry of the twin may be higher, on account of an additional symmetry plane (the twin plane) or symmetry axis, or because one of the previous axes has become of higher order (has become the twin axis).

Plane K_1 cannot be a symmetry plane of the crystal in a reflection twin, and the twin plane cannot be normal to a fourfold axis. By analogy, the η_1 direction in an axial twin cannot coincide with a twofold axis.

That is, the higher the initial symmetry the less likely are twins, and conversely; crystals of low symmetry show the greatest variety of twins. Mechanical twins of only one type occur in cubic crystals, whereas orthorhombic crystals (e.g., α-uranium) show four types of mechanical twin (§ 1.14). The active elements (K_1 and η_2) characteristically have the minimum s possible for the lattice in question (§ 1.18). In a metal, the S plane always coincides with a symmetry plane.

6. Analytic Condition for a Lattice to Twin by Simple Shear

Any twinning with change of form can be considered as deformation by simple shear (§ 1.2 and 3). An analytic study may be made (Wallerant, 1904) of the capacity of a space lattice to twin by simple shear.

All that is necessary here is to establish the indices (HKL) of K_1 and [UVW] of η_2 such that a lattice with the given unit cell may deform with mirror symmetry in simple shear (each node of the displaced lattice has a relation of mirror symmetry to the K_1 (HKL) plane with reference to the corresponding node in the undisplaced lattice). The necessary conditions here are that the pairs of nodes before shear may be joined by lines parallel to η_2 having indices [UVW] and that these lines are bisected by the (HKL) plane.

Let [mnp] and [$m_1 n_1 p_1$] be two nodes of the primitive* lattice; then the necessary condition for the line joining them to be parallel to [UVW] is

$$(m_1 - m) : (n_1 - n) : (p_1 - p) = U : V : W.$$

The condition that the two nodes are equidistant from K_1 is

$$Hm_1 + Kn_1 + Lp_1 = -(Hm + Kn + Lp).$$

These two equations are solved for m, n, and p to give

$$m = m_1 - 2U \frac{Hm + Kn + Lp}{HU + KV + LW}$$

and similarly for n and p.

The indices of lattice nodes must be integers, while U, V, and W cannot have a common factor, so the primitive lattice must have HU + KV + LW equal to ± 1 or ± 2.

Johnsen (1916) has considered the relations between the indices of K_1 and η_2 for complex lattices.

The values of the above sum for body-centered cells are ± 1, ± 2, and ± 4, but in the second case U + V + W must be even or all indices in HKL must be odd. In the third case, the HKL are odd and U + V + W is even.

A base-centered cell must have values of ± 1, ± 2, or ± 4; H + K must be even in the second case. If H + K is odd, then U and V must be odd and W must be even. In the third case, H + K must be even and W must be even, while U and V must be odd (Niggli, 1924, p. 300).

---

*A lattice whose unit cell consists of the shortest translations and which contains only one atom.

7. Transformation of Indices of Planes and Directions in Twinning

The positions of the planes and crystallographic directions are altered relative to the coordinate axes by twinning, in general. Twins of the first kind (reflection for low symmetry, any type for high symmetry) retain the same indices for crystallographic directions in the K_2 plane and for planes belonging to the η_2 zone.

Mügge (1889a) first gave an analytic form for the conversion of indices in twinning; this was extended by Pabst (1955) and by Andrews and Johnson (1955).

Let HKL be the indices of K_1 or of the initial plane K_2^0, and let UVW be those of η_2 or η_1, hkl being the indices of some plane before twinning and h'k'l' those of the plane after twinning (as referred to the initial axes); then

$$
\begin{aligned}
h' &= h(HU + KV + LW) - 2H(Uh + Vk + Wl), \\
k' &= k(HU + KV + LW) - 2K(Uh + Vk + Wl), \\
l' &= l(HU + KV + LW) \quad - 2L(Uh + Vk + Wl).
\end{aligned}
\tag{1}
$$

The conversion of [uvw] to [u'v'w'] is given by

$$
\begin{aligned}
u' &= u(HU + KV + LW) - 2U(Hu + Kv + Lw), \\
v' &= v(HU + KV + LW) - 2V(Hu + Kv + Lw), \\
w' &= w(HU + KV + LW) - 2W(Hu + Kv + Lw).
\end{aligned}
\tag{2}
$$

Equations (1) and (2) can be put in matrix form (Pabst, 1955); the form of the matrix is governed by the crystallographic system and by the twinning elements. For example, (1) gives for calcite that h - h' = k - k' = h + k; $l = l'$, the matrix for twinning on (0112) in the rhombohedral setting is given as 010/100/001. If we need to know the position of some plane with respect to the initial crystal, the procedure is as follows: We multiply the indices of this plane in turn by the elements of the first line and sum the products to obtain the first index. The second index is obtained in the same way from the second line, and so on. These formulas and matrices are used in determining twinning elements (§ 1.18).

8. Deviation of Twinning from Deformation by Simple Shear

It has been shown by analysis by reference to K_1 and η_2 or K_2 and η_1 (§ 1.3) that twinning in most space lattices cannot be described as deformation in simple shear (displacement of the basic units by amounts proportional to their distances from the twin plane).

All body-centered lattices having $K_1 = (112)$ and $\eta_2 = <111> = \eta_1$ as their twinning elements conform to the analytic conditions for simple shear. Crystals of Pb, Na, α-Fe, Cr, W, and Mo have such lattices, and they actually do show twinning on (112) with $\eta_1 = \eta_2 = <111>$ as twin direction (see § 1.19). Other crystals with body-centered cells (Li, V, K, Ta) have not been examined.

Sossinka et al. (1933) concluded from models of the atomic packing that a cubic face-centered metal can twin by simple shear if the twinning elements are $K_1 + (1\bar{1}1)$, $\eta_1 = [\bar{1}12]$, and $S = (1\bar{1}0)$; Jaswon and Dove (1956, 1957) reached the same conclusion (§ 1.20).

There is no clear-cut experimental evidence on the mechanical twinning of face-centered metals (Al, Ni, Pb, Th, Ag, Au, Rt, γ-Fe, β-CO, Sr, Rh, Pd, Ir, Ca, Tl, Pt); Mathewson et al. (1930) and Barrett (1948) made careful metallographic and x-ray studies of copper and α-brass but could not detect mechanical twinning, although the effect was later observed in copper at 4.2°K. The twinning follows the spinel law (Blewitt et al., 1954 and 1957). Elam (1928) observed reoriented regions in stretched monocrystals of aluminum; but Rybalko and Yakutovich (1948) found that these are formed by slip, not by twinning.

The reason for the absence of mechanical twins in most cubic face-centered crystals is that the possible twin plane $K_1 = (111)$ is also a slip plane T; slip is preferred to twinning, especially since s for such lattices is large (§ 1.21). Recrystallization twins are prominent in such crystals.

Grühn and Johnsen (1917a) found that a tetragonal lattice can twin by simple shear if the elements are $K_1 = (10\bar{1})$ and $K_2 = (101)$, but the analytic condition is not obeyed if the elements are $K_1 = (101)$ and $K_2 = (30\bar{1})$. Natural rutile crystals show twinning of mechanical origin in which the elements are the latter ones, but Grühn and Johnsen were able to make twins corresponding to rearrangement by simple shear. The crystal was pressed in powdered sulfur at 10,000 to 30,000 atm for 24 hr; two deformation ellipsoids, one corresponding to $K_2 = (10\bar{1})$ and the other to $K_2 = (30\bar{1})$, were observed for $K_1 = (101)$, the pressing conditions being the decisive factor. They remarked that deformation under natural conditions causes lattice reorientation that deviates from simple shear.

It has been shown that mechanical twinning is not equivalent to simple shear in other cases also: calcite and dolomite (Johnsen, 1914a); bismuth, arsenic, antimony, and millerite (Johnsen, 1916a); α-corundum and hematite (Niggli, 1924); white tin (Mügge, 1917 and 1927; Schmid and Wassermann, 1928); zinc, magnesium, beryllium, and cadmium (Schmid and Wassermann, 1928; Kolesnikov, 1933); and galena and pyrite (Niggli, 1920).

The accepted view (Johnsen, 1916a, 1916b, and 1917; Grühn and Johnsen, 1917b; Boas and Schmid, 1935, p. 69) is that the capacity for deformation by simple shear is essential to twinning, but the above survey shows that this is not so.

9. Redistribution of Basic Particles in Twinning with Change of Form

Only the initial and final orientations, or certain regions (twin bands), are observed in mechanical twinning. Twinning by simple shear has been described above (§ 1.1 and 2) by comparison of the crystallographic axes before and after twinning; the actual paths taken by the particles have not yet been examined.

The sole means of establishing the deviations of the nodes from the positions required by simple shear is to project the lattice on the S plane; this was first done for calcite [*] by Friedel (1926, p. 465), for zinc (two different methods) by Mathewson and Phillips (1928) and by Gough and Cox (1929), for magnesium by Barrett (1948) and for orthorhombic uranium by Cahn (1953a). Laves (1952b, 1954) has made the most complete recent study of this topic.

The need for deviation from the paths required by pure shear is most readily demonstrated for $CaCO_3$ (calcite; Fig. 8). All atoms here move parallel to the twin plane; the distances moved are proportional to the distances from that plane, but the group of three oxygen atoms must be turned through 52°30' around an axis which is normal to the S plane and which passes through the center of the carbon atom.[†]

Similar projections of the S plane have been made for cadmium, magnesium, α-uranium, β-tin, bismuth, α-iron (Figs. 9-14). The twinning of a diamond lattice is dealt with in § 1.15.

The trace of the K_1 plane is always shown on the projections; this trace may coincide with one of the rows of atoms (which is always more convenient) or may pass between rows. The atoms to one side of this plane show the twin position, which is found by reflection of the initial positions in this plane (Gough and Cox, 1929) or by translation of the lattice parallel to that plane (Mathewson and Phillips, 1928).

Figures 9-14 show that only some of the atoms are displaced parallel to the twin plane. Kolesnikov demonstrated this analytically in 1933.

10. Laves's Rule for Projecting the Twin Plane on the Shear Plane: Limit of Possible Deviation from Deformation by Simple Shear

Laves (1952a, 1952b) examined the motion graphically by resolving the displacement vector for each atom into components, one of which was parallel to K_1 and η_1, as in Fig. 11 for α-uranium (Cahn, 1953).

Let R_n be the displacement vector of the n-th atom and S_n the component of this parallel to the twinning elements. The size and direction of this vector are dependent on the position of atom n and on the position of the twin plane in the initial lattice (i.e., of the plane of junction). In Fig. 11 the atoms adjacent to the twin plane have their R_n (arrows) nearly normal to this plane, whereas R_n tends to be parallel to S_n for atoms far away.

[*] The construction applies to magnetite, siderite, and sodium nitrate, which have analogous lattice structures.
[†] This angle is 54°59' for dolomite (Pabst, 1955).

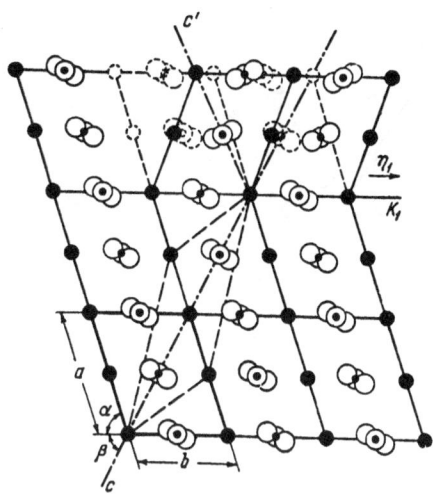

Fig. 8. Projection of the structure of calcite on the plane S = (011). The large filled circles are the calcium atoms; the open circles are the oxygen atoms; and the small filled circles are the carbon atoms. The top right corner shows part of the lattice in the twinned position. The broken lines denote the initial positions of the structure elements. Here $K_1 = (011)$, $\eta_1 = [100]$, a = 8.08 A, b = 6.41 A, $\alpha = 70°51'$, $\beta = 63°45'$ (Pabst).

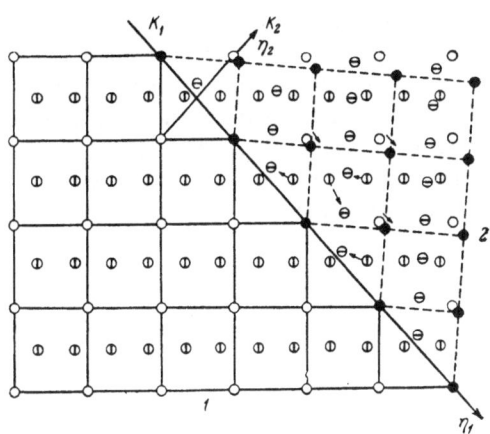

Fig. 9. Displacement of the atoms in mechanical twinning of cadmium: $K_1 = (10\bar{1}2)$, $\eta_1 = [\bar{1}011]$, $\eta_2 = [10\bar{1}1]$; 1) initial position of basal plane; 2) basal plane after twinning. The open circles with vertical lines denote atoms lying above and below the S = $(1\bar{2}10)$ plane. The filled circles denote atoms in the twin position (ones displaced parallel to the twin plane). The circles with horizontal lines denote atoms not lying at the nodes of the Bravais lattice; the displacements of these differ from those of pure shear (diagram by Hall).

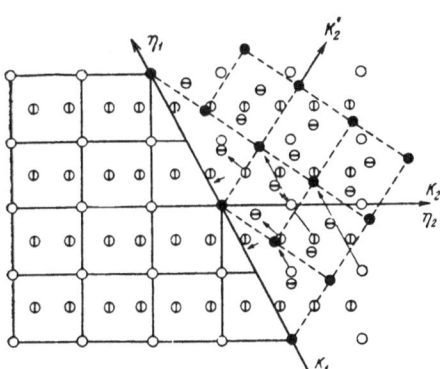

Fig. 10. Twinning of magnesium on $K_1 = (10\bar{1}1)$; $\eta_1 = [10\bar{1}2]$, projection on S = $(1\bar{2}10)$; symbols as in Fig. 9 (according to Hall).

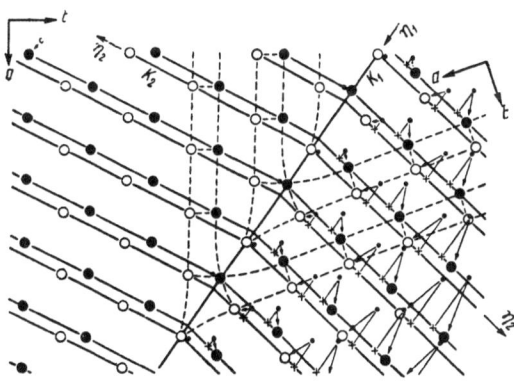

Fig. 11. Twinning of α-uranium; $K_1 = (130)$ and $\eta_1 = [3\bar{1}0]$, projection on S = (001). The filled and open circles denote atoms at different levels; the arrows denote the displacements, which are resolved into components, one of which is parallel to the twin plane and would bring the atoms to the positions shown by crosses. The broken lines indicate the distortion at the region of contact (Cahn).

Laves states that, if the interface (twin plane) passes through a row of atoms for which $\Sigma R_n \neq 0$, the vector representing the resultant of all R is directed towards the boundary or away from it (the row determines the direction). In principle, it is always possible to choose a boundary that does not coincide with any atom in either part for which $\Sigma R_n = 0$ and $\Sigma |R_n|$ is minimal. Laves considers that these relations define the best conditions for graphical analysis of twinning.

Barrett (1948) points out that the degree of distortion in the transition region is dependent on the position of the boundary.

Laves has shown that

$$Q = (\Sigma |R_n| : tn) \approx 0.2,$$

(in which t is mean interatomic distance and n is the number of the atom) for twins in tin and magnesium. Then the arithmetic mean of the displacement vectors is one-fifth of the interatomic distance; Q indicates the deviation from deformation by pure shear. Laves considers that mechanical twinning is impossible if Q exceeds a certain value. The Q for all hexagonal metals must be closely similar, for all metals having hexagonal close packing twin in the same way.

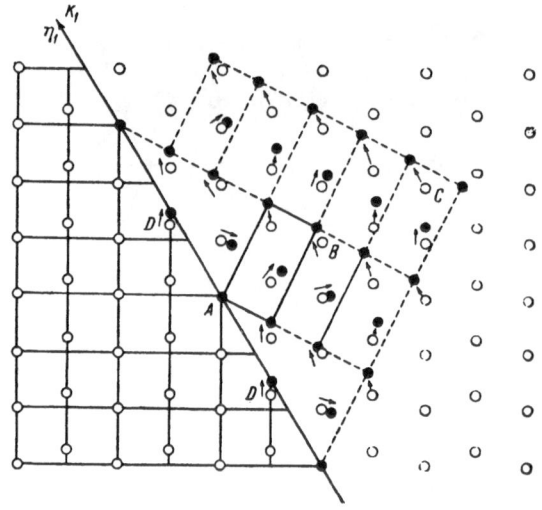

Fig. 12. Motions of atoms in the twinning of β-tin; $K_1 = (103)$, $\eta_1 = [\bar{1}03]$, $\eta_2 = [101]$, projection on S = (010). The open circles are atoms in their initial positions; the filled ones, displaced atoms. The arrows denote the paths taken. Half of the atoms are not displaced parallel to the twin plane (Hall).

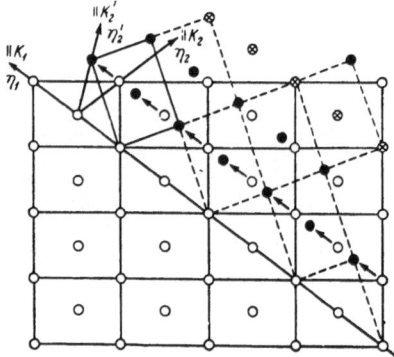

Fig. 13. Motions of atoms in the twinning of α-iron; $K_1 = (112)$, $\eta_1 = [11\bar{1}]$, $\eta_2 = [111]$, projection on S = $(\bar{1}10)$. The open circles represent atoms in their initial positions; the filled ones, atoms common to the two lattices (Hall).

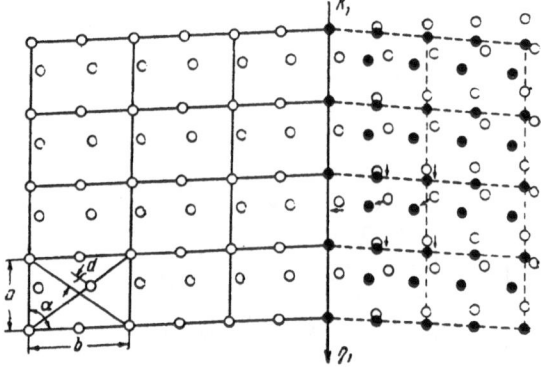

Fig. 14. Motions of atoms in the twinning of bismuth; $K_1 = (011)$, $\eta_1 = [00\bar{1}]$, $\eta_2 = [110]$, projection on S = (100); a = 6.58 A, b = 9.50 A, d = 1.41 A, α = 86°34'; symbols as in Fig. 12 (Hall).

Crystals of α-uranium can twin in two ways (§ 1.15); Cahn (1955) finds that the Q for both are 0.2 approximately.

Cahn (1954) states correctly that Laves's arguments are applicable to crystals of simple structure (tin, magnesium, uranium) but are scarcely applicable to more complex structures. We have seen (§ 1.9) that the calcium and carbon atoms in calcite are displaced as in simple shear, whereas the oxygen atoms move as a group. This makes it very difficult to evaluate Q.

The general description for calcite shows that the C–O bonds are not broken; the group rotates as a whole. Cahn (1954) considers that the lattice can allow such rotation, for the NO_3 groups in KNO_3 (whose structure resembles that of $CaCO_3$) can vibrate, as Tavonen (1947) has shown by x-ray methods.

11. Change of Length in Mechanical Twinning

If we consider twinning as a succession of shearing operations (Fig. 5), we can calculate the change in length.

Consider a stretched cylindrical monocrystal (Fig. 15); we assume that the process gives rise to a twin layer bounded by twin planes through points A and B. If the ends of the cylinder are free to move laterally, in which case the shear planes (which are parallel to the twin planes) have the same orientation throughout the twin layer, then the mutual displacements of atomic layers must be accompanied by a change in the position of the principal axis. Now the relative displacement of adjacent planes must be a fraction of the lattice parameter. The total displacement for a twin layer AB' is represented by BB'. Let AN be a perpendicular to the twin plane; then the distance AN is that between the two twin planes, so we have for s that

$$s = \frac{BB'}{AN}. \tag{1}$$

Let χ_0 and χ_1 be the angles between the pulling direction and the shear plane before and after pulling; let λ_0 and λ_1 be the angles between the pulling direction and the shear direction. Then the right-angled triangles AB'N and ABN give us for their common side AN the value

$$AN = l_0 \sin \chi_0 = l_1 \sin \chi_1. \tag{2}$$

From triangle ABB' we have

$$BB' = \frac{l_1 \sin (\lambda_0 - \lambda_1)}{\sin \lambda_0}. \tag{3}$$

We substitute from (2) and (3) for AN and BB' into (1) to get [*]

$$s = \frac{l_1}{l_0 \sin \chi_0} \cdot \frac{\sin (\lambda_0 - \lambda_1)}{\sin \lambda_0}. \tag{4}$$

Further, from triangle ABB' we have the relative change in length due to twinning:

$$\Delta = \frac{l_1}{l_0} = \frac{\sin \lambda_0}{\sin \chi_1}. \tag{5}$$

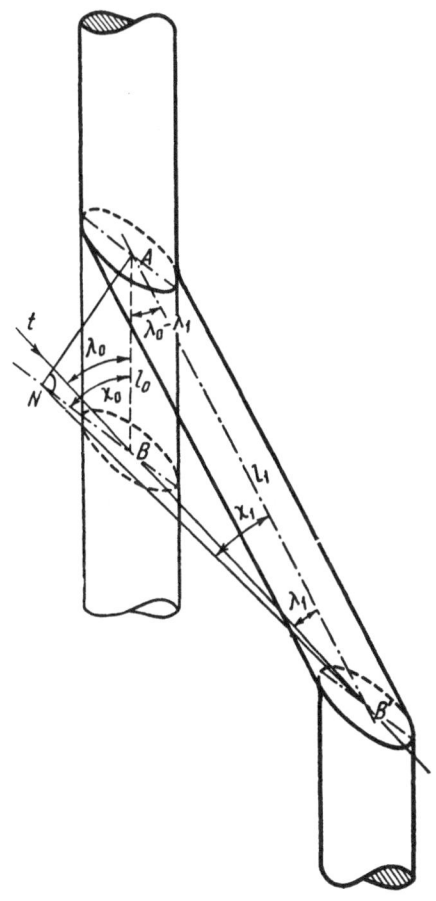

Fig. 15. Extension of a pulled monocrystal by slip of adjacent atomic layers: AB is the axis before deformation, AB' is the position of the axis after deformation, and t is the slip direction (Boas and Schmidt).

---

[*] A more convenient form for the purposes of calculation is $s = \dfrac{\cos \lambda}{\sin \chi} - \dfrac{\cos \lambda_0}{\sin \chi_0}.$

15

Then (4) and (5) give us the relation of Δ to s, χ_0, and λ_0:

$$\Delta = \sqrt{1 + 2s \cdot \sin \chi_0 \cos \lambda_0 + s^2 \sin^2 \chi_0}. \tag{6}$$

The largest deviations from the initial position occur for directions in plane S (ones specified by χ_0 and λ_0). An expression for the orientation after maximal extension (or contraction) is found by differentiating (6) and setting the derivative equal to zero. Then

$$\tan \chi_{1,2} = \frac{s}{2} \pm \sqrt{\frac{s^2}{4} + 1}. \tag{7}$$

We insert this χ_0 into (6)[*] to get the maximal extension:

$$\Delta_{max}^{e} = \frac{s}{2} + \sqrt{\frac{s^2}{4} + 1} \tag{8}$$

and the maximal contraction:

$$\Delta_{max}^{c} = -\frac{s}{2} + \sqrt{\frac{s^2}{4} + 1}. \tag{9}$$

Then (8) and (9) show that the twinning deformation is comparatively small, for the shear along the twin plane is only a fraction of the lattice parameter. Deformation by slip differs from this in that it can cause unlimited extension, as (5) (with $\chi_1 \to 0$ for extension) and (6) show. For slip we must replace s by the crystallographic shear a.

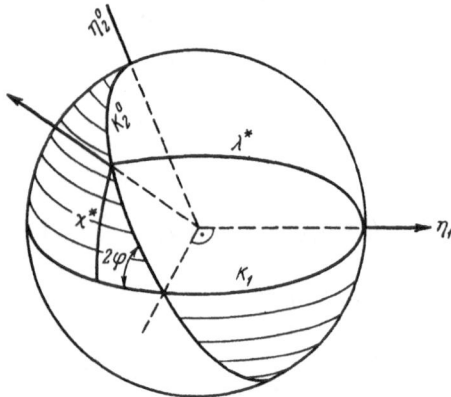

Fig. 16. Projection of the principal axes of a monocrystal to illustrate the change in length by twinning for a specimen of a given orientation. A specimen whose axis lies in the unhatched region extends during twinning; the hatched region corresponds to contraction. Symbols as in Fig. 6 (Boas and Schmidt).

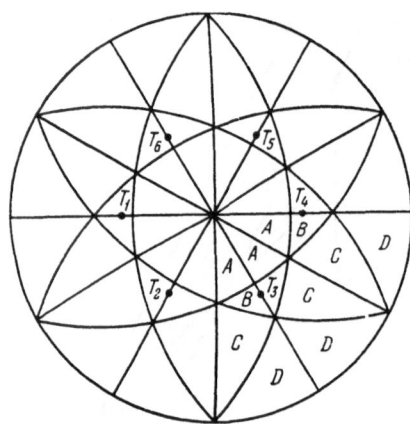

Fig. 17. Stereographic projection indicating the sign of deformation in twinning throughout the volume for zinc and cadmium. T_1 to T_6 are the normals to the twin planes. Contraction occurs upon twinning on planes 1-6 in region A; the same occurs in region B for planes 2, 3, 5, and 6, but extension for planes 1 and 4; the same also occurs in region C for planes 2 and 5, but extension occurs for planes 1, 3, 4, and 6; and extension occurs for planes 1-6 in region D.

---

[*]The sines and cosines must then be expressed in terms of the tangents.

Twinning with change of form does not necessarily cause extension or contraction; there is no change of length if the principal axis lies in the plane of K_1 or K_2. This can be demonstrated by setting the right-hand side of (6) equal to one; then,

1) $\sin \chi_0 = 0$, which corresponds to K_1 or

2) $-\sin \chi_0 / \cos \chi_0 = 2/s_0$, which corresponds to the initial position of K_2, which is denoted by K_2^0.

A projection (Fig. 16) illustrates the change in length as a function of the position of the principal axis relative to the twinning elements. This sphere is formed by the points of emergence of all possible principal axes. The surface of the sphere is divided by the traces of K_1 and K_2 into four regions; the two opposed sectors of angle 2φ (shown hatched) contain the points for axes corresponding to contraction. If the axes lie in planes K_1 and K_2^0, there is no change in length.

Figure 17 is a stereographic projection illustrating the change in length as a function of orientation for zinc twinned on $\{10\bar{1}2\}$ planes. The poles of the six possible twin planes are shown as T_1 to T_6. If the axes deviate from the hexagonal ones by not more than 50° (triangles A), contraction occurs for all six planes; if they deviate by nearly 90° (triangles D), extension always occurs. Orientations corresponding to B and C cause contraction on some planes and extension on others. Figure 17 applies also to cadmium.

If $c/a \approx 1.63$ (as for Mg and Be), the pattern is the reverse of that for Zn and Cd (extension becomes contraction, and vice versa).*

The above argument relates to the distance between points lying within the twin layer (between A and B of Fig. 15). The projections in Figs. 16 and 17 relate to a monocrystal that becomes entirely a twin.

The usual effect of deformation is to produce twin layers; Fig. 18 (Frank and Thompson, 1955) is a stereographic projection for the change in length in a specimen that produces a twin layer.

Here each principal spherical triangle is divided into nine regions (instead of the ABCD of Fig. 17); the increase is the result of the second boundary, which corresponds to great circles for the second K_2 (undeformed) plane and twin planes passing through the poles. The s for zinc is small, so the regions between the two boundaries are only 4° wide.

Figure 19 shows the principal spherical triangle for the effects of crystallographic orientation of deformation in twinning for α-iron [body-centered cubic; $K_1 = (112)$, $K_2 = (11\bar{2})$] if the specimen twins as a whole.

There are 12 twin planes in α-iron; contraction and extension must occur together if all act together, as in the case of zinc for specimens oriented as in triangles A.

The specific shear for α-iron is much larger than that for zinc, and the production of twin layers gives rise to the additional regions shown by broken lines. These regions are 19.5° wide on the stereographic projection.

12. Experimental Determination of Elements of Twinning

If a twinned crystal has even two natural faces not in the same zone, or one face and a cleavage surface, then the indices of the twinning elements can be deduced by measuring the angles between these faces in the main crystal and in the twin layer. The indices are found from transform formulas, which are solved for (HKL) and [UVW], which correspond to K_1 and η_2.

There are several methods of deducing the indices if there are no natural faces, as for the monocrystals grown from the melt by Obreimov's (1924), Bridgman's (1925), or Verneuil's (1904) method.

X-Ray Method for K_1. The indices of K_1 may be determined most precisely if the Laue pattern is recorded with the beam exactly at right angles to the twin plane (Fig. 20); this plane coincides with the boundary of a twin layer. The pattern is indexed to derive the indices (Zhdanov and Umanskii, 1937). There is sometimes a difficulty in indexing the spots, on account of asterism resulting from the deformation, if the crystals are of very high plasticity.

* Because the positions of the traces on the twin planes in Fig. 17 vary with c/a.

Metallographic Method for K_1. The monocrystal (or polycrystal) is ground and polished to give two mutually perpendicular planes, one of which lies at a known angle to one of the crystallographic axes. Slight compression or localized pressure is then used to produce twinning such that the traces of the plane are clearly visible in the two sections. Measurements are made of the angles between the traces of the twin planes and the right-angled edge formed by the ground faces (Fig. 21) as well as of the angles formed by the traces (if there are several systems of twinning).

X-Ray Method for η_1 and K_1. A pattern taken at right angles to the twin plane gives K_1, but the indices of η_1 cannot be determined in general; but some information of the indices of η_1 and S can be obtained if the crystal is of high symmetry. If there is a mirror plane normal to K_1, then this plane must coincide with S; then η_1 is defined as the line of intersection of K_1 with the plane normal to K_1.

The quickest method if the layers are reasonably broad is to use two patterns from the initial crystal and the layers (a goniometer may be used), the orientations of the two being established in a common coordinate system. A reflection twin has the lattice of one component simply a mirror image of the other in K_1; then K_1 is found from these two patterns by means of two stereographic projections on one piece of paper, the combination being examined for the normal to the face common to the two components (Zhdanov and Umanskii, 1937). Figure 22 shows this for a twin in magnesium.

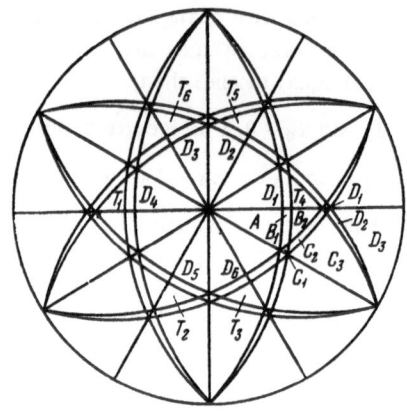

Fig. 18. Gnomostereographic projection resembling that of Fig. 17 for the formation of a twin layer in a stretched crystal of zinc or cadmium: T_1 to T_6 are the normals to the twin planes; D_1 to D_6 are the twinning directions. No system leads to contraction in region A, but contraction does occur as follows: B_2, for system 1 only; B_2 for systems 1 and 4; C_1, for 1 and 6; C_2, for 1, 4, and 6; C_3, for 1, 3, 4, and 6; D_1, for 1, 2, 4, and 6; D_2, for 1-4 and 6; D_3, for 1 and 6 (Frank and Thompson).

The η_1 for a rotation twin is found as the rotation axis of the two components; the patterns should coincide on rotation through 180° about η_1.

Determination of η_2 and K_2. Equations (1) or (2) enable us to determine [UVW] for η_2, and hence the indices of K_2, as well as s, if we know the indices of some one direction before and after twinning as well as the indices of K_1. The indices of η_2 and K_2 can be deduced for the simplest structures if those of K_1 and η_1 are known, for we may discuss the motion of the nodes in the shear plane, as for cubic face-centered crystals (Fig. 27).

13. Elements and Laws of Twinning for Crystals of High Symmetry

The twinning law specifies K_1, K_2, η_1, and η_2, as well as s (§ 1.3). The commoner laws for minerals have been given special names. The commonest law for cubic crystals is named from spinel ($MgAl_2O_4$), in which it was first observed; the commonest for orthorhombic crystals is named from calcite; the Dauphiné, Brazilian, and Japanese laws have been dealt with above for quartz (§ 3.1). and so on.

Table I gives the elements of mechanical twinning for metal crystals; Table II does the same for minerals and various compounds (these tables are at the end of the book).

Cubic System. We have $K_1 = \eta_1$ and $K_2 = \eta_2$ for this system; body-centered cubic metals (α- and β-Fe, W, Mo, Cr, Na) give mechanical twins on (112) only, the twin direction being [111]. Figure 19b (§ 1.11) shows the normals to all twelve $\{112\}$ planes. The spherical triangle of Fig. 19 can be used to establish which elements will be active in extension, and which will be active in compression for a specimen of a given orientation (Allen et al., 1956).

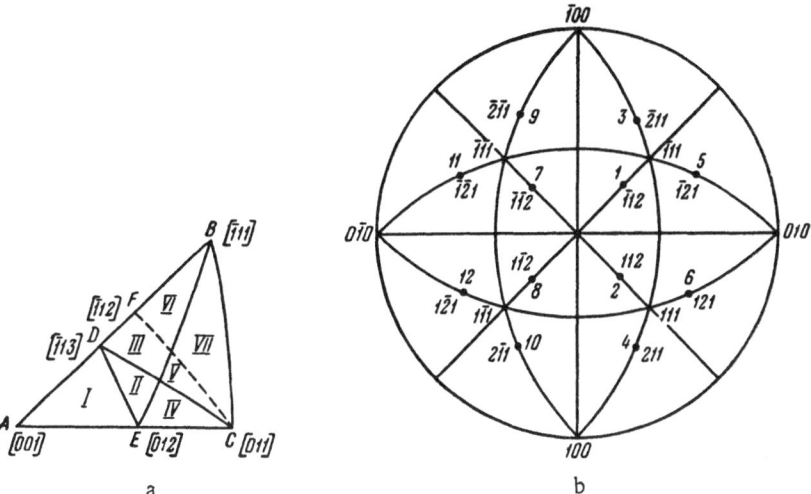

Fig. 19. Projection resembling that of Fig. 18 for twinning in α-iron: a) the full line denotes the principal spherical triangle, which has regions I and V corresponding to deformations differing in sign, as in Fig. 17; the broken line represents additional boundaries, which correspond to the formation of a twin layer, as in Fig. 18; b) stereographic projection for body-centered cubic crystals showing the 12 normals to the possible (112) twin planes. From K_1 we get extension in 1, 2, and 7 in both cases, contraction in 3 and 4 in both cases, extension in 5 and 11 in regions IV — VII for the first case and contraction in I — III and VI, extension in 6 and 12 in the first case in regions II — VII and contraction in both cases in region I, extension in 9 and 10 in the first case in regions III, V, and VII, and contraction in both cases in regions I, II, and IV (Frank and Thompson).

Systems 2, 7, and 8 can be active if the axis of extension lies within triangle ABC; systems 6 and 9 are also active if the axis of the specimen lies within triangle BCF. Only system 1 applies for orientations corresponding to triangle ACF; only 10 for BCD, only 12 for BDEC, and only 11 for BEC. Systems 3, 4, and 5 are active only in compression.

Figures 13 and 23 show the positions and possible motions of the atoms of α-Fe twinning by simple shear.

Meteorites contain crystals of pure α-iron, which show Neumann's bands in polished section (Fig. 24). The argument whether these are twin bands or not raged for many years, but it has now been shown from etch figures (Harnecker and Rassow, 1924) and x-ray patterns (Kelly, 1953) that they are twins. Bechtold (1953) observed such bands in polycrystalline molybdenum ruptured at -196°C. Cahn (1955) examined the twinning of molybdenum monocrystals in response to shock compression at liquid-oxygen temperatures; the twinning was accompanied by extensive cracking at the intersections of twin layers. Figure 25 shows x-ray patterns before and after deformation.

Metallography indicates that similar twins occur in polycrystalline chromium work-hardened (hammer) at room temperature (Carrington, 1953) and in polycrystalline tungsten ruptured in a certain temperature range (Bechtold and Shewmon, 1954).

Barrett (1954) finds that mechanical twinning cannot be produced in β-brass under any conditions, but recrystallization twins occur. Cahn considers that mechanical twinning is prevented by the substructure that occurs in β-brass at low temperatures (Laves, 1952b). Very large forces are required to disrupt this substructure.

All other body-centered cubic metals so far examined form twins in accordance with the {112} [111] law.

It is stated (Kolontsova et al., 1956) that cesium iodide and thallium bromide-iodide (CsI lattice[*]) twin on (113) in response to mechanical stress.

---

[*] Two simple cubic lattices mutually displaced by half a body diagonal.

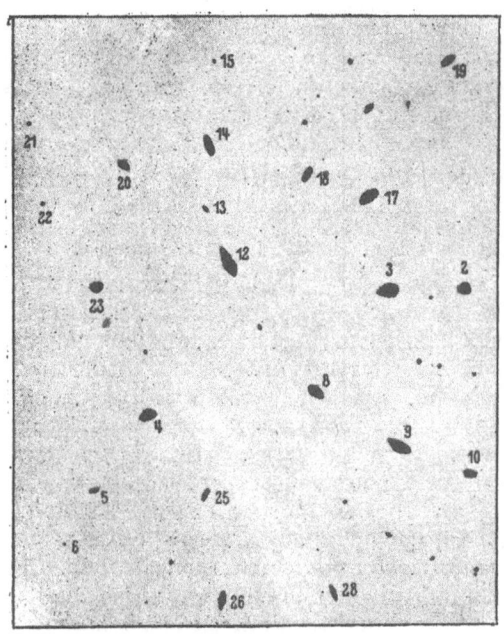

Fig. 20. Laue pattern taken at right angles to the twin plane in magnesium for the determination of the indices of this plane. The spots are numbered (Boas and Schmidt).

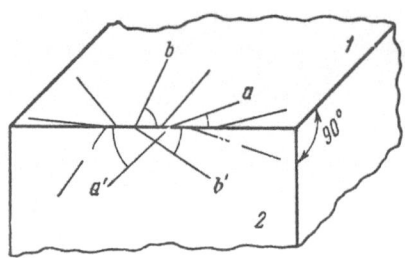

Fig. 21. Metallographic determination of K_1 from the traces a, b, a', and b' of the boundaries of twin layers on polished surfaces 1 and 2 of a crystal of beryllium (Boas and Schmidt).

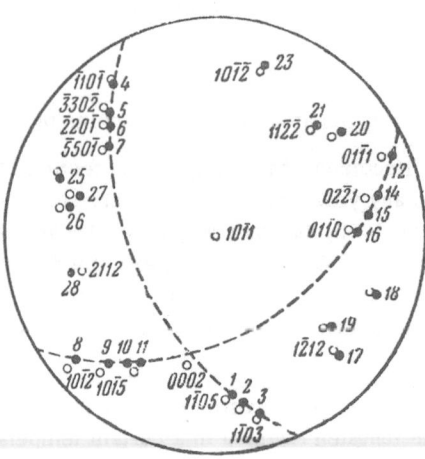

Fig. 22. Projection for the magnesium crystal of Fig. 20; the numbers from 1 to 28 denote projections (filled circles) corresponding to the Laue spots, and the open circles denote projections of the nearest rational crystallographic planes (with indices). The pole of the twin plane is shown at the center (Boas and Schmidt).

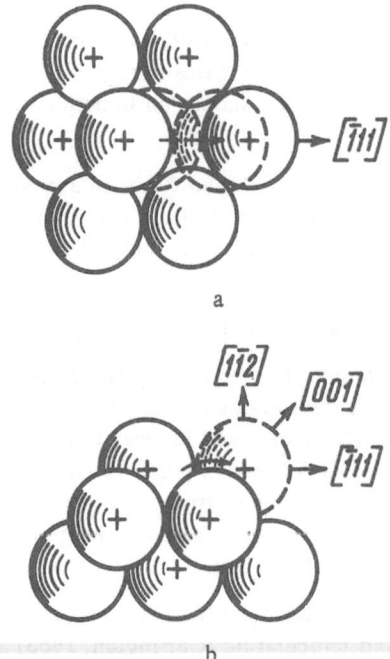

Fig. 23. Motion of atoms in the twinning of α-iron: a) projection on (1$\bar{1}$2), here acting as twin plane, with [$\bar{1}$11] as twin direction, the circle shown broken being an atom in a twin position; b) projection on the (110) plane of shear (Barrett).

Almost all cubic minerals and metals show growth (recrystallization) twins of spinel type: $K_1 = K_2 = \{111\}$; Fig. 26a shows the disposition of the components. The two interpenetrating cubes have a common [111] axis (twin axis); one component is derived from the other by rotation through 180° on [111] or through 74°31' on [110]. Figure 26b shows the disposition of the principal crystallographic planes and directions.

Figure 27 shows the projection on the S plane of a spinel twin in a face-centered cubic crystal, for which it is assumed that $K_1 = K_2 = (111)$, $S = (0\bar{1}0)$, and $\eta_1 = [112]$; the displacements in the shear plane give $K_2 = (11\bar{1})$, $\eta_2 = [11\bar{2}]$, and $s = \sqrt{2}/2 = 0.7071$ (these hypothetical elements satisfy the analytic condition for deformation by simple shear).

Repeated attempts (Barrett, 1947) failed to produce mechanical twinning in face-centered crystals, but the effect was finally observed (Blewitt et al., 1954 and 1957) for copper deformed at 4.2°K. Deformation by slip interferes at higher temperatures.

Cubic crystals with lattices of diamond type show twinning in response to indentation with a diamond pyramid at high temperatures under vacuum (Churchman et al., 1956). Germanium, silicon, gallium antimonide, and indium antimonide show twinning on $\{111\}$ and $\{123\}$; the first plane has $\eta_1 = [11\bar{2}]$, $s = 0.4084a$ (a is the cell parameter), and $S = (\bar{1}10)$, while the second has $\eta_1 = [412]$, $s = 0.6552$, and $S = (\bar{1}20)$. Sphalerite twins only on $\{111\}$. Germanium shows this response at 300-600°C, silicon at 500-900°C, indium antimonide at 80-320°C, gallium antimonide at 100-500°C, and sphalerite at 300-400°C.

Hexagonal System. Crystals with hexagonal close packing (Zn, Cd, Mg, Be, Ti) all have the same law; $K_1 = (10\bar{1}2)$, a pyramid face, in all cases. There are six such planes, as against only one slip plane, (0001), at normal temperatures. Zinc is found to have $K_2 = (10\bar{1}\bar{2})$, and this is assumed (by analogy) for all other hexagonal metals. Magnesium has been found to have $(10\bar{1}1)$ as a second twin plane. For zinc,

$$s = \frac{\left(\frac{c}{a}\right)^2 - 3}{\frac{c}{a} - \sqrt{3}} = 0.143,$$

in which c/a applies to the unit cell (Mathewson and Phillips, 1928; Schmid and Wassermann, 1928). Table I gives the values for other hexagonal metals.

The $(10\bar{1}2)$ system occurs in beryllium together with rarer twin bands parallel to $(10\bar{1}1)$ and $(10\bar{1}3)$ (Garber et al., 1955). The entire specimen can be converted to the twinned orientation at elevated temperatures;

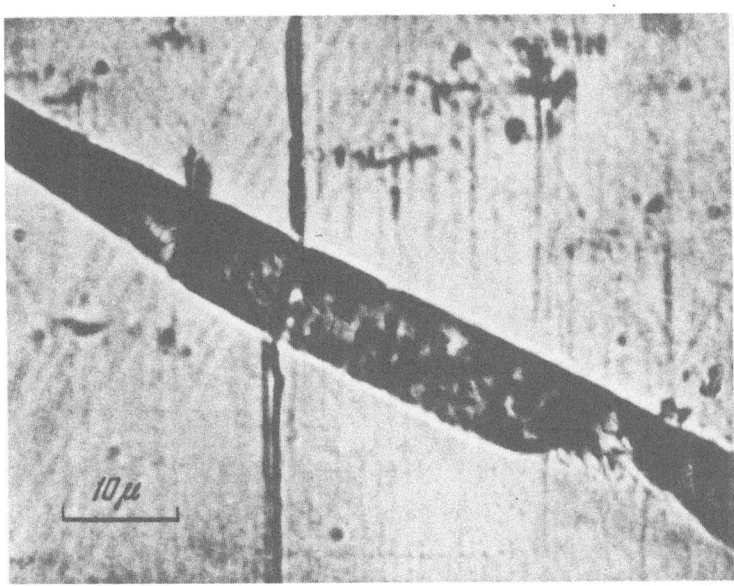

Fig. 24. Neumann band in iron containing 3.7% silicon (Hall).

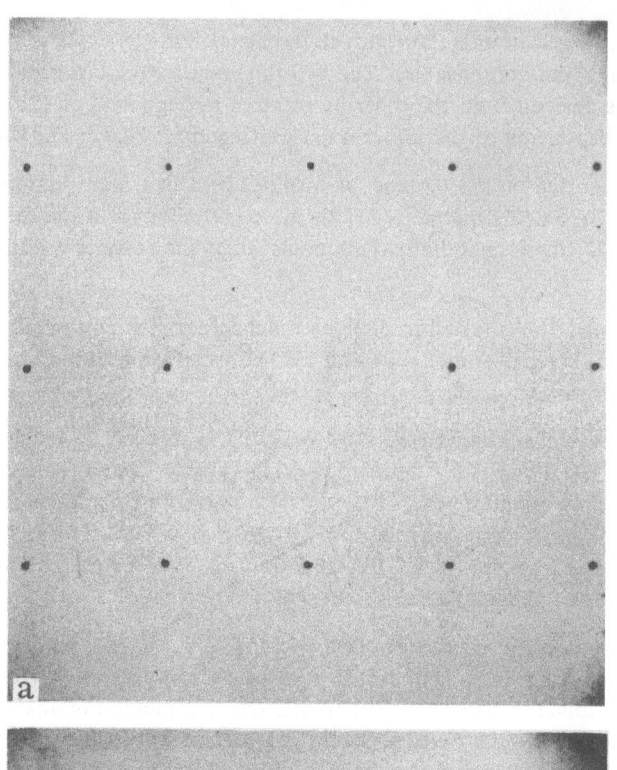

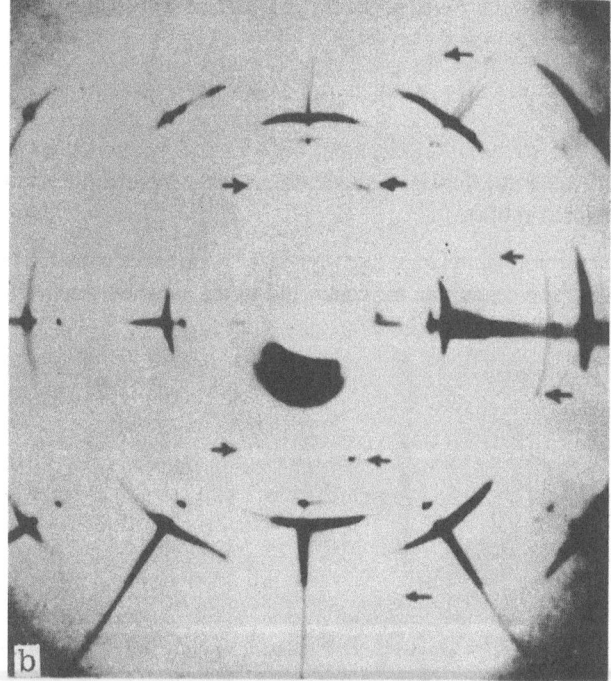

Fig. 25. X-ray patterns of a molybdenum monocrystal: a) undeformed (Mo K_α radiation, 45 kV, 60 mA-hr); b) deformed (45 kV, 900 mA-hr). The plane of the figure coincides with the plane perpendicular to (110) in the reciprocal lattice.

Fig. 29, parts a and b, shows Laue patterns recorded along sixfold and twofold axes for a monocrystal slightly compressed at 200°C; twins had not appeared at that point. Figure 29c is for the same specimen after twinning at 200°C.

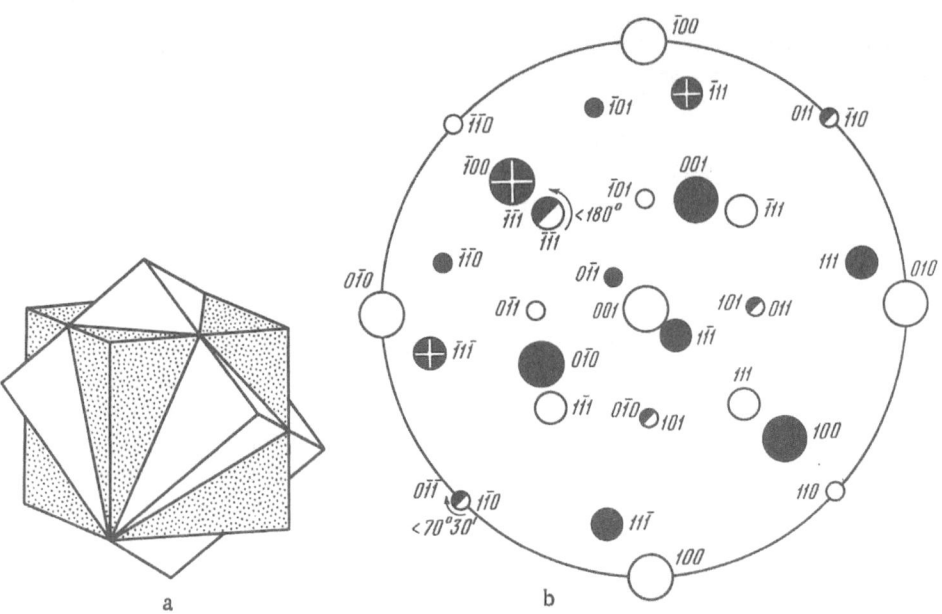

Fig. 26. a) Disposition of the components in a spinel twin; b) stereographic projection on the cube plane for such a twin. The open circles denote normals to the main faces of the initial curve; the filled ones are the same for the second (twin) component. The half-filled ones represent faces common to the two; black with a white cross represents a projection for faces whose normals emerge in the upper and lower hemispheres. The arrows denote twin axes.

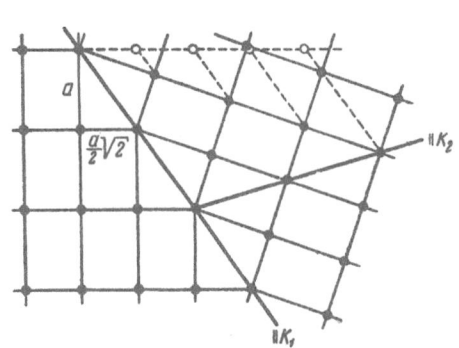

Fig. 27. Projection on $S = (0\bar{1}0)$ of the lattice of a twin on $K_1 = (111)$ and $\eta_1 = [11\bar{2}]$ for a face-centered cubic crystal. The twin position is produced by simple shear parallel to K_1.

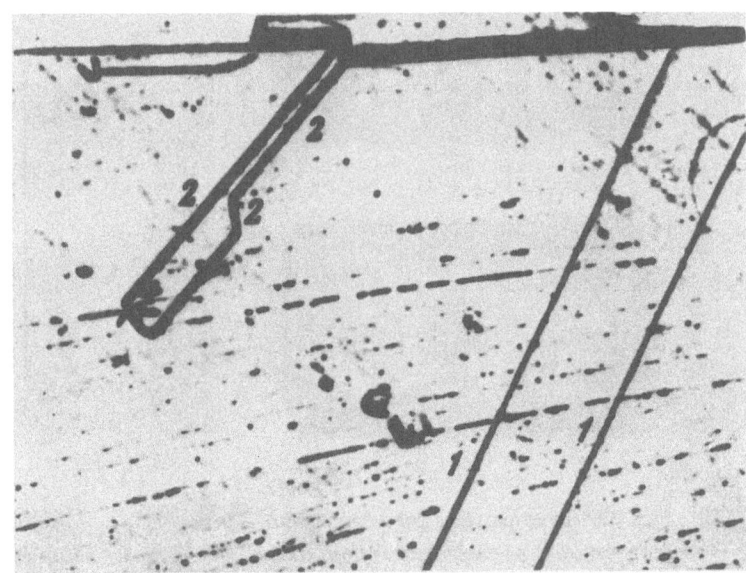

Fig. 28. Twins in a polished and etched (111) plane in silicon: 1) traces of twin bands parallel to (111); 2) the same, parallel to (123) (Churchman et al.).

Titanium monocrystals show five systems of mechanical twins (Seidle et al., 1953; Table I).

Figure 30 shows the disposition of the components in a hexagonal crystal with close packing; the $(10\bar{1}2)$ twin plane passes through vertices A and C of the hexagonal prism. Vertex B moves parallel to the AC plane to take up position B'; the angle BAB' is only 4°, but this small displacement turns the principal (L^6) axis through 84°, because the (0001) basal plane (AB in figure) is transferred to position BB'. This large change in (0001), which is the slip plane in a hexagonal crystal, with respect to the force (which acts along the axis of the mono-crystal) should cause a change in the plasticity, as is found.

Deformation by slip starts within the twin layer; this is called secondary slip (Fig. 31). Figure 32 shows how the crystallographic orientation alters within the twin layer (ϑ_0 denotes the initial orientation). Twinning on plane I causes a stepwise change in the orientation within the layer; the new orientation along the geometrical axis of the rod is defined by ϑ_1, which is reflected in plane I. Plane II corresponds to ϑ_2, and so on. The initial (ϑ_0) orientation is such that II and V lead to contraction, while I, III, IV, and VI should cause extension. The change in length as a direct result of twinning is slight, but the change in orientation within the twin layers favors secondary slip and so can cause increased plasticity.

Graphite (Table II) forms mechanical twins on $(11\bar{2}1)$ planes (Laves and Baskin, 1956; Platt, 1957; Palache, 1941).

In 1898 Veit (1922) produced twins on (0001) and on $(10\bar{1}1)$ in natural rubies (the latter are planes of the rhombohedron) by compressing the crystals mounted in powdered sulfur in a hollow steel cylinder (13,000 to

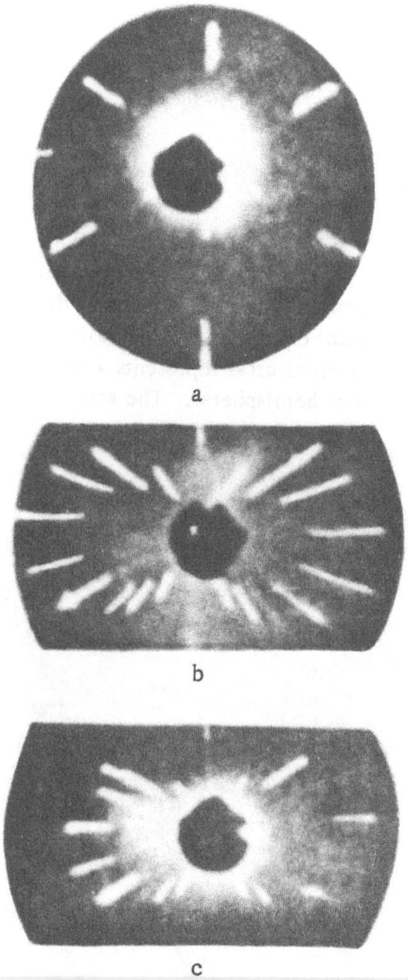

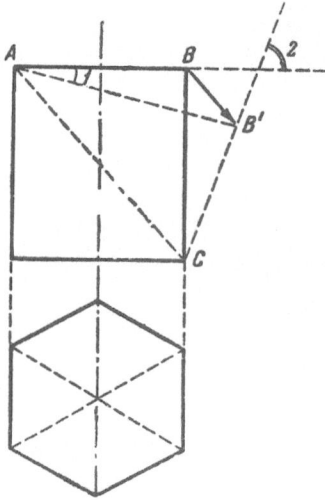

Fig. 29. Laue patterns for a beryllium monocrystal: a) along the sixfold axis; b) along a twofold axis (in both cases slightly deformed at 200°C; no twinning); c) along the original sixfold axis after mechanical twinning at 200°C. A twofold axis has been produced along this direction.

Fig. 30. Reorientation of the (0001) plane in the twinning of a hexagonal crystal: AB is the trace of the basal plane on the hexagonal prism for the undeformed crystal; AC is the trace of the $(10\bar{1}2)$ twin plane; BB' is the shear in twinning; ∠1 is the shear angle (4°); B'C is the position of the basal plane after twinning; and ∠2 is 84°.

18,000 atm).[*] He found that $K_1 = (0001)$ and $K_2 = (02\bar{2}1)$ had $S = (2\bar{1}\bar{1}0)$, $\eta_1 = [0\bar{1}10]$, $\eta_2 = [01\bar{1}2]$, $s = 0.685$, and the angle $K_1K_2 = 72°22.5'$. The expected η_1 for twins on $(10\bar{1}1)$ is $[\bar{1}011]$. I have produced twins in synthetic colorless corundum (Al_2O_3) by rapid cooling from 600°C to room temperature (Klassen-Neklyudova, 1942a). Figure 33 shows the disposition of the components. Twinning is one reason why artificial stones for bearings are scrapped; cracking and cleaving are liable to occur along the twin plane.

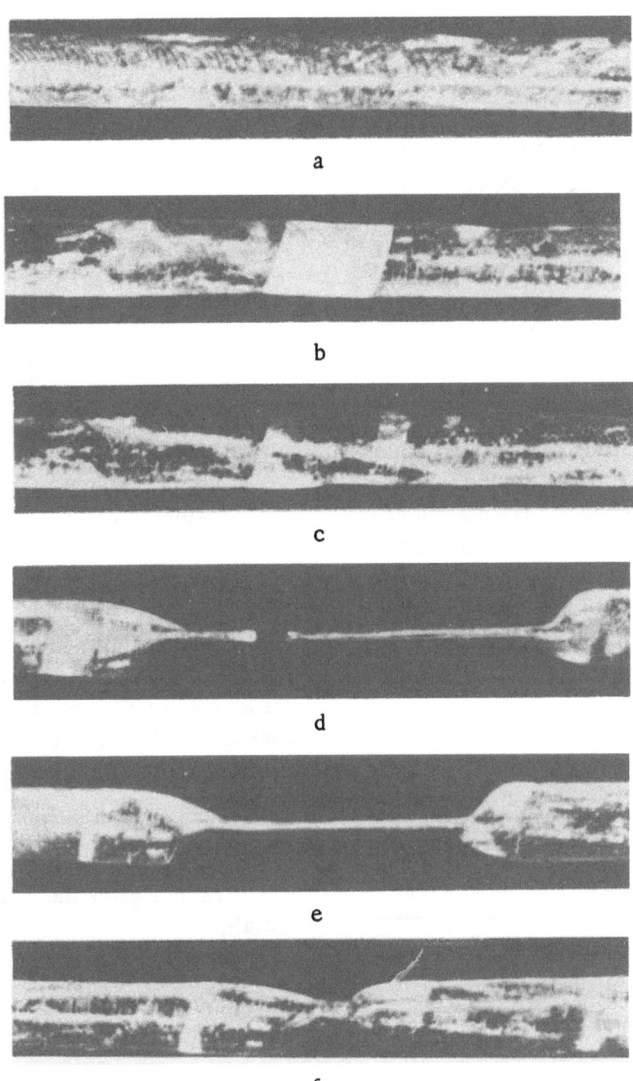

a

b

c

d

e

f

Fig. 31. Successive stages in the pulling of a cadmium monocrystal oriented favorably for twinning: a) undeformed crystal; b) twin layer increasing in width; c) slip on basal plane in twin layer; d) and e) part of specimen showing slip drawn out into a band, with secondary slip; f) specimen as of e) in another orientation (Boas and Schmidt).

[*] Kronberg (1957) discusses the mechanical twinning of corundum on the basal plane.

Trigonal System. Calcite ($CaCO_3$), sodium nitrate ($NaNO_3$), magnesite ($MgCO_3$), and siderite ($FeCO_3$) all twin in the same way (Table II); rhombohedron planes serve as the twin planes. Table I gives the indices of the twin planes and directions for the calcite law in the hexagonal and trigonal settings.

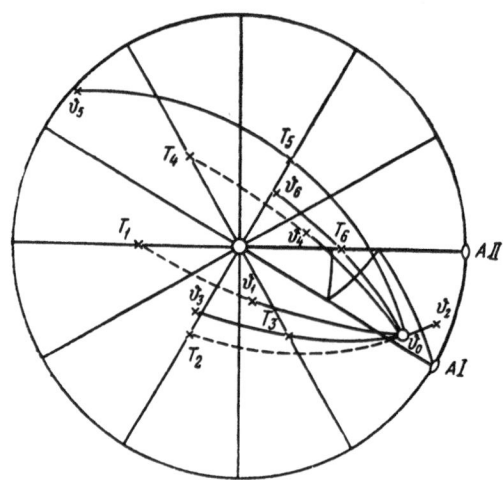

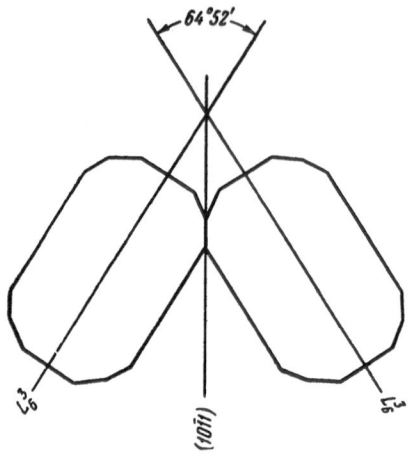

Fig. 32. Change in lattice orientation within a twin band in a monocrystal of zinc (or cadmium): ϑ_0, pole defining the orientation before deformation; T_1 to T_6, poles (normal to K_1) of possible twin planes; ϑ_1 to ϑ_6, possible orientations of specimen in twin layer when the corresponding twin planes are active (Boas and Schmidt).

Fig. 33. Disposition of the principal crystallographic axes and faces in the twinning of α-Al_2O_3 (corundum) with $K_1 = (10\bar{1}1)$.

TABLE 1

Twinning Elements for Calcite in Two Settings

Hexagonal		Trigonal	
K_1	η_1	K_1	η_1
$(\bar{1}012)$	$[10\bar{1}1]$	(011)	$[100]$
$(1\bar{1}02)$	$[\bar{1}101]$	(101)	$[010]$
$(01\bar{1}2)$	$[0\bar{1}11]$	(110)	$[001]$

Calcite can twin in three distinct systems; the K_1 planes of the twin layers intersect if several systems are active simultaneously, and these give rise to Rose channels (§ 1.18). The movement of the atoms in the twinning of calcite is considered in § 1.9.

Millerite (NiS) shows additional twinning on $(2\bar{1}\bar{1}0)$ (Bravais).

Table 1 gives the twinning elements for trigonal metals.

Andrade and Hutchings (1935) observed twinning in mercury monocrystals stretched at temperatures below -38.8°C (the melting point of mercury).

14. Twinning in Crystals of Low Symmetry

The probability and number of modes of twinning increase as the symmetry falls (§ 1.4). The subject has been examined reasonably completely only for quartz, Rochelle salt, and α-uranium. The only mechanical twinning in quartz is that involving no change of form, so this considered separately (§ 3).

Twinning of α-Uranium. The α-form is stable up to 670°C; it is orthorhombic. The crystal consists of giant molecules (Fig. 34) held together by covalent bonds (the bonds within the molecules are metallic). Cahn (1953a) has shown that four twinning laws apply to deformed coarse-grained α-uranium; one of these has an irrational twin plane approximating to $(1\bar{7}2)$ and a rational twin axis $\eta_1 = [312]$, so it can be described only as a rotation twin.

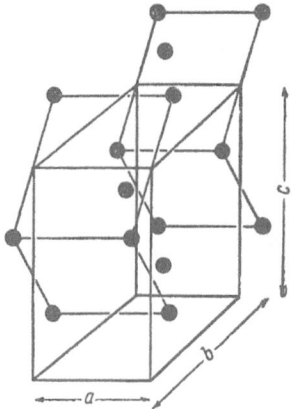

Fig. 34. Unit cell of α-uranium: a = 2.852 A, b = 5.865 A, c = 4.945 A (Hall).

Twins on $K_1 = (130)$ have all other elements also rational; they can be considered as ones of reflection or rotation, as for crystals of higher symmetry. Twins on $K_1 = (112)$ have $\eta_2 = [312]$, and ones on $K_1 = (121)$ have $\eta_2 = [311]$; K_2 and η_1 are irrational, so these are ordinary twins of the second kind.

Figure 11 shows the structure of a twin with $K_1 = (130)$ in α-uranium projected on $S = (001)$; § 1.10 deals with the motion of the atoms for this case. The broken lines in Fig. 11 illustrate the disturbance to the regular lattice structure near the junction of the two components.

Twinning of Rochelle Salt. This salt ($KNaC_4H_4O_6 \cdot 4H_2O$) is orthorhombic below -18°C (lower Curie point) and above 24°C (upper Curie point); the symmetry class is $3L^2$. It is monoclinic (L^2) between the Curie points, the angle between the b and c axes differing from a right angle by about 3' (Jaffe, 1937; Ubbelohde and Woodward, 1946). A second-order phase transition occurs at the Curie point.

The transition from orthorhombic to monoclinic causes the crystal to split up into a system of polysynthetic twins (Fig. 35; Klassen-Neklyudova, Chernysheva, and Shternberg, 1948), which are produced by the shear deformation in the bc plane.

An external force alters the initial polysynthetic twin; a plate cut normal to a, or at a small angle to it, responds to the localized load from a needle or glass ball by producing larger

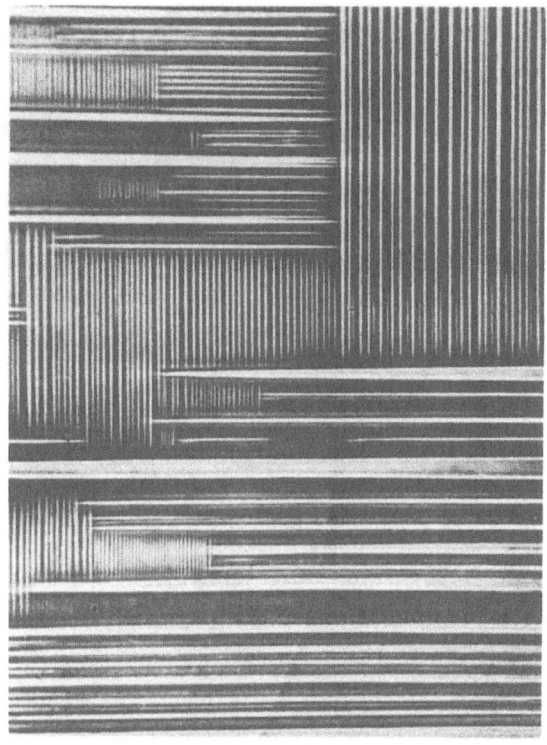

a 200μ

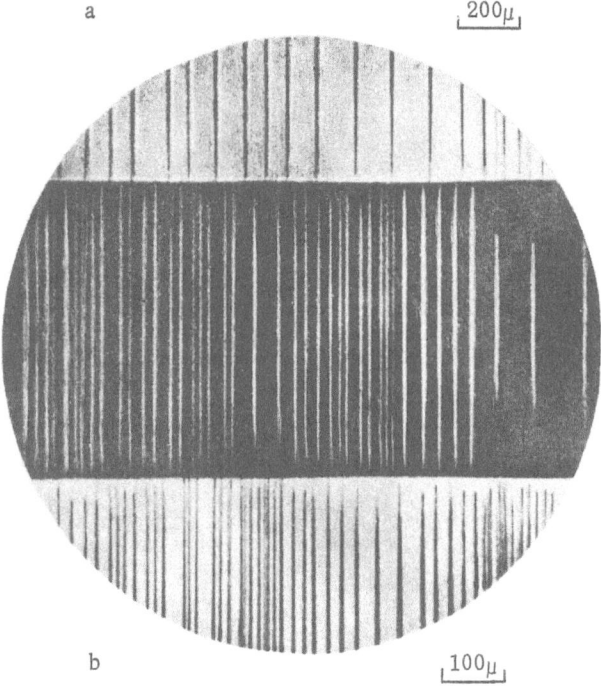

b 100μ

Fig. 35. Twin (domain) structure of Rochelle salt; plate cut normal to a, with combinations of twin components extending along b and c: a) one of the possible combinations; b) broad component containing narrow wedge-shaped components (Chernysheva).

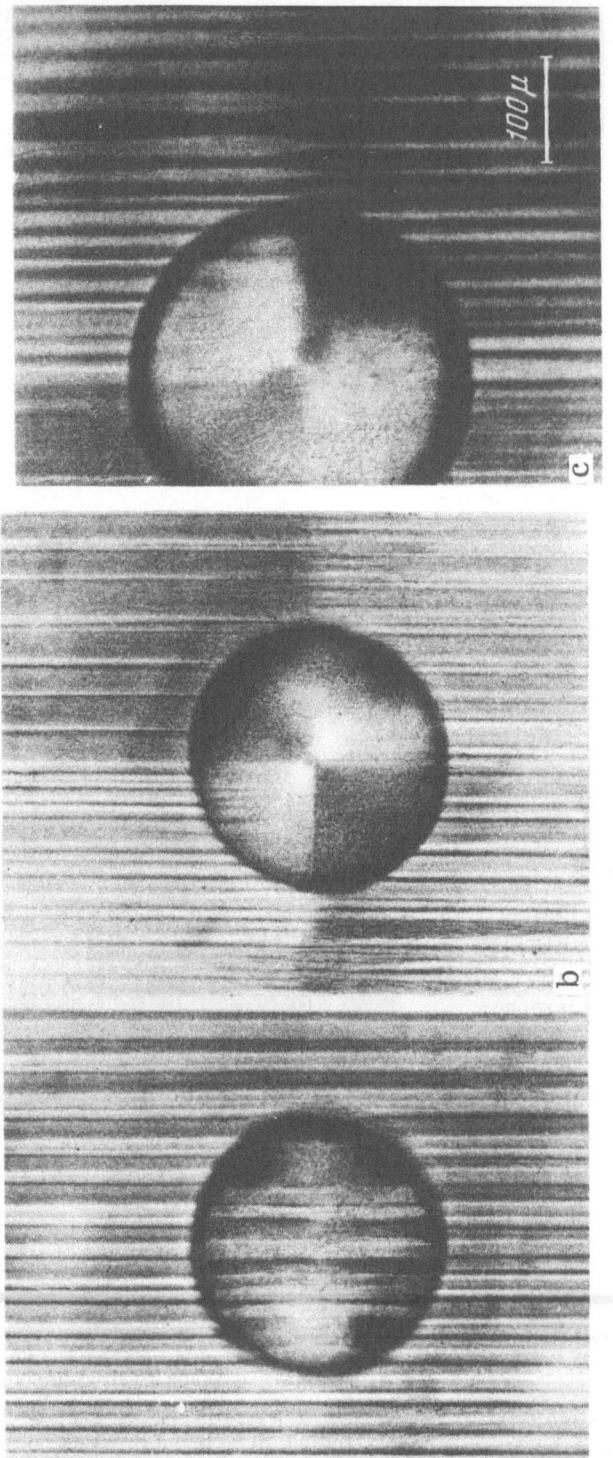

Fig. 36. Formation of an enlarged twin by reorientation of the components of a polysynthetic twin in response to pressure from a glass ball: a) before; b) during compression; c) initial stage of reorientation (Chernysheva).

twin components (Fig. 36, a and b; Chernysheva, 1950), on account of the displacement of boundaries (Fig. 36c). Both components develop simultaneously from the point of loading. Figure 37 shows a mechanical twin produced by internal stresses localized at the end of a crack. The two dark regions are almost monocrystalline; the two light ones are also, these pairs being the components of a twin. Comparison of Fig. 36c with Fig. 37 shows that the right upper (light) monocrystalline part is formed by extension of the light components in the polysynthetic twin, while the right lower (dark) part derives from the dark ones (narrowing occurs in the bands at the point of transition from one quadrant to another.

Fig. 37. Tapering crack twin in Rochelle salt; section normal to a axis
(Chernysheva).

The components of the polysynthetic twin (bands in Fig. 36, which enlarge to form the mechanical twins) extend along c or along b. Each component of the twin may be considered as a crystal showing oblique extinction. Figure 38a shows the optical indicatrix in the bc plane for an orthorhombic crystal; the shear in the bc plane between the Curie points causes the indicatrix to rotate sufficiently to reveal the twin structure from the differences in extinction. Figure 38, b and c, shows sections of the indicatrix for the monoclinic components in the bc plane. These mechanical twins are formed by combination of twin layers of the same orientation, so Fig. 38b applies also to mechanical twins.[*]

The polysynthetic twins are not seen in sections normal to b and c; the plate appears as a monocrystal (Fig. 38e).

Figure 39 is an idealized representation of a twin in Rochelle salt (Chernysheva, 1950). The habit corresponds to orthorhombic symmetry at all temperatures, but this is only the pseudosymmetry between -18 and +24°C, for the crystal is actually twinned and is not a monocrystal; the two components are related by rotation through 180°, which can be performed on c or on b (more precisely, about a normal to a (010) face of the orthorhombic crystal).

This law for Rochelle salt is analogous to the Dauphiné law for quartz (Chernysheva, 1950).

Figure 38, b and c, shows that K_1 = (010) for polysynthetic twins along c, with η_1 = [c] = [001]; again, K_1 = (001) and η_1 = [b] = [010] for those along b. The twin plane is also the mirror plane of the twin in a calcite or

---

[*] Strictly speaking, mechanical stress produces a further rotation of the indicatrix.

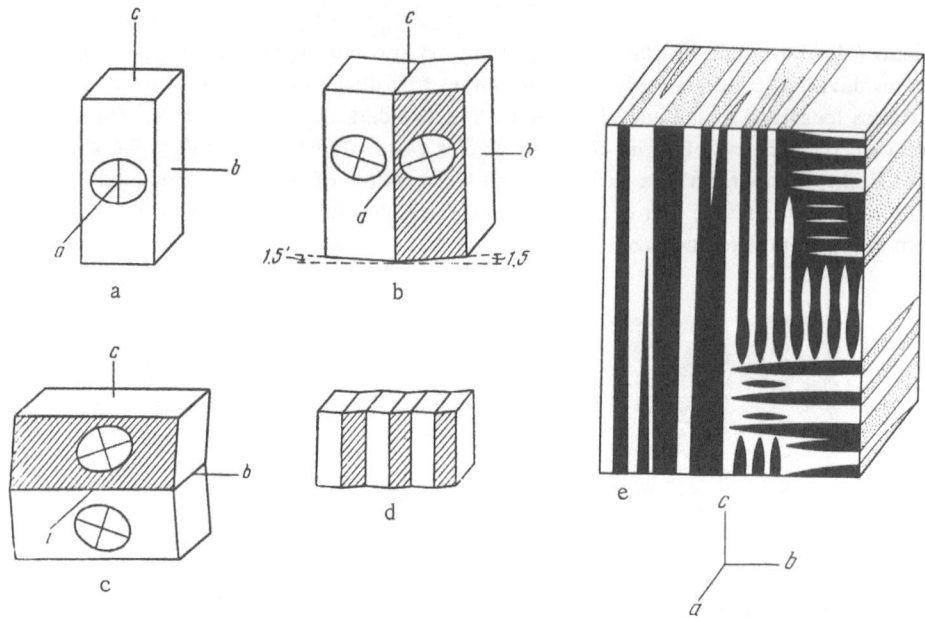

Fig. 38. Disposition of optical indicatrices in Rochelle salt: a) in an orthorhombic crystal (above Curie point); b) and c) in twin components along c and b in a monoclinic crystal (between Curie points); d) monoclinic polysynthetic twin; e) scheme of real polysynthetic twin (Chernysheva).

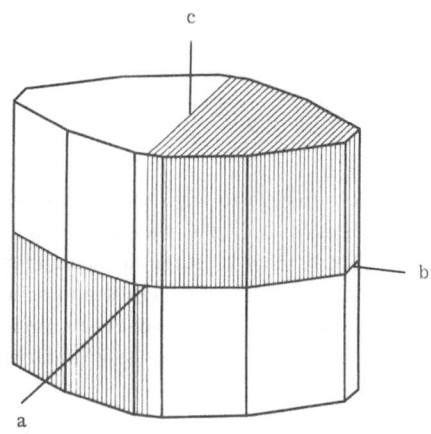

Fig. 39. Conversion of the b and c axes of orthorhombic Rochelle salt to the twin axes in the monoclinic form; rotation through 180° on c produces the same result as that on b (Chernysheva).

metal crystal, but this is not so for Rochelle salt, because the left component cannot be brought into coincidence with the right one for several reasons. In particular, the crystal as a whole is optically active between the Curie points, so all components must be active (right-handed). The antiparallel electric axes in adjacent domains also make the reflection operation inapplicable.

The large twin components are layers of variable thickness normal to b or to c. The small twin components enclosed in the body of the crystal take the form of oblate triaxial ellipsoids, which are also normal to b and c (Fig. 40).

The polysynthetic components in this ferroelectric and in others are electrical domains (regions of spontaneous polarization; § 2.20).

Twinning of Tetragonal Crystals. Simple shear on $K_1 = (101)$ along $\eta_1 = [\bar{1}01]$ produces twinning here; $K_2 = (\bar{1}01)$ and $\eta_2 = [101]$:

$$2\varphi = 2\tan^{-1}\frac{a}{c}\ ;\quad s = \frac{c^2 - a^2}{ac}\ ,\quad 3e = 2\tan^{-1}\frac{c}{a} - \frac{\pi}{2}\ ,$$

in which Φ is the angle between [001] in the initial lattice and [001] in the twin layer (Cahn, 1953a). For indium-thallium alloys, for example,

$$\left|\left(\frac{c}{a} - 1\right)\right| \ll 1 \text{ and } s = 2\left(\frac{c}{a} - 1\right) = 6e\ ;\quad \frac{c}{a} \approx 1.04\ ;\quad s = 0.08\ .$$

15. Disposition of Atoms in the Twin Boundary When This is Parallel to the Twin Plane

The boundaries of a twin layer (twin sutures) often act as points of weakness; such boundaries may be coherent or incoherent. Coherent ones are plane and are parallel to the twin plane; the lattices are conjugate along the boundary, and they do not give rise to macroscopic stresses (first-order stresses),[*] although the short-range order may be disturbed near the plane of contact (this disturbance may be ascribed to third-order stresses). Incoherent ones may be plane, stepped, or curvilinear; they cause macroscopic stresses.

Incoherent boundaries undoubtedly act as points of weakness, for the energy of the elastic distortions may be very large. It is found, though, that mirror cleavages occur on twin planes (and also on slip planes); this is called parting in mineralogy, to distinguish the effect from that for cleavage planes. (The easy cleavage on cleavage planes is a result of pronounced anisotropy in the lattice.)

Mathewson (1928) was the first to discuss the atomic distribution at twin boundaries. The following concerns the structure of twin boundaries.

The projection on the S plane for cadmium (§ 1.9) shows that the atoms in the first row are arranged very differently from those in the initial lattice. The degree of distortion along the boundary varies with the choice of position for the trace of K_1 on S (§ 1.9); the short-range order is disturbed along a coherent boundary, so the energy must be somewhat increased.

Parting can occur along a coherent boundary if its energy is comparable with the surface energy; such parting is therefore determined by the structure. The projection on S for calcite (Gogoberidze, 1952; Pabst, 1955) shows that the first row differs from the initial one (Fig. 8); calcite shows perfect parting on $\{01\bar{1}2\}$ twin planes, but cadmium does not give perfect parting on these planes.

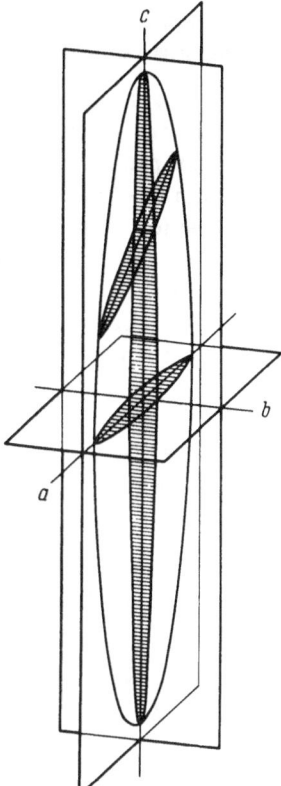

Fig. 40. Possible form of a twin component (domain) in Rochelle salt; sections by the ab, ac, and bc planes, and also by a plane at 45° to a and c (Chernysheva).

Twinning does not disrupt the regular lattice structure in cadmium, zinc, or calcite if we consider only the nodes of the Bravais lattice; the distortion merely displaces the atoms lying within the cells of the Bravais lattice, and such distortion may be treated as being of third order. First-order stresses occur only at the first stage of twinning; the completion of the plane-parallel twin layer should leave no first-order stresses, for parting along the lines of junction should cause no change in volume in the adjacent lattices.

Aminoff and Broomé (1931) examined the geometry of twin structures for zinc blende ZnS, galena PbS, and diamond, as well as for gold and copper.

The rules they demonstrated may be formulated as follows:

1.   If two crystals form a twin, then either a) the two lattices are directly in contact (have a common atomic layer) or b) the lattices are separated by a transition region (there are two atomic layers along the boundaries of this region that are common to the two lattices).

---

[*] First-order stresses are residual stresses within a macroscopic specimen, which are relieved when the specimen is broken up; second-order ones are confined to individual grains and are relieved only by breaking up the grains; third-order ones occur within volumes of atomic dimensions.

2.   The array along the boundary in case a) corresponds precisely to the initial structure (e. g., diamond); in case b) the transition region has a structure corresponding to that of one of the polymorphic forms found in nature or to that of one of the possible modifications of the initial structure.

Twins in diamond (Fig. 41) and zinc blende (Fig. 42) illustrate this; Fig. 41 is the projection on the (110) plane (Ellis and Treuting, 1951). The twin plane is (111), which is also the plane of contact in this case; the trace of this plane passes between the rows in Fig. 41. The projections of atoms in the first row above the interface (shown by circles with lines) are common to the two lattices,* so the initial structure is not disturbed at the site of junction. The normal array in the projection is preserved, and the tetrahedra formed by the atoms near the interface are still of the same size. The projections of atoms in the second and subsequent rows lie in the positions required by twinning, but they may also be derived by translation of the atoms in the initial lattice. Twinning on (111) thus should cause no disturbance in diamond, but some disturbance in the distribution does in fact occur if we consider the rows lying further from the plane of contact, for the twin consists of two components differing in orientation.

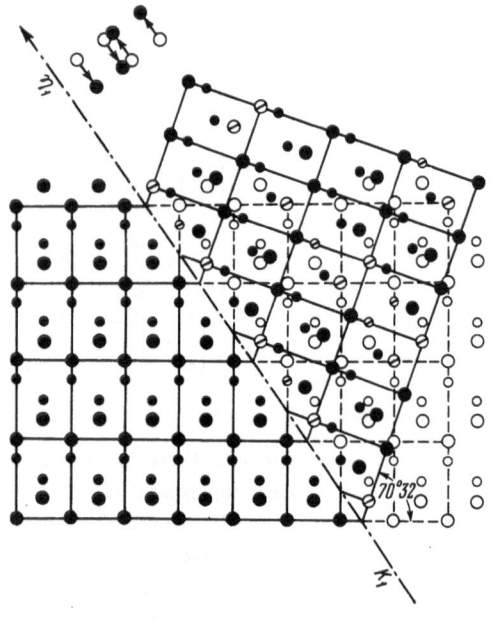

Fig. 41. Projection of the structure of a twinned diamond on S = (110); K_1 = (111) and η_1 = [$\bar{1}\bar{1}2$]. The large filled circles are atoms in the (110) plane passing through the center of the unit cell; the small filled ones are atoms in a (110) plane separated from the first by a quarter of a diagonal; the open circles are initial positions; and the circles with lines represent atoms after the twinning displacement (Ellis and Treuting).

Figure 42 shows a structure model for ZnS (cubic close-packed lattice) based on the (111) plane; a and c are the lattice repeat distances parallel to [110] and [111] respectively. This ZnS structure has no center of symmetry (not all vertices of the tetrahedra have the same orientation). The arrow denotes the direction of the [111] (polar) axis. Consider a ZnS crystal growing layer by layer; it may be assumed to develop a defect at some point, as if a zinc atom in the unit cell were deposited on a (111) plane as shown in Fig. 42c (instead of in the proper place). The error is not repeated in the subsequent growth, so the result is a crystal whose lattice is turned through 180° about [111]. These growth twins do occur in zinc blende; Fig. 42b shows that such a twin gives rise to no distortion at the plane of contact, but the structure in the region shown by full lines is not that of zinc blende; it is rather a unit cell (with hexagonal packing) of another form of ZnS (wurzite). The polar axes of the components are parallel, so this is an axial twin, not a reflection one. Figure 42 shows two other growth errors; in c) we have a reflection-rotation twin, which is not found in nature, while in   d) we have a reflection twin, also unknown.

Aminoff and Broomé showed that there are 14 possible modes of incorrect deposition when ZnS grows on a (111) face, but the only growth twins known are ones in which the transition region has the wurzite structure.

These same workers examined growth errors for gold, silver, and aluminum, which have cubic lattices and centered faces. Only one mode of anomalous deposition on (111) is possible here, and this corresponds to twinning by the spinel law (the components are turned through 180° on [111]). Here the transition region (which is bounded by two rows) has hexagonal close packing, which does not occur in these metals but is found in cobalt.†

---

* See Whitwham and Lacombe (1961) on "coincidence lattice" and Friedel (1926) on "macle par merihedrie reticulaire."

† The structure of β-cobalt is the same as that in gold.

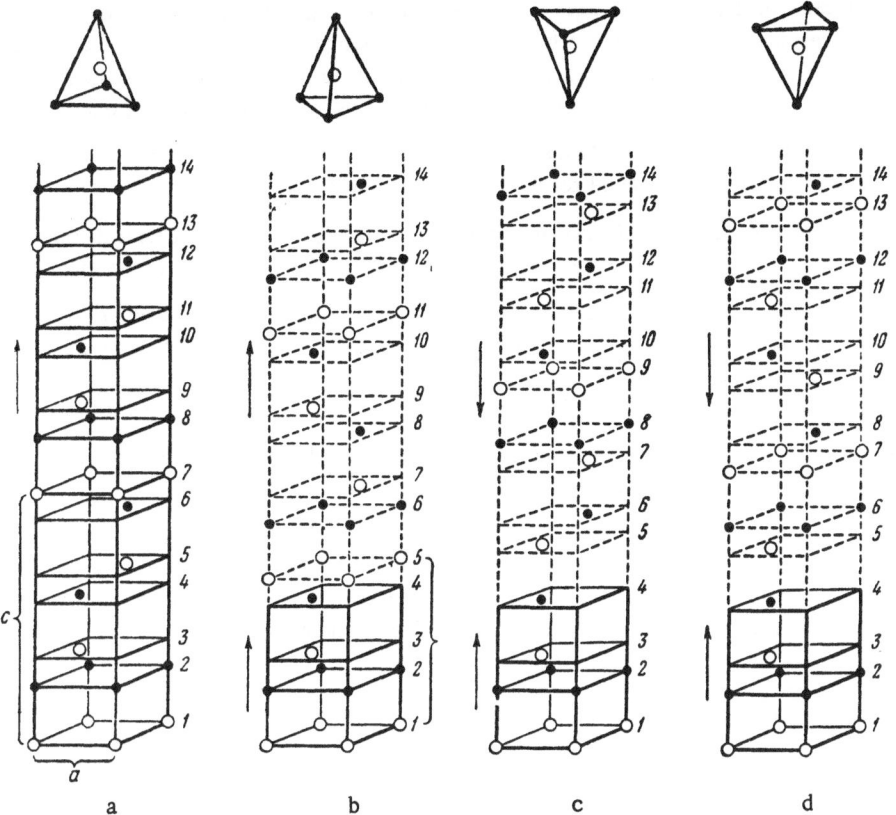

Fig. 42. Lattice of zinc blende, ZnS; the horizontal rows are parallel to (111) planes:
a) normal structure, no twinning, in which a and c are the vectors for translation along
[110] and [111]; b) twin generated by rotation about [111], as found in nature; c) hypothetical twin derived by combined reflection and rotation; d) hypothetical reflection
twin. The arrows denote the direction of the polar axis (Aminoff and Broomé).

There are four types of error that give rise to twins for deposition on (111) in lattices of NaCl type. The
second row adjacent to the interface is slightly distorted, but the region near this boundary has the structure of
the unit cell of NiAs. Although NaCl and PbS (galena) do not exist in this modification, it is known for FeS.
Mechanical twins on numerous faces have been demonstrated for PbS (Seifert, 1928; Table II). All known growth
twins in galena have (111) as their twin (contact) plane, though.

Aminoff and Broomé based their deductions on an analysis of growth conditions, but the results are of
great value to the theory of mechanical twinning, for they show that some structures can give rise to twins
without distortion along the interface.

16. Structure of a Twin Boundary Deviating from a Twin Plane: Formation of Accommodation Bands and Shape of Twin Layers

A twin layer does not always have its boundaries parallel to the twin plane; a layer with plane-parallel
boundaries (a plate) occurs only if the twinning extends throughout the monocrystal. The components of a twin
often have a tapering or scale-type cross section; in the first case, one or both boundaries may not be parallel
to the twin plane (may correspond to the planes with high or irrational indices). In the second case, the boundaries deviate from crystallographic planes nearly everywhere, so the lattices cannot meet coherently. First-
order stresses are bound to occur if the boundaries are incoherent; if these are large, there may be an additional
deformed region on the side representing the initial crystal.

Garber's geometric scheme (1947a, 1947 b) implies that a wedge-shaped twin layer can give rise to elastic distortion or even to residual deformation nearby (Fig. 69). A twin step on the surface of a crystal must inevitably imply transition(reoriented) regions, which correspond to features of the relief such as CD', D'E', or AD, DG. These regions may appear on the coherent side or on the other side, or on both sides at once (Fig. 69).

These regions must be elastic if the twins are elastic; Garber has observed this (see Fig. 68). However, residual transition regions can occur around residual twin components.

Jillson (1950) observed that several rows of bands may occur along the boundaries of mechanical twins in zinc monocrystals cleaved on the basal plane; these could be fairly broad relative to the twin layers (Fig. 43).

These bands lie at smaller angles to the basal plane (they are seen as dark bands in the photograph), so the lattice here must be turned through angles much smaller than those involved in twinning. Moore (1952, 1955) measured the inclination of the surfaces of these relative to the basal cleavage; the angle was 47' ± 2.5' for one band. The angle near the band was 21' (14' farther away) if there were two bands adjacent to the twin.

Jillson called these accommodation bands; he considered that they arose by adaptation of the initial lattice to the lattice of the twin component.

Figure 44 shows a geometrical scheme for the most general type of junction between twin components in zinc for a section normal to the basal plane (Urusovskaya, 1956a and 1956b). Figure 45 shows Moore's photograph of a zinc monocrystal containing a wedge component together with the adjacent transition zone (plane of section normal to basal plane); this confirms Jillson's supposition and also the scheme of Fig. 44.

Fig. 43. Twin layer (dark band) and adjacent accommodation bands (shown by arrows); basal cleavage in zinc (Moore).

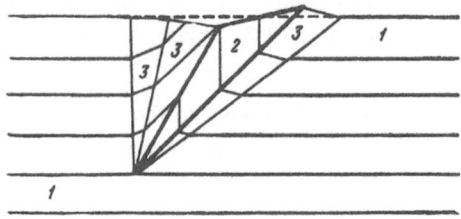

Fig. 44. Junction of twin boundaries with main crystal via accommodation bands in zinc: 1) initial crystal; 2) twin layer; 3) two accommodation bands on incoherent side and one on coherent side. The parallel lines that change direction in the layer and bands represent traces of the basal planes (Urusovskaya).

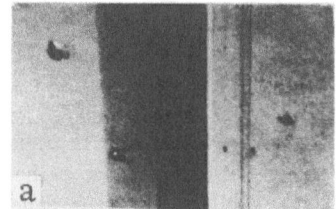

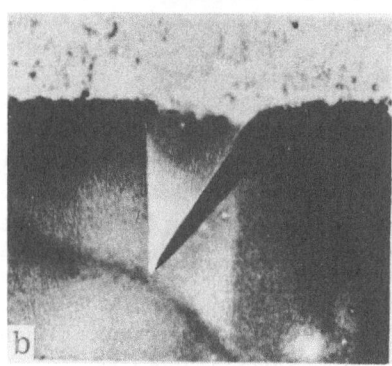

Fig. 45. Accommodation bands in zinc seen in two projections by reflected polarized light: a) on basal cleavage; b) in section normal to basal plane. The dark wedge is the twin component, while the light area is the accommodation region or transition zone (Moore).

Pratt and Pugh (1952) applied Berg and Barrett's x-ray method to mechanical twins in zinc; they found that the transition zones were accompanied by accommodation bands parallel to the boundaries of the twin layers and by analogous transition regions, which were normal or inclined to these boundaries. They also found that the accommodation bands taper off and do not occur at the end of the twin layer if this tapers off.

Jillson found similar bands at grain boundaries in coarse-grained zinc; Fig. 46 shows a set of twin lobes and their associated accommodation zones in Moore's representation, which is for the particular case of accommodation zones formed at a coherent boundary.

It is stated (Churchman et al., 1956; Fig. 28) that twin components in silicon meet on (123) with pronounced disturbance to the short-range order and (apparently) without a transition region. Figure 47 shows the atomic array in a lattice of diamond type for this mode of junction.

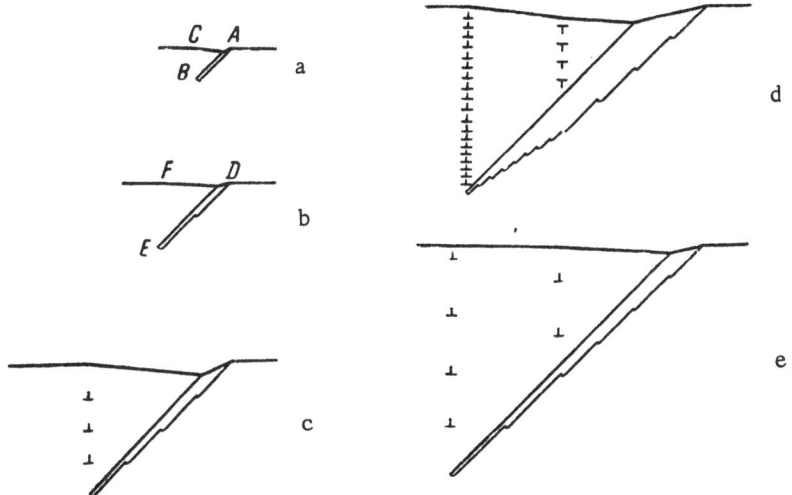

Fig. 46. Development of a twin layer and formation of a transition zone (Moore's scheme): a) initial layer propagating along CB with parallel faces and with a transition zone; b) and c) further development, with an incoherent boundary to the plane-parallel layers, whose length increases by one step at each stage, the length of a step corresponding to several rows of atoms, and the lengthening being accompanied by widening of the transition region; d) and e) two possible types of structure for the boundaries of the transition region. In the first case, the boundaries are formed by rows of edge dislocations differing in sign, while in the second they are formed by ones of the same sign. The relief of the cleaved surface differs in the two cases.

Shape of Layers. Yakovleva and Yakutovich (1939, 1940) examined the residual twins produced in zinc monocrystals by pressure on the basal plane (the plane of slip and cleavage). Thin twin layers (rays) radiated from the point of contact with the steel sphere; these lay in the (10$\bar{1}$2) planes. A ray usually consisted of a set of narrow twin layers, which were examined by grinding and etching the plate. Figures 48 and 49 show some results; the twins take the form of petals lying in the twin planes and extending along the twin directions. The thickness normal to the twin plane was very small.

The pronounced differences in size in the various directions are to be ascribed to the additional stresses set up in the adjacent material by the growth of the rays. "In areas in which these stresses have the same sign as the external ones, they aid conversion; and where they are in opposition, they oppose extension" (Yakovleva and Yakutovich, 1939).

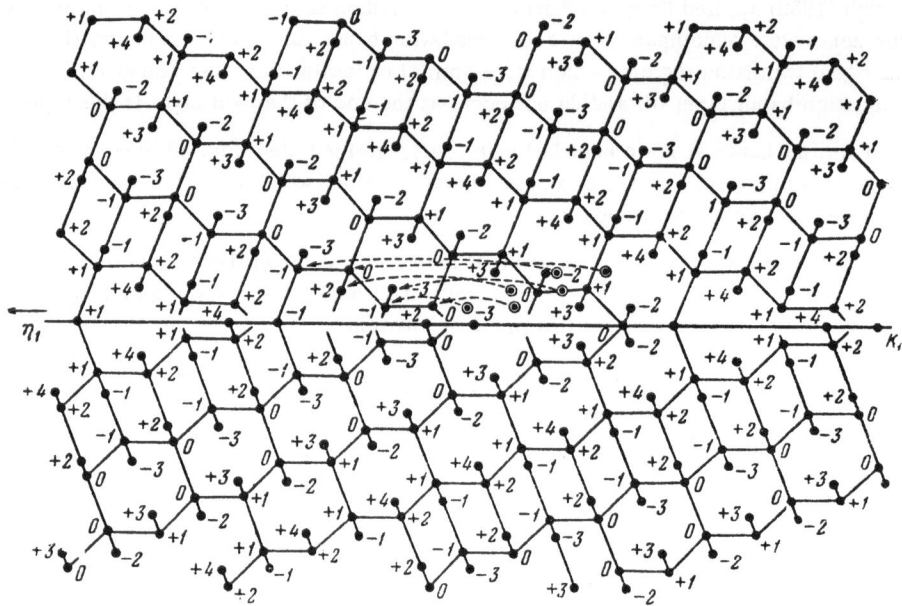

Fig. 47. Disposition of atoms in a lattice of diamond type upon twinning on $K_1 = (123)$ along $[\overline{4}1\overline{2}]$; projection on $S = (1\overline{2}1)$. See Fig. 41 for the case of a coherent twin on (111) for this lattice (Churchman, Geach, and Winton).

Figure 49a shows sections of a twin ray; the twin layers within the crystal are often lens-shaped (Fig. 49b). It may be supposed that accommodation bands join these twin layers to the initial lattice. The sizes of these rays are dependent on the loading rate; dynamic loading gives longer rays.

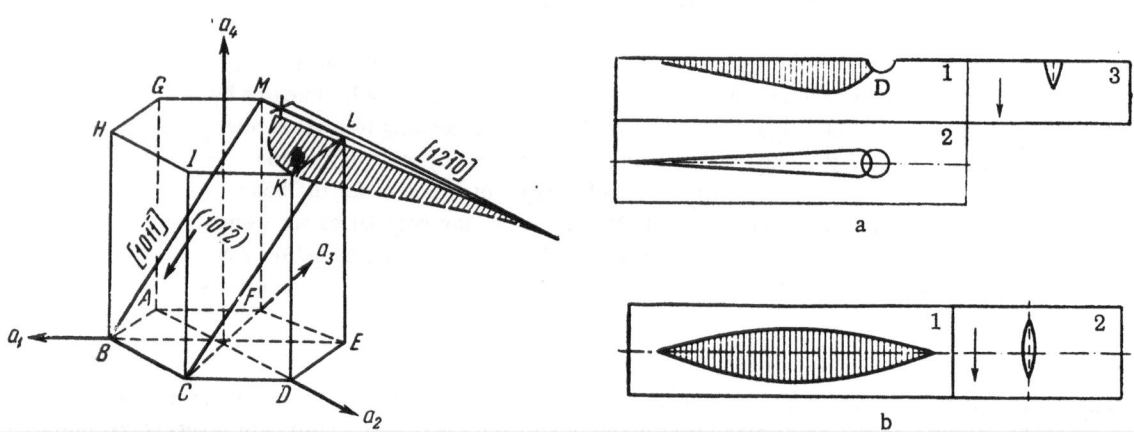

Fig. 48. Disposition of a twin ray with respect to the crystallographic planes in zinc. The heavy lines show the $(10\overline{1}2)$ twin plane; the arrow denotes the $[10\overline{1}\overline{1}]$ direction (that of twinning) in this plane. Twinning propagates along $[10\overline{1}\overline{1}]$ and at right angles to this (Yakutovich and Yakovleva).

Fig. 49. a) Form of twin ray produced by pressure of a steel sphere on the basal plane in zinc (D denotes the dent): 1) section by the twin plane; 2) section by the $(1\overline{2}10)$ plane; 3) section by the plane normal to $(10\overline{1}2)$ and $(1\overline{2}10)$. The arrow denotes the $[10\overline{1}\overline{1}]$ (twinning) direction. b): 1) shape of twin ray within crystal; 2) section by the plane of Fig. 49 (Yakutovich and Yakovleva).

Layers on several systems of intersecting planes can be produced; the state of stress is decisive here. Three such planes occur in bismuth, calcite, and antimony, six in cadmium and zinc, and so on. A channel of rhombic form is produced if two twin layers meet (Fig. 50), as in calcite. In addition, at the points of intersection we find also Rose channels, which may run throughout the crystal and cause what is called the optical anomaly of calcite.

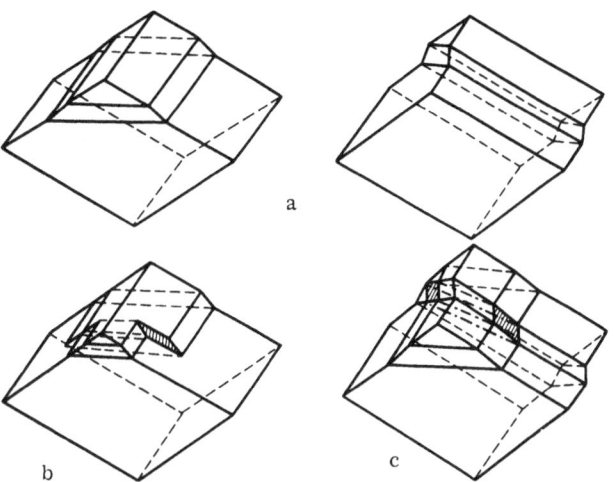

a

b   c

Fig. 50. Formation of Rose channels in calcite at the intersection of two twin layers: a) twin layers parallel to different edges of cleaved crystal; b) hollow channels of the
first kind at region of encounter; c) hollow channels of the
second kind formed by intersection of two layers. The points
of emergence of the channels are shown hatched (Rose).

Networks of very thin channels of this type increase the volume of the crystal. There are 12 possible {112} twin planes in iron and other body-centered cubic metals. The volume of iron increases a fraction of a percent on twinning (Czochralski, 1924). Smith et al. (1928) have described the intersection of twins in meteoritic iron; Pratt (1953) has observed the effect in zinc. Cahn (1953a) made a detailed study of the geometry of twin intersection for α-uranium, and also for zinc monocrystals (Bell and Cahn, 1958).

Figure 51a (Cahn) shows the intersection of twin layers; Cahn proposed the term crossing twin for primary layer A and crossed twin for primary layer B, C being the secondary twin layer. Cahn demonstrated the following rules.

1. The traces of A and C on the plane of B either coincide or are parallel.

2. The magnitude, direction, and sign of the shear during twinning must be the same in A and C; if the first condition is not obeyed, layers A and C are not conjugate; and if the second is not, the displacement caused by the twinning is so large that the layer cannot develop. If B is a reflection twin (twin of the first kind), the two rules become one, which is illustrated by the projection of Fig. 51b.

The plane of Fig. 51b is that of K_{1B}; K_1C and K_{1A} may be found by simple reflection in K_{1B}, so their traces are parallel to XY. This corresponds to condition 1. Further, if η_{1A} lies in K_{1B}, then $\eta_{1C} = \eta_{1A}$, for A and C have the same indices and so the shear is the same for A and C. This corresponds to condition 2. The intersection is therefore possible if η_{1A} is parallel to K_{1B}.

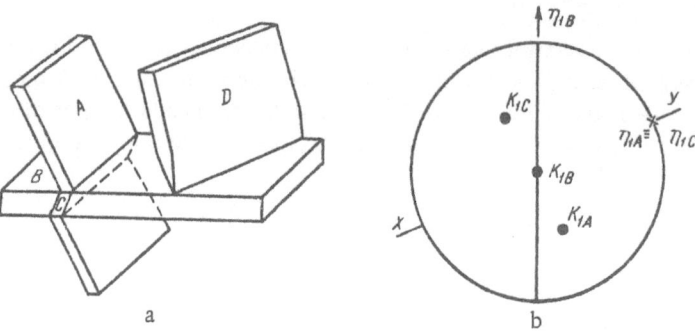

Fig. 51. Intersection of twin layers in α-uranium: a) scheme of
intersection; b) stereographic projection of intersection. The K
are the twin planes and the η are the twinning directions (Cahn).

As K_1 is irrational for twins of the second kind, the relation does not apply, so the two conditions given above should be used. Twins in uranium (§ 1.14) having $K_1 \approx (1\bar{7}2)$ may intersect (130) twins or other twins of the $(1\bar{7}2)$ type. Cahn used these conditions as a convenient means of deducing the twin indices.

It is usual to find that A is blocked when it meets B if one of the above conditions is not obeyed; Fig. 51a shows this case (for D), and Fig. 52 shows the effect in tin.

No intersection should occur in tin, because $\eta_1 = [\bar{1}03]$ is parallel only to the $K_1 = (301)$ plane; the same applies to hexagonal metals, because $\eta_1 = [10\bar{1}\bar{1}]$ is parallel to $K_1 = (10\bar{1}2)$ only. In α-iron, though, $\eta_1 = [11\bar{1}]$ is parallel to $K_1 = (112)$, $(\bar{1}21)$, and $(2\bar{1}\bar{1})$, so intersection is possible.

Localized stresses often occur at the end of a blocked twin; these can give rise to a similar twin on the other side without the formation of a secondary twin (§ 2.20). Twin A may also cause localized slip on the other side of the blocking twin (Hall, 1954).

18. Prediction of Twin Elements for Metal Crystals[*]

Twins in crystals are described geometrically in terms of K_1, K_2, η_1, and η_2 (§ 1.3). A method enabling one to predict these has been developed; this is based on the geometrical and physical features of the crystals (Jaswon, 1956; Jaswon and Dove, 1957). The method was designed for twins having K_1 rational (reflection twins). The test for elements active in mechanical twinning is the magnitude of the displacement, not the energy of distortion at the interface between the initial and twinned parts.

$\lfloor$ 100μ $\rfloor$

Fig. 52. Intersection of twin layers in tin (Hall).

[*]This section was compiled by A.A. Urusovskaya.

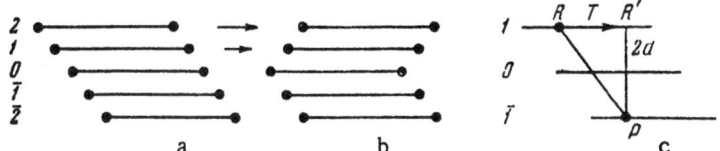

Fig. 53. Twinning of a simple lattice by simple shear: a) initial state; b) twin; c) relation between shear T and the parameters $\mathscr{E}$ and d for simple lattices.

The twinning in a crystal with a simple lattice* (body- or face-centered cubic, tetragonal, or trigonal) can be represented as the result of uniform shear of the atomic planes parallel to K_1 along η_1 (Fig. 53). This scheme is correct for the process on any scale. The diagram shows the traces of the mutually parallel lattice planes denoted by . . ., $\bar{2}, \bar{1}, 0, 1, 2, \ldots$. Plane K_1 coincides with the 0 plane. We now displace plane 1 parallel to itself through a distance such that it becomes a mirror image of plane $\bar{1}$ in plane 0. Jaswon and Dove suppose that the best mode of displacement is that in which the movement of the atoms is minimal. The translation vector T for this is the least of all possible vectors giving rise to the twinned state. The direction of T is that of η_1, while the magnitude is given by $s = T/d$, in which d is the distance between adjacent lattice planes.

We may (Fig. 53c) express T in terms of $\mathscr{E}$ and d: $T = \mathscr{E} - 2\,d$; $T^2 = \mathscr{E}^2 - 4\,d^2$.

Then for some hypothetical shear we have

$$G^2 = \frac{T^2}{d^2} = \frac{\mathscr{E}^2 - 4d^2}{d^2},$$

$$G^2 = \frac{\mathscr{E}^2}{d^2} - 4. \tag{1}$$

This equation is solved for $1/d^2$:

$$\frac{1}{d^2} = \frac{G^2 + 4}{\mathscr{E}^2}.$$

We replace $\mathscr{E}$ by the least vector E to get

$$\frac{1}{d^2} \leqslant \frac{G^2 + 4}{E^2}. \tag{2}$$

This inequality restricts the indices of K_1 and K_2 for a simple lattice.

Twinning on the microscopic scale cannot be treated as a process of simple shear for atomic planes if the lattice is more complex; here displacements additional to uniform ones along η_1 parallel to K_1 are needed (Fig. 11), and these give rise to a twin without altering the macroscopic form of the crystal (§ 1.9).

Jaswon and Dove considered any complex lattice as a combination of two simple lattices, one displaced relative to the other. Here pairs of planes are considered instead of single planes. Let the sequence of parallel planes in one lattice be . . ., $\bar{2}a, \bar{1}a, 0a, 1a, 2a, \ldots$, and let the other sequence be . . ., $\bar{2}b, \bar{1}b, 0b, 1b, 2b, \ldots$; the traces of these planes are shown in Figs. 54a and 55a, in which the broken line denotes the trace of the twin plane D. Planes 0a and 0b, and also D, can be chosen in several ways; those shown in Figs. 54 and 55 enable one to demonstrate the displacements most readily.

---

*By this is meant a lattice of simple type that can twin by simple shear. The term is sometimes used with the meaning of the primitive lattice in crystallography.

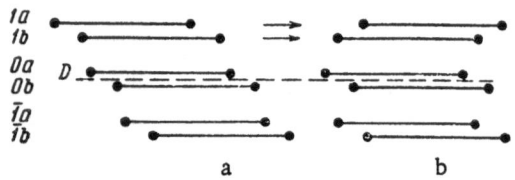

Fig. 54. Twinning of a complex lattice by the X
mechanism: a) initial state; b) twin.

The vector of semihomogeneous shear is used to describe the twin shear in complex lattices. This vector has homogeneous and inhomogeneous components; the first coincides in magnitude and direction with the macroscopic twin shear, while the second causes no macroscopic change but is necessary to produce the twin array in the deformed part. This latter may be parallel or perpendicular to the twin plane; the mechanisms are respectively denoted by X and Y. The component is

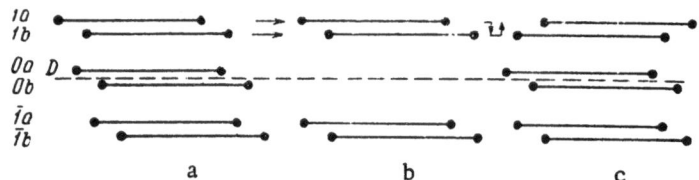

Fig. 55. Twinning of a complex lattice by the Y mechanism: a)
initial state; b) and c) twin state.

chosen to be the least of all possible displacements giving rise to the twin; if it is parallel to the twin plane, it does not exceed the interatomic distance in the most closely packed row of atoms in the twin plane. If it is perpendicular to this plane, it does not exceed the distance between adjacent a and b planes.

The X mechanism brings plane 1a above plane $\bar{1}$b, plane 2a above $\bar{2}$b, and so on. Planes 1a and 1b are displaced by T (homogeneous component) and are relatively displaced by t (inhomogeneous component), which makes them mirror images of $\bar{1}$b and $\bar{1}$a, respectively (Fig. 55b). Vector T can be used to determine the macroscopic shear s = T/ d.

The Y mechanism involves a common displacement T' (homogeneous component) of planes 1a and 1b; then these two exchange planes by displacement perpendicular to the twin plane, the displacement being the inhomogeneous component (Fig. 55, b and c). The more probable of the two mechanisms can be established by comparing the possible T and T'; the smallest of these determines the mechanism. Energy considerations must be employed if T and T' are nearly equal.

Plane 2a comes above $\bar{2}$a and 2b above $\bar{2}$b if the displacements are $2T$ and $2T'$; the hypothetical shear is described by the vector Q (Fig. 56):

$$G = \frac{Q}{2d}.$$

This Q can be expressed in terms of the lattice vectors F and d:

$$-Q = F - 4d; \quad Q^2 = F^2 - 16d^2.$$

Then

$$G^2 = \frac{F^2}{4d^2} - 4. \tag{3}$$

We replace F by E; then we have

$$\frac{1}{d^2} \leqslant \frac{4(G^2 + 4)}{E^2}. \tag{4}$$

We are now in a position to predict the twin elements.

<u>Determination of K_1.</u> We use (2) or (4) to determine the indices of possible twin planes; first of all, we determine the right-hand side of the inequality, in which G is equated to one, on the assumption that the angular displacement does not exceed 45°. Vector E (and hence $E^2 = \sqrt{E_x^2 + E_y^2 + E_z^2}$) is determined directly by examination of the structure. Once the right-hand side has been determined, we express $1/d^2$ in terms of hk$l$. Table 2 gives the corresponding quadratic forms. The right-hand part of the form must be multiplied by four for face-centered cubic and tetragonal lattices if the hk$l$ differ (some even and some odd, as in 112 and 100); the same must be done for body-centered cubic lattices if the sum of the hk$l$ is odd (as for hk$l$ = 111, sum 3). This need to multiply by four is a result of the additional atomic planes in centered cubic and tetragonal lattices; the interplanar distances are thereby halved. The quadratic forms are used to choose hk$l$ satisfying (2) and (4).

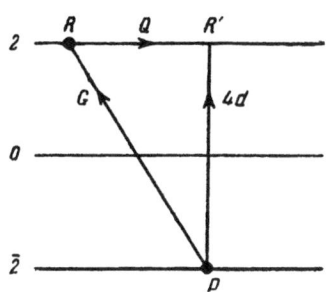

Fig. 56. Relation between Q and the parameters d and G of complex lattices.

<u>Magnitude and Direction of Twin Shear.</u> The direction of η_1 is parallel to the homogeneous component of T or T' (η_1 is parallel to T in a simple lattice); s = T/ d. The direction and magnitude of T are deduced from the projection of the atoms on the twin plane; the atomic planes parallel to the twin plane and lying immediately above and below it are the ones required. The projection of some one atom in the upper plane is displaced to bring it into coincidence with that of one in the lower plane, the smallest displacements being sought. The difference between the coordinates of atoms in a mutual-twin position gives the components of T (η_1); $1/d_{hkl}$ is calculated from the quadratic form (Table 2).

TABLE 2

Lattice	Quadratic Form
Cubic	$\dfrac{1}{d^2} = \dfrac{1}{a^2} (h^2 + k^2 + l^2)$
Tetragonal	$\dfrac{1}{d^2} = \dfrac{1}{a^2}\left[h^2 + k^2 + \left(\dfrac{a}{c}\right)^2 l^2 \right]$
Hexagonal	$\dfrac{1}{d^2} = \dfrac{1}{a^2}\left[\dfrac{4}{3}(h^2 + hk + k^2) + \left(\dfrac{a}{c}\right)^2 l^2 \right]$
Trigonal	$\dfrac{1}{d^2} = \dfrac{1}{a^2} \dfrac{\cos^2 \frac{\alpha}{2}}{\sin \frac{\alpha}{2}\cdot\sin\frac{3\alpha}{2}} - \left[(h^2+k^2+l^2) - \left(1-\tan^2\frac{\alpha}{2}\right)(hk+kl+hl) \right]$
Orthorhombic	$\dfrac{1}{d^2} = \dfrac{h^2}{a^2} + \dfrac{k^2}{b^2} + \dfrac{l^2}{c^2}$

Note: a, b, and c are the lattice parameters.

The direction and magnitude of the shear can be determined in another way. The direction is given by

$$T = e - 2d \quad \text{for simple lattices,}$$
$$-Q = F - 4d \quad \text{for complex lattices.}$$

Vectors e and F are determined geometrically as the least vectors between nodes lying in parallel hk$l$ planes; vector d can also be determined geometrically from the structure, but we may also use the analytic expression

$$d = d^2 [uvw].$$

Here u, v, and w are found from equations in terms of h, k, l , and the normal to the hk$l$ plane:

$$\frac{a^2}{h} u + \left(-\frac{b^2}{k} w\right) + \left(-\frac{bc}{k} w \cos \alpha + \frac{ac}{h} w \cos \beta + \frac{ab}{h} v \cos \gamma\right) + \left(-\frac{ab}{k} u \cos \gamma\right) = 0 ,$$

$$-\frac{b^2}{k} v + \frac{c^2}{l} w + \left(-\frac{bc}{k} w \cos \alpha\right) + \frac{bc}{l} v \cos \alpha + \frac{ac}{l} u \cos \gamma + \left(-\frac{ab}{k} u \cos \gamma\right) = 0 .$$

Here a, b, and c are the lattice parameters, while α, β, and δ are the angles between the unit vectors a, b, and c. The expression for d simplifies for cubic lattices to

$$d_{hkl} = d^2 \, [hkl].$$

The components of vectors $e(\boldsymbol{F})$ and d enable us to find s as

$$s = \frac{\sqrt{T_x^2 + T_y^2 + T_z^2}}{\sqrt{d_x^2 + d_y^2 + d_z^2}}$$

or from (1) and (3).

As regards K_2 and η_2 we may say as follows: we assume that several sets of K_1 and η_1 have been found as above; two of these give the same s, so one is taken as K_1, η_1 and the other as K_2, η_2. This method for K_1, K_2, η_1, and η_2 is applicable to crystals having conjugate twin systems (§ 1.3); the K_2 and η_2 corresponding to K_1 and η_1 are found geometrically from the definitions of K_2 and η_2 (§ 1.3) for other crystals.

19. Examples of Prediction of Twin Elements

Body-Centered Cubic Lattice. This is a simple type; the indices of K_1 and K_2 are found from (2):

$$\frac{1}{d^2} \leqslant \frac{G^2 + 4}{E^2} .$$

For body-centered cubic lattices

$$\boldsymbol{E} = \left[\frac{1}{2} \frac{1}{2} \frac{1}{2}\right], \quad E^2 = \frac{3}{4} a^2 .$$

We substitute $E^2 = \frac{3}{4} a^2$, G = 1, to get

$$\frac{a^2}{d^2} \leqslant \frac{5 \times 4}{3} \approx 6 .$$

The quadratic form is

$$\frac{a^2}{d^2} = h^2 + k^2 + l^2 \qquad \text{even } hkl \text{ ,}$$

$$\frac{a^2}{d^2} = 4\,(h^2 + k^2 + l^2) \qquad \text{odd } hkl \text{ ,}$$

or

$$h^2 + k^2 + l^2 \leqslant 6 \qquad \text{even } hkl \text{ ,}$$
$$h^2 + k^2 + l^2 \leqslant 1 \qquad \text{odd } hkl \text{ .}$$

The possible twin planes are then (100), (110), and (112).

Planes (100) and (110) are symmetry planes in these lattices, so they cannot be twin planes; therefore K_1 must be (112) and K_2 must be $(11\bar{2})$.

Now we determine s and the indices of η_1; we project the nodes on the (112) plane for this purpose. Figure 57 shows the projection for one of these planes and also the projections for two nodes, one of which lies in the plane directly above the plane in question and the other of which lies in the next (112) plane. These three planes are denoted by $\underline{1}$, 0, 1; we displace plane 1 to bring the projection of its nodes into coincidence with that of the nodes of $\bar{1}$, and use the smallest displacement, which here is clearly that to bring node 8 (in projection) into coincidence with the projection of node $\left[\frac{3}{2} \ \frac{3}{2} \ \frac{3}{2}\right]$. This displacement is specified by $\boldsymbol{T}$:

$$\boldsymbol{T} = \left[\frac{3}{2} \ \frac{3}{2} \ \frac{\bar{3}}{2}\right] - \left[\frac{5}{3} \ \frac{5}{3} \ \frac{\bar{5}}{3}\right] = \left[\frac{\bar{1}}{6} \ \frac{\bar{1}}{6} \ \frac{1}{6}\right] \quad ;$$

$$s = \frac{T}{d}; \qquad T = \frac{1}{\sqrt{12}} \; ; \quad d_{112} = \frac{1}{\sqrt{6}} \; ;$$

$$s = \frac{1}{\sqrt{2}} \; ; \quad \eta_1 = [\bar{1}\bar{1}1] \; ; \quad \eta_2 = [111] \; .$$

Face-Centered Cubic Lattice. The indices of K_1 and K_2 care found from

$$\frac{1}{d^2} \leqslant \frac{G^2 + 4}{E^2}.$$

$\boldsymbol{E} = \left[\frac{1}{2} \ \frac{1}{2} \ 0\right]$; $E^2 = \frac{1}{2}$; G is expressed as one.

We replace $1/d^2$ by hkl to get

$$h^2 + k^2 + l^2 \leqslant 2 \qquad hkl \; \text{ not all odd,}$$
$$h^2 + k^2 + l^2 \leqslant 10 \qquad hkl \; \text{ all odd.}$$

These indicate that the only possible result is

$$K_1 = (111); \quad K_2 = (11\bar{1}).$$

The direction of η_1 is found from $\boldsymbol{T} = \boldsymbol{E} - 2\boldsymbol{d}$.

The structure readily shows that the smallest $\boldsymbol{E}$ between nodes of two (111) planes separated by $2\boldsymbol{d}$ is a vector whose components are $\left[\frac{1}{2} \ \frac{1}{2} \ 1\right]$:

$$\boldsymbol{d}_{hkl} = d^2 \, [hkl] \; ; \quad d_{111} = \frac{1}{3} \, [111] \; ;$$

$$T = \left[\frac{1}{2}\ \frac{1}{2}\ 1\right] - \left[\frac{2}{3}\ \frac{2}{3}\ \frac{2}{3}\right] = \left[\frac{1}{6}\ \frac{1}{6}\ \frac{1}{3}\right].$$

Then $\eta_1 = [11\bar{2}]$; $\eta_2 = [112]$.

From (1)

$$s^2 = \frac{E^2}{d^2} - 4 = \frac{6-3}{4} - 4 = \frac{1}{2};\quad s = \frac{1}{\sqrt{2}}.$$

Face-centered cubic metals show twins with $K_1 = (111)$; $K_2 = (11\bar{1})$; $\eta_1 = [11\bar{2}]$; $\eta_2 = [112]$, which are formed during growth. Twins have been produced mechanically in these metals only in the case of copper at 4.2°K.

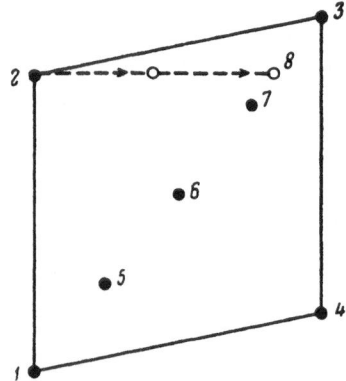

Fig. 57. Projection of the nodes of a body-centered cubic lattice on the (112) plane; the numbers and nodes correspond as follows:

1 - [000]; 2 - [02$\bar{1}$]; 3 - [22$\bar{2}$];

4 - [20$\bar{1}$]; 5 - $\left[\frac{1}{2}\ \frac{1}{2}\ \frac{\bar{1}}{2}\right]$; 6 - [111];

7 - $\left[\frac{3}{2}\ \frac{3}{2}\ \frac{\bar{3}}{2}\right]$; 8 - $\left[\frac{5}{3}\ \frac{5}{3}\ \frac{\bar{5}}{3}\right]$.

<u>Tetragonal Face-Centered Lattice.</u> This is the only case to be considered, for a body-centered tetragonal lattice can be treated as a face-centered one by suitable choice of the elements of the unit cell. Twinning on the microscopic scale can be represented here as homogeneous shear, so the indices of K_1 and K_2 are found from (2):

$$E = \left[\frac{1}{2}\ \frac{1}{2}\ 0\right];\quad E^2 = \frac{1}{2};$$

G is expressed as one.

The quadratic forms are

$$\frac{1}{d^2} = \frac{4}{a^2}\left[h^2 + k^2 + \left(\frac{a}{c}\right)^2 l^2\right] \quad hkl\ \text{not all odd},$$
$$\frac{1}{d^2} = \frac{1}{a^2}\left[h^2 + k^2 + \left(\frac{a}{c}\right)^2 l^2\right] \quad hkl\ \text{all odd}.$$

With $c/a = q$ we have

$$h^2 + k^2 + \frac{1}{q^2} l^2 \leqslant 2 \quad hkl\ \text{not all odd},$$
$$h^2 + k^2 + \frac{1}{q^2} l^2 \leqslant 10 \quad hkl\ \text{all odd}.$$

If we assume that the shortest translation here is that from a lattice node at a corner of the cell to a node at the center of a face (as is the case for metals), q^2 must lie between 1/3 and 3. Then for crystals with $q = 1$ (e.g., indium) we have (101) as a possible twin plane. Figure 58 illustrates the calculation of the shear along this plane; the plane is a (010) prism plane, and there are shown the traces of three (101) planes, which are denoted by $\bar{1}$, 0, 1. We project nodes B and C on plane 0 to get B' and C'. To form a twin with plane 0 as its symmetry plane we must displace plane 1 parallel to plane 0 through a distance AB such that B' coincides with C'; then the coordinates of B are

$$\frac{a}{2} \cdot \frac{q^2 - 1}{1 + q},\ 0,\ \frac{aq}{2} \cdot \frac{3 + q^2}{1 + q^2};$$

and those of C' are

$$-\frac{a}{2}\cdot\frac{q^2-1}{1+q},\ 0,\ \frac{aq}{2}\cdot\frac{1+3q^2}{1+q^2}.$$

$$\mathbf{T} = \mathbf{AB} = \left[\frac{a(q^2-1)}{(1+q^2)},\ 0,\ \frac{aq(1-q^2)}{1+q^2}\right];\quad \eta_1 = [ao\bar{a}];\quad \eta_1 = [10\bar{1}];$$

$$s = \frac{T}{a} = \frac{OC\cos\alpha}{OC\sin\alpha}\cdot 2 = 2\cot\ \alpha.$$

But α is related to β, and $\cot\beta = q$, so

$$\cot\alpha = \cot 2\beta = \frac{1}{2}(\cot\beta - \tan\beta) = \frac{1}{2}\frac{q^2-1}{q}\ ;\ s = \frac{q^2-1}{q}\ .$$

(The expression for s implies that $K_1 = (101)$, $\eta_1 = [10\bar{1}]$ is impossible for face-centered cubic crystals, for which $q = 1$.)

We take $(10\bar{1})$ as K_2, with $\eta_2 = [101]$.

Now we consider an example of a complex lattice.

Diamond Lattice. This consists of two face-centered cubic lattices relatively displaced by 1/4 of a body diagonal; we use (4) to determine the twin plane:

$$\frac{1}{d^2} \leqslant \frac{4(G^2+4)}{E^2}.$$

$$E = \left[\frac{1}{2}\ \frac{1}{2}\ 0\right];\quad E^2 = \frac{1}{2}a^2.$$

We put $E^2 = a^2/2$ and $G = 1$ in the inequality and use

$$\frac{1}{d^2} = \frac{1}{a^2}(h^2+k^2+l^2)\quad hkl\ \text{all odd},$$

$$\frac{1}{d^2} = \frac{4}{a^2}(h^2+k^2+l^2)\quad hkl\ \text{some odd, some even}.$$

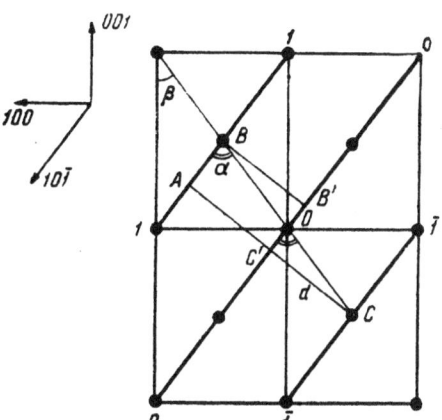

Fig. 58. Disposition of the nodes in a (101) plane of a face-centered tetragonal lattice.

Then

$$h^2+k^2+l^2\leqslant 10\quad hkl\ \text{all odd},$$

$$h^2+k^2+l^2\leqslant 2\quad hkl\ \text{some odd, some even}.$$

Table 3 lists the indices of possible K_1 and K_2 planes so selected; it also gives the F, η, and G for each plane.

These results indicate that the most probable elements are $K_1 = (111)$, $K_2 = (113)$, $\eta_1 = [\bar{1}\bar{1}2]$, $N_2 = [332]$, for the shear is then least. The X mechanism can occur in diamond lattices, the contact between the conjugate lattices being an exact geometrical plane, so there is no transition layer.

Mechanical twinning has been observed (Churchman et al., 1956) on $\{123\}$ planes in some crystals with a diamond structure (§ 1.13); here we must have the very high value $G = \sqrt{3}$. Tables 3 and 4 given by Jaswon and Dove (1957) do not list these twins, for they considered that twinning was improbable for $G > 1$. Bullough (see part III) discussed the twinning of diamond lattices on $\{123\}$ in 1957.

Finally, Table 4 gives the shear and twin elements for some metals; these have been derived in the following way (see p. 46).

TABLE 3

Possible Twin Elements in Crystals with a Diamond Lattice

hkl	$h^2 + k^2 + l^2$	F	η	G	hkl	$h^2 + k^2 + l^2$	F	η	G
111	8	$\frac{3}{2}, \frac{3}{2}, 1$	[332]	$\frac{1}{2\sqrt{2}}$	100	1	2, 0, 0	[100]	0
113	11	$\frac{1}{2}, \frac{1}{2}, 1$	[112]	$\frac{1}{2\sqrt{2}}$	110	2	1, 1, 0	[110]	0
133	19	$1, \frac{1}{2}, \frac{1}{2}$	[211]	$\frac{5}{2\sqrt{2}}$	120	5	0, 1, 0	[010]	1
115	27	$1, \frac{1}{2}, \frac{1}{2}$	[211]	$\frac{7}{2\sqrt{2}}$	112	6	0, 0, 1	[001]	$\sqrt{2}$
135	35	$0, \frac{1}{2}, \frac{1}{2}$	[011]	$\frac{\sqrt{3}}{2\sqrt{2}}$	122	9	$0, \frac{1}{2}, \frac{1}{2}$	[011]	$\frac{1}{\sqrt{2}}$

TABLE 4

Twin Elements for Metals, Calculated by Jawson and Dove's Method

Metal	K_1	K_2	$[\eta_1]$	$[\eta_2]$	s	Unit Cell	Mechanism
Bismuth	(110)	$(00\bar{1})$	$[00\bar{1}]$	[110]	$\dfrac{2\sqrt{2w}}{\sqrt{(1+2w)}\,\sqrt{(1-w)}}$	Face-centered trigonal, $w = \cos\alpha$	Y
Mercury	(110)	$(00\bar{1})$	$[00\bar{1}]$	[110]	$\dfrac{2\sqrt{2w}}{\sqrt{(1+2w)}\,\sqrt{(1-w)}}$	Face-centered trigonal, $w = \cos\alpha$	Y
Close-packed hexagonal	$(10\bar{1}2)$	$(10\bar{1}\bar{2})$	$[\bar{1}011]$	$[10\bar{1}1]$	$\dfrac{q^2-3}{\sqrt{3}q}$	Hexagonal, $q = \dfrac{c}{a}$	X
Diamond lattice	(111)	$(11\bar{3})$	$[\bar{1}\bar{1}2]$	[332]	$\dfrac{1}{2\sqrt{2}}$	Face-centered cubic, $q = 1$	X
Face-centered cubic	(111)	$(11\bar{1})$	$[11\bar{2}]$	[112]	$\dfrac{1}{\sqrt{2}}$	Face-centered cubic, $q = 1$	
β-tin	(331)	$(\bar{1}\bar{1}1)$	$[\bar{1}\bar{1}6]$	[112]	$\dfrac{1-6q^2}{2\sqrt{2}q}$	Face-centered tetragonal, $q = 0.383$	X
Indium	(101)	$(10\bar{1})$	$[10\bar{1}]$	[101]	$\dfrac{q^2-1}{q}$	Face-centered tetragonal, $q = 1.078$	X
Body-centered cubic	(112)	$(11\bar{2})$	$[\bar{1}\bar{1}1]$	[111]	$\dfrac{1}{\sqrt{2}}$	Body-centered cubic	X
α-uranium	(111)	$(1\bar{7}6)$	$[12\bar{3}]$	$[5\bar{1}\bar{2}]$	0.214	Orthorhombic	Y
	(112)	$(1\bar{7}2)$	$[3\bar{7}2]$	[312]	0.227		X
	(121)	(111,4)	[100]	[132]	0.286		X
	(130)	$(1\bar{1}0)$	$[3\bar{1}0]$	[110]	0.298		Y

46

§ 2. Production and Evolution of Twins in Response to Mechanical Stress

1. Methods of Producing and Detecting Mechanical Twins

It is simplest to produce mechanical twinning in soft metals by the use of a localized load (blow or pressure from a steel point or ball, pressure from a blunt knife-edge). Heating followed by rapid cooling is best for hard crystals.

It is common practice in mineralogy to produce twinning by electrical breakdown or by the application of a hot point. Tension, compression, or bending can also be used, but the desired result can be ensured only if the orientation is selected to prevent or hinder deformation by slip. Low temperatures are useful also, for they reduce the plasticity.

Pressing while embedded in a powder may be used with brittle materials or ones difficult to twin; pressures of hundreds of atmospheres may be used.

Twinning with change of form is readily observed in opaque crystals by examination in reflection; twinning in transparent optically anisotropic crystals is easily detected by the unaided eye from the reflection at the twin planes. If the optical anisotropy is low, or if the rotation of the lattice is slight (as for Rochelle salt), the reflection is very weak. The twins in Rochelle salt are seen well by polarized light, because the components differ in extinction position (§ 1.14).

It is more difficult to observe twinning without change of form; for example, Dauphiné twins in quartz are not optically detectable at all (§ 4), and etching is required to reveal them. Etch or impact figures are useful here; in addition, twinning is always detectable by x-ray methods, for the twins raise the symmetry of the Laue pattern on account of their symmetry plane (reflection twins) or higher-order symmetry axis (rotation twins).

2. Formation of Mechanical Twins in Calcite: Polysynthetic Twins, Detwinning, Parting Planes, Channels

Calcite is orthorhombic; it is still the sole material for Nicol (polarizing) prisms. The birefringence of calcite was examined by Bartholin in 1670; in 1678 Huygens observed the presence of thin twin platelets, which are readily visible in reflection and often have interference colors. Reusch (1867) showed that compression could produce these in a perfectly homogeneous crystal. Mechanical twinning in calcite was afterwards examined by Baumhauer (1879), Mügge (1883), Vernadskii (1897), and Garber (1938-1953).

Baumhauer's method is the simplest to use; a cleaved rhomb is supported by an obtuse edge in wood or plaster, and the opposite edge is indented with the slightly blunt edge of a knife (Fig. 59a). The knife displaces part of the crystal into the twin position (Fig. 59b); the size of the resulting twin is dependent on the distance from the corner through which the optical axis passes (Fig. 59c). The knife should be located a few millimeters from this. The depth of insertion is dependent on the size and quality of the crystal; the more perfect it is, the larger the region that can be converted to the twin orientation. Considerable force is needed to produce twinning.

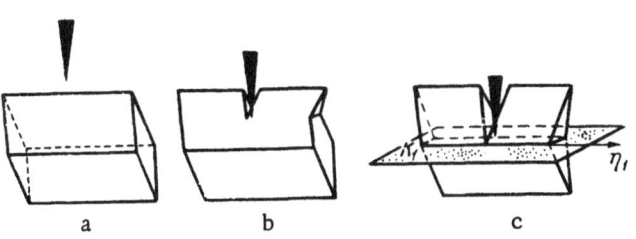

Fig. 59. Mechanical twinning in calcite by Baumhauer's method (Tertsch).

The faces normal to the plane of the knife are unaffected and remain smooth (Fig. 60), but the third face that meets the other two usually becomes covered in lines parallel to the twin plane in the reoriented region. These lines are caused by the layer-by-layer reorientation of the crystal. This layering gives rise to what is called a polysynthetic twin. If great care is taken, it is sometimes possible to produce a twin free from these lines. Large reoriented regions cannot be produced by Baumhauer's method.

Mügge ground two parallel areas normal to the body diagonal* through the acute-angled corners (the diagonal lying in the same plane as the optic axis). Compression (Fig. 61) produced reorientation without the production of layers (polysynthetic twins).

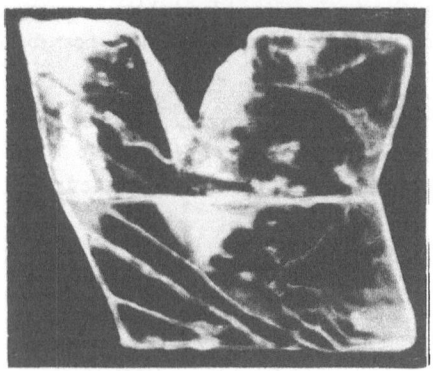

Fig. 60. Calcite crystal twinned by the knife method (Tertsch).

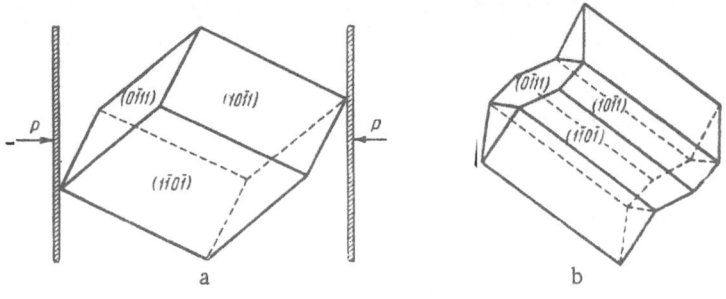

Fig. 61. Mechanical twinning of calcite by Mügge's method: a) force applied along a body diagonal (this sometimes produces twinning throughout the volume simultaneously); b) crystal twinned in upper and lower parts, with middle part unchanged (Mügge).

Twinning throughout the volume interchanges the body diagonal (force direction) and the optic axis (other body diagonal); the crystal shortens along the first direction and extends along the other.

Garber cut rectangular prisms having a possible twin plane as their cross section (§ 1.13), one pair of side-faces being normal to η_1 and the other pair normal to S (Fig. 62a). The crystal was cut (wire saw), ground, and polished with great care; the specimen, protected by cardboard inserts, was gripped as in Fig. 62, b and c. The load was applied via a metal prism working into a grooved plate resting on paper, the force P being parallel to the twin plane and along the twin direction.

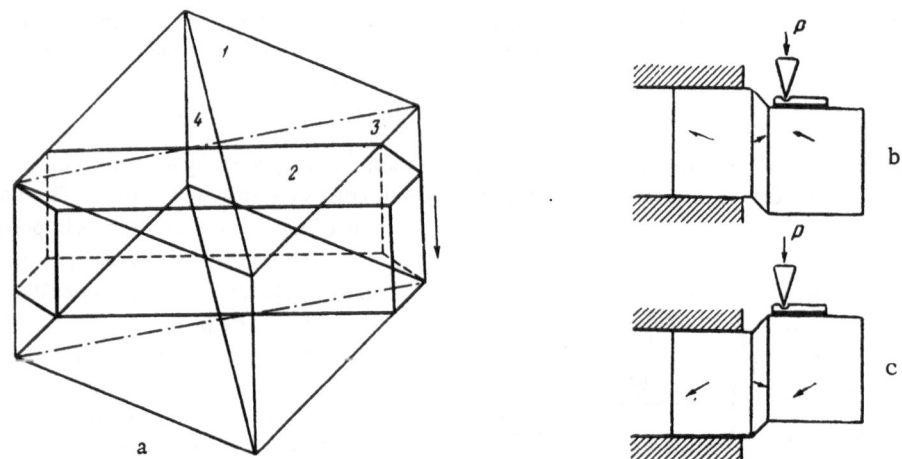

Fig. 62. Mechanical twinning of calcite by Garber's method: a) mode of cutting from cleaved crystal: 1) cleavage rhombohedron; 2) specimen; 3) S plane; 4) K_1 plane (the arrow denotes the direction of η_1); b) mode of twinning, with force acting in the K_1 plane parallel to η_1; c) method of eliminating twin, the arrows indicating the directions of the principal axes (Garber).

---

*Mügge modified an experiment by Reusch, who ground flats perpendicular to the edges of obtuse corners.

The specimens produced by Reusch's, Baumhauer's, Garber's, and Mügge's methods are optically homogeneous.

The Laue patterns (Gogoberidze and Ananiashvili, 1935 and 1936; Gogoberidze, 1936 and 1938) of twin layers are the same as those of undeformed calcite, so the lattice is not disrupted by the twinning operation. Figure 63 shows such patterns, part b being for a region containing a polysynthetic twin. The latter clearly consists of a sequence of plates of a normal calcite crystal separated by twin layers, the lattices of the latter being turned through a fixed angle. A mechanically twinned crystal of quartz similarly shows no change in its Laue pattern (Tsinzerling and Shubnikov, 1933).

A twin structure can be eliminated by the application of a force in the reverse direction, as in Fig. 62c for Garber's method. This is sometimes called detwinning and has often been used for practical purposes. Calcite crystals rejected on account of twin layers can be converted to usable material by compression in a screw press (Shternberg, 1944). Two narrow lead strips are used to localized the loading in the twin layer; one is placed on top on one side of the layer, and the other is placed underneath on the other side. Pressures up to 150 kg/cm^2 are needed to detwin large crystals; the force is normally larger than that needed to produce twinning.

Any attempt to repeat the twinning after the crystal has been twinned and detwinned causes it to cleave along the twin plane. The surfaces are perfect.

These parting surfaces differ from primary cleavage ones on rhombohedron planes; they have been called second cleavage or parting surfaces.

Twinning may cause cracking along the cleavage, as Fig. 64 shows.

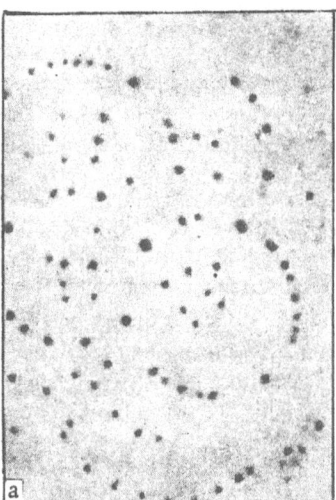 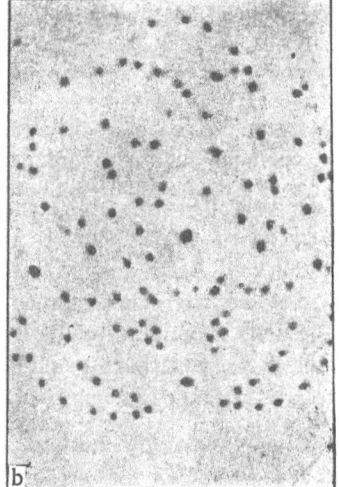

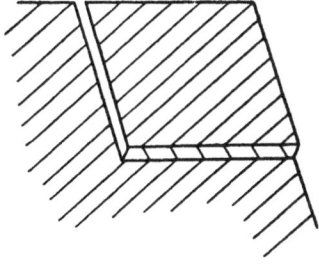

Fig. 64. Crack on cleavage in calcite resulting from production of a twin layer in calcite (Cahn).

Fig. 63. Laue patterns of calcite taken at right angles to the plane of the rhombohedron: a) undeformed crystal; b) deformed crystal with twins (Gogoberidze).

Local surface disturbances (grooves of isosceles-triangle form) are produced on the side surfaces near the line of junction of the two components.

Further, small crystalline fragments (powder) occur along the boundary between the components; the reason for this is that residual stresses occur along the boundaries (§ 1.15 and 16). The distortion is removed by detwinning, for the resulting crystal no longer parts along the twin plane (Gogoberidze, 1952).

Channels first described by Rose (1868) occur where twin layers or twins meet, if the two systems are not parallel. Figure 50 illustrates the origin of these. These channels reduce the density somewhat, whereas the production of polysynthetic twins does not.

3. Elastic Twinning and Wedge-Type Elastic Twins

So far we have considered mechanical twinning as one of the forms of residual deformation. Reusch(1867), Mügge (1889a), Vernadskii (1897), and Johnsen (1907a) remarked on the spontaneous loss of mechanical twinning; Reusch and Vernadskii observed the effect in calcite. Mügge found that the pressure of a needle on monoclinic $BaCl_2 \cdot H_2O$ could produce twins, which were seen as very narrow streaks on the surfaces; these streaks vanished when the force was removed but persisted if the force was such as to cause cracking. Johnsen observed a similar effect in $Ta(NiNa)(UO_2)_3 \cdot (CH_3COO)_9 \cdot 9H_2O$.

Garber (1938) made the first systematic study of this spontaneous reversal. He found that calcite or sodium nitrate would show a wedge-shaped twin layer when pressure was applied from a blunt needle along a twin direction in a possible twin plane. This layer appeared directly under the needle and extended downwards and laterally as the force increased. He believed that the layer did not become thicker (did not develop at right angles to the twin plane), but a more detailed study by Obreimov and Startsev (1958) showed that this is not so, all three dimensions being proportional to the load (§ 2.8). Figure 65 (from Garber's paper) shows such a wedge produced in calcite by pressure from a glass rod. The wedge vanished completely when the force was removed.

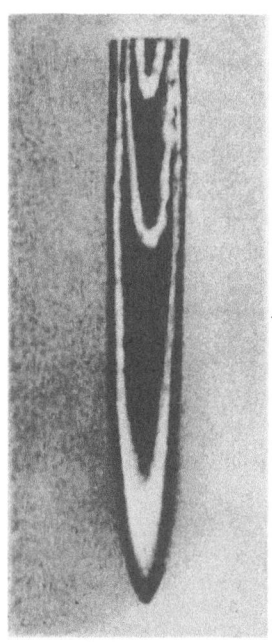

Fig. 65. Elastic twin in calcite seen by polarized light. Lateral illumination (under the microscope) reveals interference fringes, which show that the twin varies in thickness. Room temperature (Garber).

A few cycles of loading and unloading enable one to estimate the force at which the effect becomes irreversible. Garber called the spontaneous reversal "elastic twinning" (the lattice constants change in proportion to the load in ordinary elastic deformation; elastic twinning does not obey Hooke's law, because reversible lattice reorientation occurs). This "elastic twinning" is, in fact, reversible plastic deformation; the process is reversible because the twin layer is ejected by elastic stresses remaining in the crystal, so there is no essential difference between "elastic" and residual twin layers.

Calcite and sodium nitrate sometimes showed wedge twins that only shortened when the load was removed. Garber called this "reducing elastic twins," on the assumption that complete reversal was prevented by microcracks or by plastic deformation (slip) accompanying the indentation under the needle.

Laves and Baskin (1956) found that graphite gives reversible mechanical twinning; Laves (1952, a and b) observed elastic twins in heat-treated albite. Cahn (1954) stated that crystals of uranyl acetate on compression give twins parallel to two crystallographic planes; the twins did not vanish at once when the load was removed, though they did after a few seconds. The resistance to twinning in uranyl acetate is greater than that in albite; a localized load (needle) is needed. Elastic twins thus occur in a number of substances.

Attempts to produce elastic twins in metals were at first unsuccessful (Gindin and Startsev, 1950; Cahn, 1954). Cahn and Startsev supposed that twins in zinc and bismuth are lost by slip accompanying twinning.

Startsev and Kosevich (1955) finally produced elastic twins in antimony, which is less plastic than zinc or bismuth and has a higher melting point. These twins have a general resemblance to those in calcite; Fig. 66 shows photomicrographs of cleaved surfaces, on which are visible intersecting twin systems produced by slight bending of the plate. Parts b and d show the right-hand elastic twin being transformed to a residual twin by increasing loads; the twin does not vanish completely when the load is removed, as in $NaNO_3$ and calcite at high temperatures.

Elastic twins in calcite and sodium nitrate vanish entirely on unloading in the early stages, but they are suddenly transformed to residual twins at a certain load. Antimony behaves as does calcite at first, but the elastic twins are converted to residual ones only gradually. The twins remain partly elastic up to the highest loads; they always contract to a certain extent when the load is removed.

A more careful study for zinc and bismuth showed that their behavior is intermediate between that of calcite and that of antimony (Startsev and Kosevich, 1955). Elastic twinning does not occur; residual twins are produced stepwise, but these layers are always partly elastic (the width is reduced on unloading, but the length rarely alters). Table 5 lists crystals for which elastic twinning has been observed.

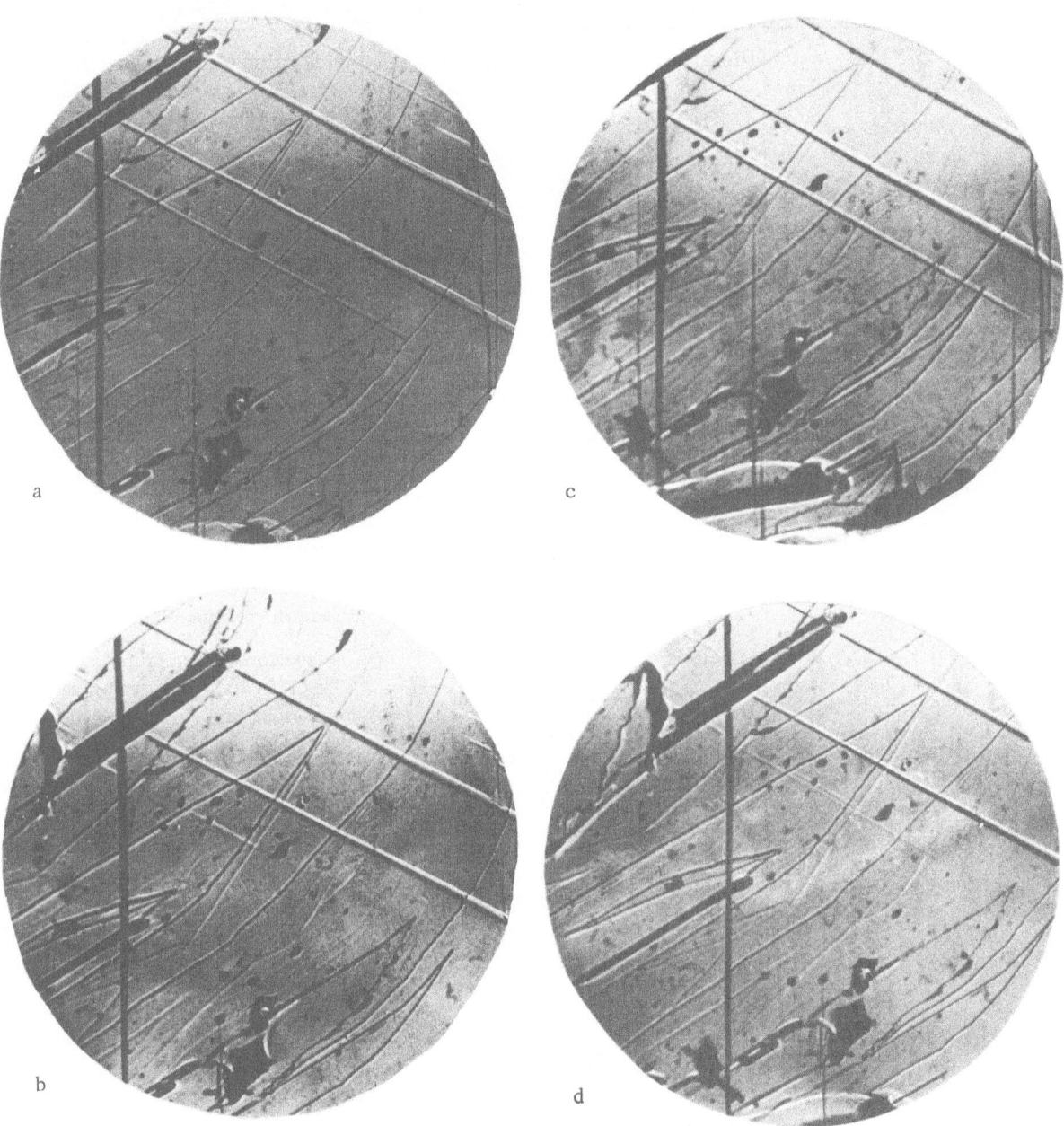

Fig. 66. Elastic twinning in antimony at room temperature: a) crystal under load, face coincident with cleavage plane, elastic twins produced by bending are denoted by arrows; b) load removed, one layer lost, two others reduced; c) second load, somewhat larger, production of larger elastic layers; d) load removed, one layer lost, two others shortened (Startsev et al.).

TABLE 5

Elastic Mechanical Twinning

Crystal	Symmetry	Literature reference	Notes
Zn	Hexagonal	Startsev and Kosevich, 1955	Production of twin layers that vanish on unloading, residual twinning possible
Sb	Trigonal	Startsev and Kosevich, 1955	Twin layers partly lost on unloading
Bi	Trigonal	Startsev and Kosevich, 1955	Twin layers partly lost on unloading
$CaCO_3$	Trigonal	Mügge, 1889a Johnsen, 1907b Vernadskii, 1897 Garber, 1938	Elastic and residual layers possible
$NaNO_3$	Trigonal	Garber, 1940	Elastic and residual layers possible
Graphite	Trigonal	Laves and Baskin, 1950; Platt, 1957	Production of layers, lost on unloading
Albite $Na(AlSi_3O_8)$	Triclinic	Laves, 1952b	Production of layers, which vanish slowly on unloading
$NiNa(UO_2)_3 \cdot (CH_3COO_9) \cdot 9H_2O$	Triclinic	Johnsen, 1907a; Cahn, 1954	Production of twin layers parallel to (110) and (310), which vanish on unloading
$Na_2Pt(CN)_4 \cdot 3HO_2$	Triclinic	Baumhauer, 1911; Mügge, 1930	Elastic and residual layers possible
$BaCl_2 \cdot 2H_2O$	Monoclinic	Vyrubov, 1886	Elastic, reducing, and residual layers can occur
See Table 7	Monoclinic Triclinic Ortho- rhombic	Mügge, 1888	Elastic monocrystallization

Friedel (1926, p. 493) described some unusual effects for $(NH_4)_3H(SO_4)_2$ and $Al_2O_3 \cdot 3CaO \cdot 3CaCl_2 \cdot 10H_2O$. These compounds have the structures of polysynthetic twins at room temperature; twinning occurs as a result of a polymorphic transition when the growing crystal is cooled. One of the components changes its orientation under stress and ejects the other, so the crystal becomes partly or completely monocrystalline. Similar effects have been observed in my laboratory (Chernysheva, 1950) in Rochelle salt in the region between the Curie points (-18 to +24°C).

Detwinning can be spontaneously reversible (§ 2.20 and 21) or residual; the twinned state is gradually regained on unloading in the first case. Table 7 (§ 2.21) collects the experimental evidence on elastic monocrystallization.

4. Surface Relief of Calcite in Elastic Twinning

The colors of the interference fringes (Fig. 65) indicate the thickness of the layer. The extreme point gives only a gray color; the higher-order colors occur nearer the surface (see § 2.8 for details). Calculations (Garber, 1947a) show that the thickness is about a micron when the length and width are a few millimeters. The width as measured under the microscope on a side face is also of the order of a micron. Garber (1947) examined the interference pattern at the site of contact of a calcite crystal with a glass lens (spherical or cylindrical) in order to eludicate the causes of these very thin twin layers (Fig. 67). The force was applied along a twin plane in the twin direction, the observations being made in reflection (Fig. 68).

Figure 68a shows that homogeneous elastic deformation occurs at first; the size of the central spot increases with the load. The shape of the rings alters when the deformation is sufficient to give rise to an elastic twin; the rings show streaks parallel to the twin plane (Fig. 68b). Further increase in load gives rise to discontinuities (Fig. 68c).

These changes show that the surface rises on one side of the twin and falls on the other (Fig. 69).

Garber gives the shearing stress τ around the point of contact C (Fig. 70) as

$$\tau = c \left(x \sqrt{r^2 - x^2} + r^2 \arcsin \frac{x}{2} - 2hx \right),$$

in which h is a constant dependent on the elastic properties of the crystal, r is the radius of curvature of the lens, and x is the distance of C from AD.

This equation shows that τ is always zero at D (Fig. 70), so twinning cannot start there; it takes opposite signs on the two sides of D. In the region where the stresses act along the crystallographic direction of shear (arrow M) there develops a twin layer; the twin direction is polar, so twinning should not occur on the other side of D, as is found.

The depth H_{max} of the pit is given by $2H_{max} = \delta \tan 38°$, in which δ is the width of the elastic twin (this may be measured under the microscope).

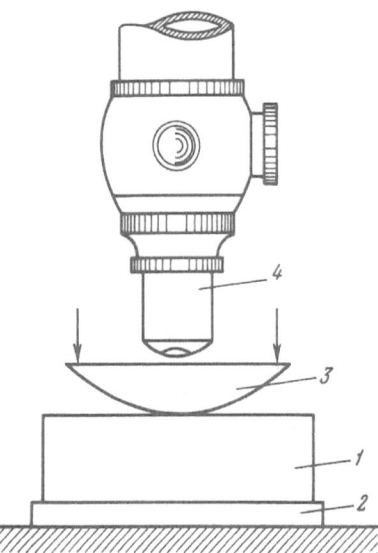

Fig. 67. Interference studies on pressure figures. The crystal 1 lies on the soft support 2 and is loaded via the lens 3, which is also used to observe the crystal in reflection (Garber).

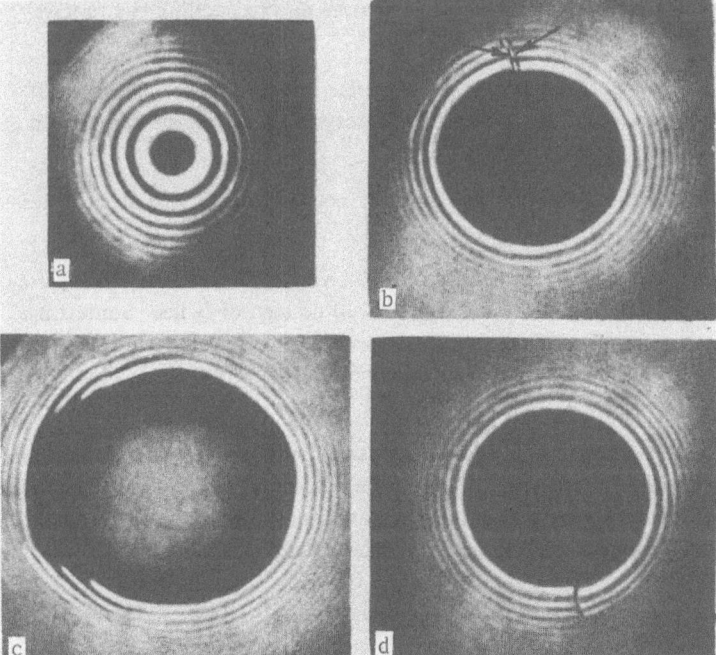

Fig. 68. Newton's rings in a pressure figure: a) lens in contact with surface; b) elastic twin produced under increasing load, with arrows denoting streaks parallel to twin plane; c) distorted rings corresponding to wedge-shaped surface; d) reduction of shear by partial unloading (Garber).

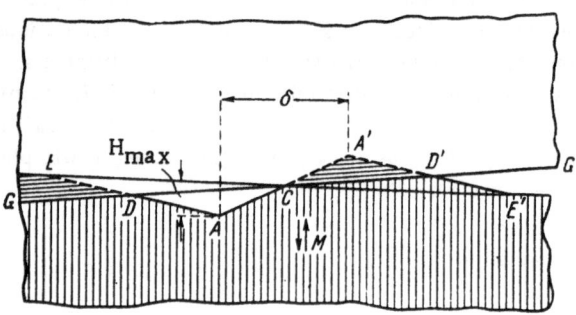

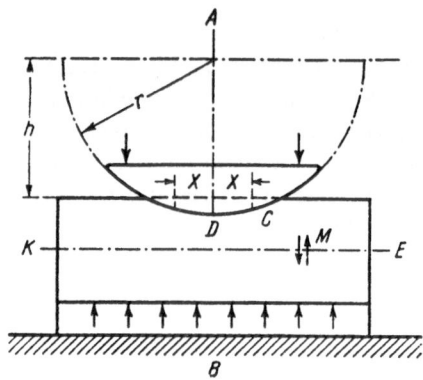

Fig. 69. Redistribution of load by elastic deformation in twinning at point C (contact between lens and crystal): EE' and GG' are the surfaces of crystal and lens before deformation (center of lens to left, off figure); AA' is the width δ of the freely developing elastic twin; CD' and D'E' show the relief corresponding to regions of elastic accommodation; H_{max} is the depth of the pit produced by elastic twinning; and M denotes the shear direction (Garber).

Fig. 70. Contact of a cylindrical lens with a crystal: KE is the axis of the crystal, which is normal to the twin plane; D is a point on the lens; C is a point on the surface of contact between lens and crystal; and A is the axis of the cylinder, which lies in the twin plane (Garber).

Garber did not observe elastic twins thicker than one micron, which he explained by saying that an elastic twin layer with the shape shown in Fig. 69 caused a redistribution of the stress, which became less than that needed to displace the boundary (to cause the twin to grow). This effect is the more pronounced the thicker the layer. Increase in stress makes the layer wider, not thicker, until the limits of the specimen are reached. Success in tests with calcite is largely dependent on the correct choice of orientation with respect to the plane and direction of twinning.

5. Role of Localized Loads

Twinning in calcite should be produced very cautiously and slowly (Lifshits and Obreimov, 1948); rapid loading may break the crystal without producing twinning, which shows that stresses approaching the rupture stress are needed to start twinning.

A localized load should be used for this purpose; the twin component arises directly under the point of application (Garber, 1938). The point or knife-edge must be harder than the crystal; insufficient hardness in the point may result in difficulty in producing twinning (Garber, 1946).

A distributed load does not cause twinning if the surface is free from defects; conversely, dust particles facilitate twinning.

Bell and Cahn (1953) have shown that local overstress is needed to produce twins in zinc; no twins were found up to the moment of rupture on stretching a zinc monocrystal that showed no sign of twins. Sometimes twins were found directly at the fracture surface, where overstress naturally occurs. Twinning during stretching occurs if a nucleus is present at the start, or if the surface is scratched during stretching. Twinning then takes an avalanche form.

Ancker (1953) evaluated the order of the local stresses needed to nucleate a twin layer. His very precise x-ray measurements of lattice constants demonstrated that there are considerable residual stresses in the surface layers of zinc monocrystals grown in glass tubes; twin layers suddenly appeared after a few days, whereupon repeat measurements showed that the lattice constants had become normal. The changes in the lattice constants point to surface stresses of the order of 100 kg/mm^2.*

---

* See p. 63 on the work of Price.

There had been many previous attempts to evaluate these nucleation stresses; ordinary deformation (stretching), with the load distributed over the cross section, had been used. The values obtained for the critical stress for cadmium monocrystals of about the same degree of purity (§ 2.7) are 140 ± 4 g/mm^2, 420 ± 50 g/mm^2, and 50 ± 25 g/mm^2. Bell and Cahn considered that this lack of agreement is the result of local stresses produced during growth, during removal from the tube, during mounting for test, and so on, but not of stress produced by the pulling. Their later work (1954) contains the assertion that slip is usually a necessary forerunner to nucleation of a twin; homogeneous slip causes sets of dislocations to accumulate, and these cause local stress concentrations. First, a twin nucleus about 250 A in size is formed, which then develops into a lens, a wedge twin, and a twin plate. The nucleation step demands the largest stress.

6. Stages of Twinning: Elastic Limits and Yield Point

Garber (1947a) concluded from his study of calcite and sodium nitrate that the lattice reorientation in twinning occurs in several stages:

1. The deformation is elastic (Hooke's law is obeyed).

2. An elastic twin is formed at a certain critical stress (the first elastic limit).

3. The twin layer becomes a stable residual or reducing twin (no longer vanishes when the load is removed); reducing layers occur if complete ejection is prevented by internal defects (cracks, inclusions). Residual layers occur if the reserve of elastic stresses (related to an incoherent twin boundary) is insufficient for ejection.

4. The residual twin layer becomes thicker by displacement of twin boundaries. *

It is assumed that the transition from elastic deformation to residual occurs at a critical stress. The assumption of four stages implies that there are three limits (Garber, 1947a), which correspond to the production of an elastic twin, the formation of thin residual twin layers, and thickening of these layers by displacement of twin boundaries. This last limit is better called the yield point.

Garber (1938) stated in his first paper on the elastic twins of calcite that small tangential stresses (about 26 g/mm^2) are sufficient to produce elastic twins. Later (1947) he stated that a nucleus is needed, a very high stress being required to produce this; the very high values are produced in practice by the use of localized loads and do not allow of precise determination. There are also no reliable methods of determining the second elastic limit (Garber, 1947a); the stress corresponding to it for a localized load is dependent on the size of the specimen, on the stress distribution, and the time allowed. The third limit (the yield point) can be determined the most reliably, for the twinning at this state "is localized at the plane bounding surfaces of the twin layers and is easily produced by homogeneous stresses" (Garber, 1947a). Garber's rough measurements on a single calcite crystal imply that the first limit is about 8 times the third.

Garber determined the yield point from the stress needed to increase the thickness of a residual twin layer by 1μ; the experiment was done as described in § 2.2 and is shown in Fig. 62b.

The deformation was observed with a microscope fitted with an eyepiece micrometer. The yield point was taken as the stress at which the boundary of the residual twin had moved by 0.5 scale division (0.7μ). The yield rate was 2 to 18μ per minute.

Detwinning (§ 2.2) can be observed if the force is appropriately applied to a crystal containing a twin layer (Fig. 62). The yield points for the two processes do not differ appreciably (Garber, 1946). The usual range for calcite is 90 to 110 g/mm^2, but values of 40-50 g/mm^2 are sometimes found. Crystals hardened by previous twinning or by heat treatment (§ 2.10) give values of 300-400 g/mm^2.

The concept of stages is fully applicable to twinning in metals. Bismuth, for example, gives rise to residual twins at stresses of 215-225 g/mm^2, while displacement of the boundaries (fourth stage) requires 140-150 g/mm^2.

*Creep on account of displacement of the boundaries of the twin (§ 2.11).

A further discussion of displacement at stresses below the second limit is to be found in § 2.11 in relation to creep in crystalline bodies.

7. The Law of Critical Shear Stresses

We resolve the force P into components P_n and P_s, respectively, normal and parallel to the twin plane, the latter acting along the projection of the loading axis (Fig. 71). Let the twin plane 1 make an angle χ_0 with the axis of the specimen; let the twin direction t (in plane 1) make an angle λ_0 with this axis; let the area of the section of 1 be ω and let the area of the normal section 2 be ω_0. Here λ_0 and χ_0 specify the initial crystallographic orientation, so Fig. 71 shows that the normal stress is

$$\sigma_n = \frac{P_n}{\omega} = p \sin^2 \chi_0, \tag{1}$$

in which $p = P/\omega_0$. The shearing stress along the projection of the loading axis is

$$\sigma_{s_1} = \frac{P_s}{\omega} = p \cos \chi_0 \sin \chi_0. \tag{2}$$

The stress σ_s along the shear direction in the initial stage* of twinning is

$$\sigma_s = \frac{P_s \cos \alpha}{\omega} = p \cos \lambda_0 \sin \chi_0. \tag{3}$$

It is assumed (Boas and Schmidt, 1938, p. 100) that a monocrystal starts to twin or to deform by slip when σ_s in the appropriate direction exceeds sone limit σ_{sc}, which is not the same for crystallographically nonequivalent planes; we may say that there is an indication of critical shearing stress (Regel' and Zemtsov, 1955). The σ_{sc} for slip is dependent on the temperature, on the previous deformation, on the deformation rate, on the composition, and on the normal stress (at high hydrostatic pressures).

If a deformed crystal has several equivalent planes and directions of slip (as in a cubic crystal), then the slip occurs on the system for which σ_s is largest.

The law of critical stress is confirmed very precisely for slip in monocrystals of readily fusible metals (Boas and Schmidt, 1938).

It is not entirely clear whether the law applies to deformation by twinning; in cadmium (Thompson and Millard, 1952), calcite (Griggs, 1938), zinc and molybdenum (Cahn, 1954), and beryllium (Brick, 1953; Garber et al., 1955) the twinning actually does start in this way, but the σ_{sc} show a range of 100:1 (Cahn, 1954; see § 2.5). This enormous spread occurs partly because some σ_{sc} are determined for the second stage (nucleation of an elastic twin), in which case σ_{sc} for cadmium is 3000 g/mm² (Bell and Cahn, 1953), while others are for the fourth stage (displacement of boundaries), when the σ_{sc} are naturally very low (50 ± 25 g/mm² for cadmium; Yakovleva and Yakutovich, 1940).

The critical stress for slip or twinning is very much dependent on the state of the surface. King (1952) found that a film of oxide on a monocrystal of cadmium increased σ_{sc}, while Gilman and Read (1952) found a considerable increase in resistance to twinning when a zinc monocrystal was coated with copper.

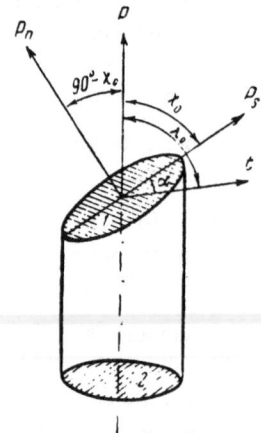

Fig. 71. Resolution of an applied force into components normal and tangential to the twin plane.

---

* If the deformation is large, $\sigma_s = p \sin \chi_0 \sqrt{1 - \frac{\sin^2 \lambda_0}{d^2}}$, in which d is the ratio of the lengths of the working part after and before stretching.

The evidence on the twinning of iron monocrystals is conflicting. Vogel and Brick (1953) state that the law does not apply, while Krafft et al. (1953) state the converse. Siems and Haasen(1958) determined the critical stress for zinc (99.995%) from the first sound and from the first layer visible under microscope. The σ_{sc} for zinc having an orientation favoring slip on the basal plane is 300 to 500 g/mm^2, whereas specimens of unfavorable orientation gave 1000 g/mm^2 or more. The reason for the large spread in σ_{sc} is that slip precedes twinning, which ultimately gives rise to the local stress concentrations necessary to initiate twinning.

8. Work of Formation for an Elastic-Twin Layer: Invariant of Deformation in Twinning

Garber (1946) measured the tangential stress corresponding to the third elastic limit. We have already noted (§ 2.5 and 7) the difficulties in determining the stress needed to generate and extend an elastic twin.

Obreimov and Startsev (1958) measured the dimensions of an elastic twin layer generated by localized loading in calcite; the force was applied along the twin direction. The length and width were measured under the microscope, while the thickness was calculated from the interference fringes (Figs. 72 and 73). All three dimensions were found to increase in proportion to the load (Figs. 74 and 75).

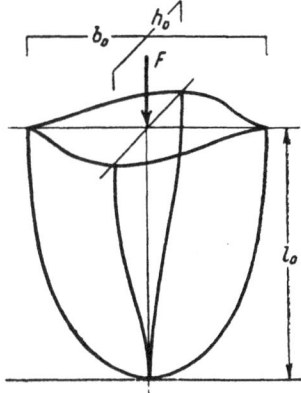

Fig. 72. Model of an elastic twin: l_0 is the over-all length, b_0 is the maximal width, and h_0 is the maximal thickness (Obreimov and Startsev).

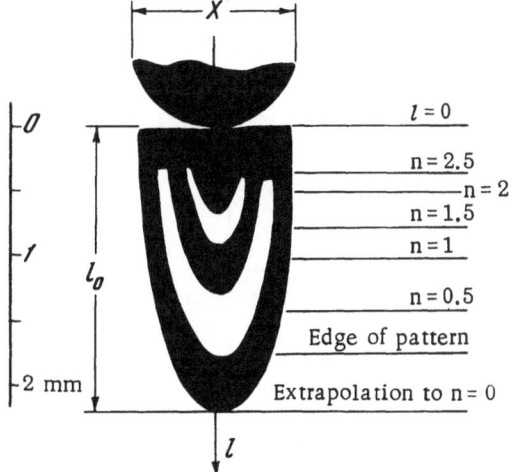

Fig. 73. Elastic twin as seen under the microscope; the scale of sizes is seen on the left. The ball end of the steel plunger is shown at the top. X' is the apparent width, $b_0 = 3.945X'$ being the true width; n is the order of the interference fringe. The thickness $h = n \times 6.505 \times 10^{-5}$cm (Obreimov and Startsev).

From these dimensions they evaluated the work of twinning as the product of the force by the shear (observed thickness of the twin). This work goes to produce the fresh surfaces and to generate stresses nearby. Obreimov and Startsev assumed that the proportion absorbed by elastic deformation was small (that nearly all the energy went to produce fresh surfaces). This gave values of 1200 ergs/cm^2 or more, which are very high. It is not clear why so large an energy should be needed if the lattices remain in contact along the boundaries.

Obreimov and Startsev supposed that the work was taken up in the formation of stepped boundaries and that the steps represent discontinuities on cleavage planes. If one of the surfaces is parallel to the twin plane, then the energy relates to the other surface only; if both deviate from the twin plane, the energy is that of the whole surface area. Electron microscopy (Whelan et al., 1957) has shown that the boundaries are stepped in the twins produced by annealing of stainless steel (Fig. 76).

Reproducible values are not obtained if the force of rupture is used as a measure of the strength of a crystal, for this is dependent on how the force is applied. There is also an enormous spread in the elastic limit for

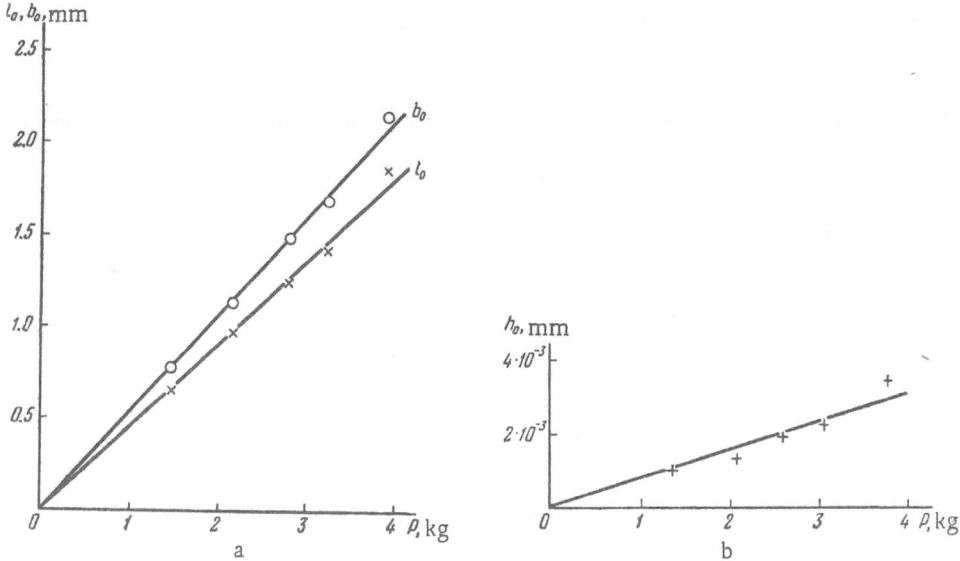

Fig. 74. Linear relation of dimensions to load for an elastic twin: a) length l_0 and width b_0 as functions of load; b) thickness as a function of load (Obreimov and Startsev).

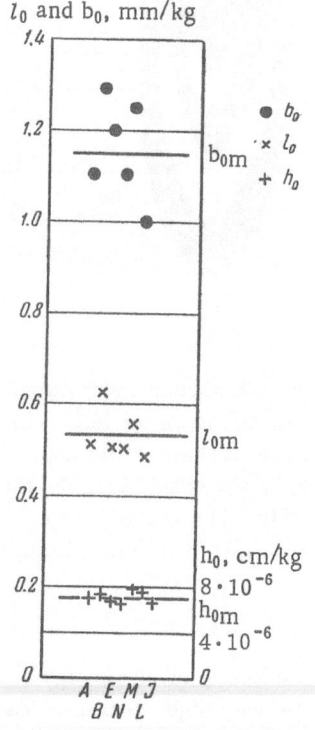

Fig. 75. Constancy of dimensions for an elastic twin in repeat experiments for various points in a crystal: b_0 is the width, l_0 is the length, and h_0 is the thickness (on another scale). The individual tests are indicated by letters; the mean values b_{0m}, l_{0m}, and h_{0m} are shown. b_0, l_0, and h_0 are for loads of 1 kg (Obreimov and Startsev).

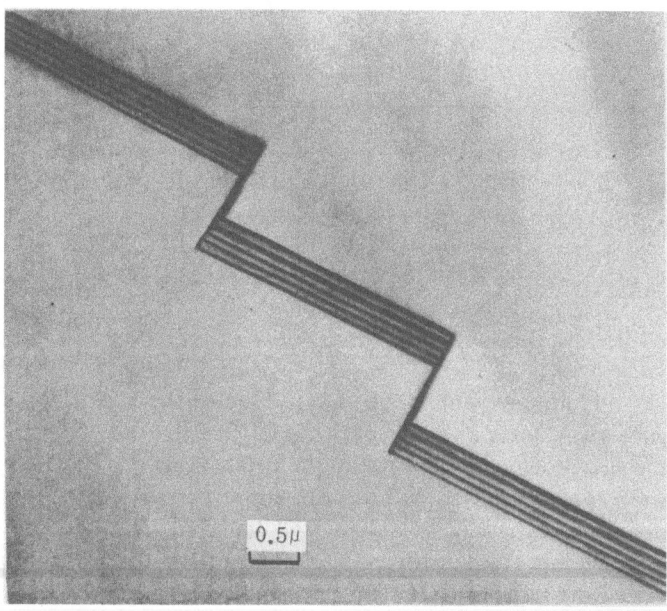

Fig. 76. Steps on the boundary of a twin in a crystal of stainless steel having its surface roughly parallel to the (110) plane. One face of the step is parallel to (111) and the other is parallel to (110) × 60,000 (Whelan, Hirsch, Horne, and Bollman).

twinning. Obreimov's experiments (1930) on the cleavage of mica under vacuum showed that the work needed to produce a unit area of fresh surface (the surface energy) is a constant for any material; he assumed that this work is the invariant of rupture.

The work of formation of the boundary may be taken as the invariant for elastic twinning, although Obreimov and Startsev obtained very high values for it. Direct measurement of the work of deformation is needed.

9. Rate of a Twinning Process: Jumps in Twinning, Deformation Curves, and Jumps in Deformation

We must distinguish the rate of twinning parallel to the twin plane (the rate of tangential displacement of the twin dislocation, i.e., of the steps at the boundary) from the rate normal to the twin plane (rate of displacement of the boundary).

Only rough values are available for the first rate. Sometimes (as for tin and zinc) the twinning is accompanied by a sound which is audible out to several meters away for tin (the 'cry of tin'). Forster and Scheil (1940) recorded the increase in specific resistance associated with twinning for bismuth. The individual pulses lasted for not more than 10^{-4} sec. Mason et al. (1948) used an ultrasonic method to record the pulses associated with the stretching of polycrystalline tin; the durations were found to be about 30 μ sec, and oscillograms revealed a fine structure of scale 1 to 3 μsec. The time needed to transmit sound through the specimen was 1 μsec, so they concluded that the twin dislocations propagate with a speed approaching that of sound.

Thompson and Millard (1952) used a similar method to detect steps associated with waves of period 10^{-4} sec and decay time 10^{-2} sec when thin (5μ) twin layers were formed in zinc. The precise form of the wave was dependent on the elasticity, but it was clear that an elementary act of twinning takes less than 10^{-4} sec. Kontorova and Frenkel's (1938) theory would indicate that dislocations cannot propagate at a rate higher than the speed of sound (see § 20, Chapter 8); in fact, it is found (Gilman and Johnston, 1959) that the dislocations move at a far lower speed during deformation. The experimental results can be explained only if we assume that twin dislocations move much more rapidly than ones causing slip (with less loss of energy), the rate approaching 1/10 of the speed of sound (van Bueren, 1960, p. 216).

The rate normal to the plane (to the direction of motion of the twin dislocation) may be much less. Mason et al. found that the mean rate of change of the length was about 0.1 mm/sec, so the lengthening during the period corresponding to the fine structure was 1-3 A (ten atomic layers or so). The rate of thickening (on the assumption that twinning could occur throughout the cross section) was 1 mm/sec. Garber (§ 2.6) observed a very low rate of displacement (0.03 μ/sec) for calcite.

High-speed cinematography was first used by Jillson (1950; 3000 frames per sec) to observe the growth of twin layers; Siems and Haasen (1958; 2000 frames per sec) later repeated the work. Jillson observed the broadening and loss (detwinning) of thick (125 μ) layers in zinc. The initial state of uniaxial compression involved a rate of 4 mm/sec, the subsequent stepwise growth having a mean rate of 0.2 mm/sec.

Siems and Haasen found a very wide spread in the second rate for 99.995% zinc; the stage in the process, or the presence of coherent or incoherent boundaries has a marked effect (Fig. 77). The rate for a coherent boundary should be lower than that for an incoherent one. The observed rates are 0.1 mm/sec, 15 mm/sec, and 20-40 mm/sec. Figure 78 shows the two most typical types of curve.

Their explanation for the slow movement is that the movement of dislocations is hindered by foreign atoms, intersection with other dislocations, and so on.

Either twinning occurs very rapidly (stepwise) in the first stage, subsequent thickening being slow and gradual, or the actual thickening is rapid but occurs in steps, with halts in-between. As yet, we cannot say for certain which actually occurs.

Deformation Curves. Figure 79 shows the stress as a function of the deformation for cadmium monocrystals (Boas and Schmidt, 1938). In case a, the orientation was such that deformation began at once with twinning.* In case b, the twinning was preceded by slip on (0001) planes. The sudden change in length produces a pronounced fall in the stress (the pulling was done on a Polyani machine at a fixed deformation rate). The formation of Neumann bands in iron has a similar effect.

---

*Figure 81a shows a similar curve for a dislocation-free 0.5 μ monocrystalline platelet of zinc parallel to the basal plane.

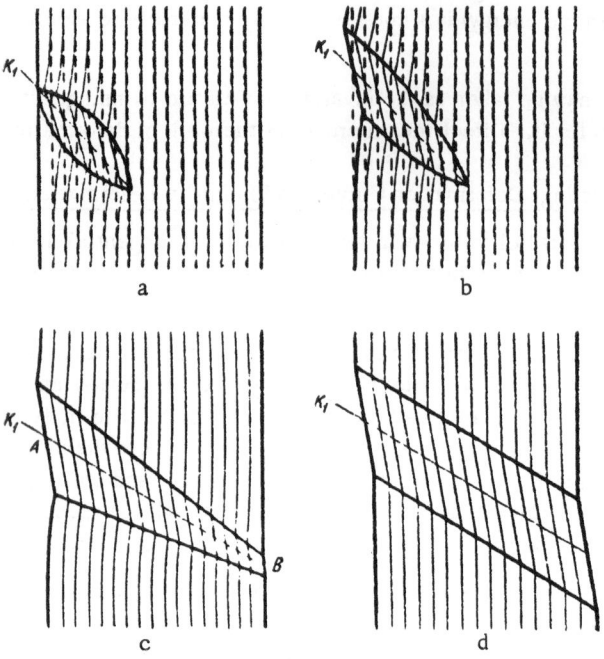

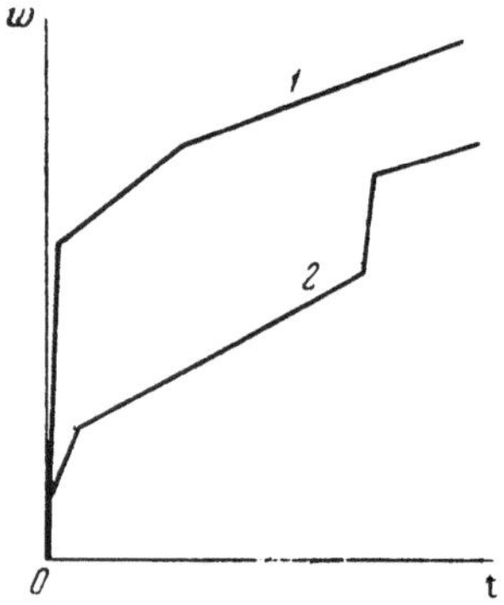

Fig. 77. Stages in the development of a twin layer (axis AB) in the body of a crystal: a, b, and c show incoherent boundaries, while d shows coherent ones. The twinning rate as observed on the surface is dependent on the stage (Siems and Haasen).

Fig. 78. Curves 1 and 2 represent the thickness of twin layers in zinc as a function of time t (Siems and Haasen).

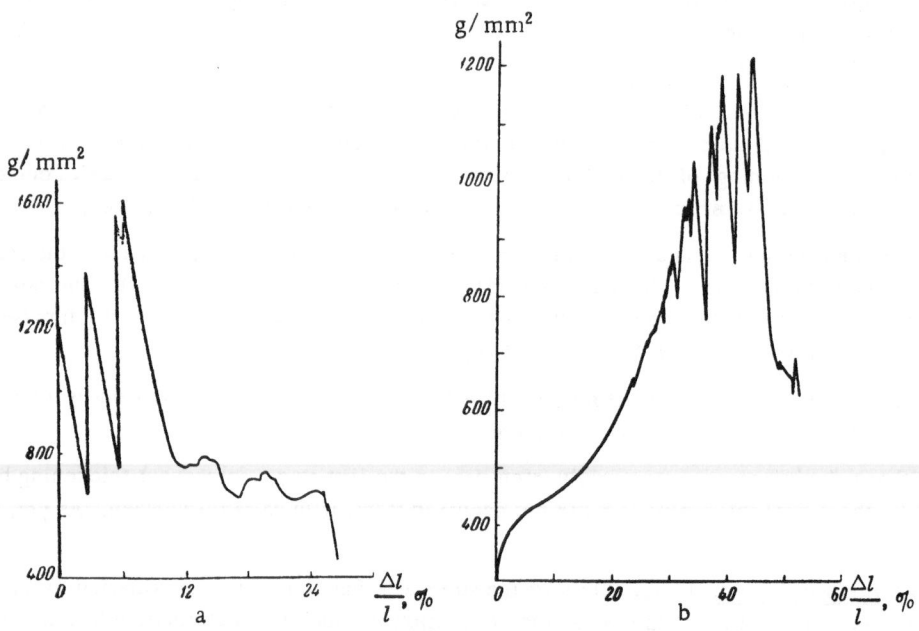

Fig. 79. Stretching curves for cadmium: a) sudden relief of stress by mechanical twinning, X = 3.2° and λ = 10.4°; b) twinning preceded by slip, X = 7.0° and λ = 17.3°. The stress is shown as g/mm² of the original cross section.

Stepwise deformation is not always related to twinning; the effect is sometimes caused by the production of a series of shears in the deformed crystal (Klassen-Neklyudova, 1927; Regel' and Zemtsov, 1955; Klassen-Neklyudova and Urusovskaya, 1956b; Rozhanskii, 1958). Figure 80 shows how the deformation curves are affected if secondary slip occurs during twinning (Fig. 31). Stepwise deformation can also occur in a monocrystal if kink bands (§ 15.4) are produced.

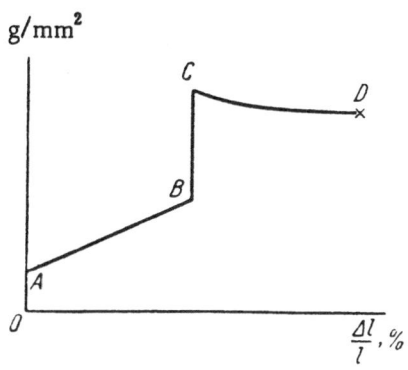

Fig. 80. Slip resistance in g/mm² for the basal plane in a monocrystal of a hexagonal metal as a function of $\Delta l/l$ (in %). AB represents primary basal slip; BC, mechanical twinning; and CD, secondary slip in the twin layer (Boas and Schmidt).

10. Work-hardening as a Result of Mechanical Twinning

A stress of 90-110 g/mm² is usually needed to transform an elastic twin to a residual one in calcite (see § 2.6 on yield points). Some calcite crystals (the most homogeneous ones, without trace of yellow color) require only 40-50 g/mm², though. If a twin layer is produced (Fig. 62b) and then is removed (Fig. 62c), no work-hardening occurs; the yield points in the two processes are roughly equal (Garber, 1946). On the other hand, bismuth (Gindin and Startsev, 1950) requires a larger force to remove the twin layer.

Increased resistance is found if a calcite crystal treated once as above is deformed again at the same site; the yield point rises to 300-400 g/mm² (work-hardening has occurred).

This hardening is associated with the distortion of the structure at the boundaries (Garber, 1946); vacancies can accumulate along the boundaries, and this can hinder further extension (can increase the resistance to deformation). The dislocation theory indicates that vacancies are formed when opposed dislocations meet (Read, 1957), so we would expect them to be formed when the twin dislocations meet any other dislocations. The mechanism of work-hardening by twinning for metals should not differ essentially from that for calcite.

The work-hardening is reduced if the experiment is done at high temperatures; thicker twin layers are formed in beryllium at elevated temperatures, and the reduced resistance to twinning enables one to convert the entire specimen to the twin orientation (Garber et al., 1955).

11. Displacement of a Twin Boundary Under Stress

A metal or alloy will flow slowly (creep) at high temperatures at loads below the elastic limit; the creep rate increases with the temperature. A similar effect occurs in rocks (Bridgman, 1925). Creep in an assembly of crystals occurs mainly by flow at grain boundaries; creep in a monocrystal under a fixed load may be the result of slip or of growth of twin layers (or kink layers, § 15.4). Slip results in the largest plastic deformation of course.

A crystal that twins with change of form (calcite, Rochelle salt, metals) may show either broadening or narrowing in the twin layers, the sign of the stress being decisive here. Any change in the dimensions of a monocrystal is dependent on the disposition of the twin elements in the stress field (§ 1.12).

Garber (§ 2.6) has observed the displacement (creep) of twin boundaries in calcite under a fixed load not exceeding the critical twinning stress; Thompson and Millard (1952) observed the same for crystals of cadmium (Fig. 81). In both cases, the creep occurred at room temperature.

There are six possible twin planes in cadmium; thickening occurs on layers whose twin planes receive the highest stress. Creep is halted at -183°C; low-temperatures stabilize the boundaries.

Tsinzerling (1940) has observed the displacement of twin boundaries in quartz under fixed load; this process is considered in some detail below (§ 4.6). Lifshits (1948) had discussed some theoretical aspects of creep in twinning.

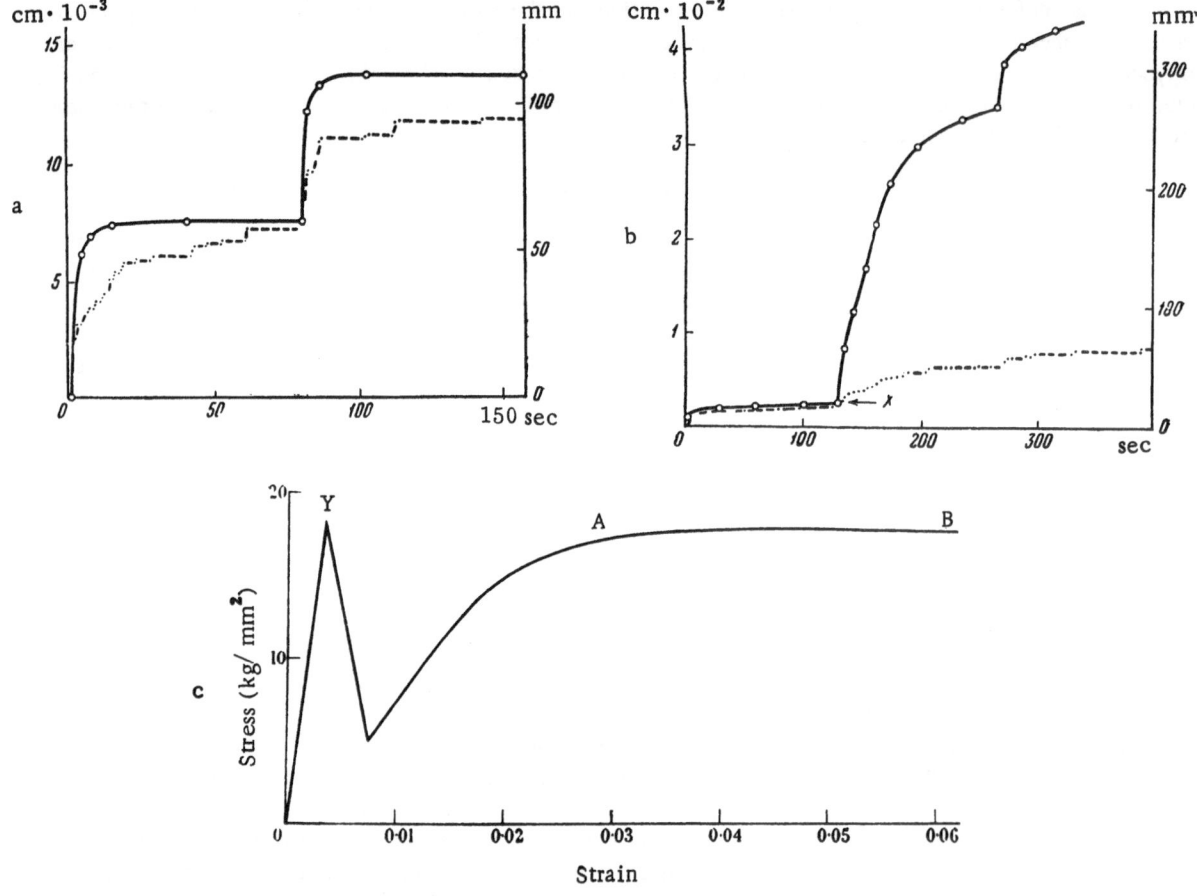

Fig. 81. Creep curves for cadmium monocrystals at room temperature: a) creep by displacement of twin boundaries (full line, strain as a function of time in creep; broken line, strain associated with twinning only); b) creep by slip and in part by twinning (full line, creep curve; broken line, twinning curve recorded as above by oscillograph). The left vertical axis shows the strain; the right, the oscillograph reading (Thompson and Millard); c) typical stress-strain curve for a platelet which deformed by twinning (Price).

12. Twinning and Slip Along a Common Crystallographic Plane

The two effects sometimes occur on the same plane; for example, iron and molybedenum twin on $\{112\}$ planes, which are also possible slip planes (Chen and Maddin, 1951; Brick and Steijn, 1954). Monoclinic $BaBr_2 \cdot 2H_2O$ twins on (110) planes (Mügge, 1889a), and slip occurs on these planes also if the stress is then applied in the reverse direction, whereas loading in several other ways produces kink bands (§ 12.3). The directions of shear for slip and twinning are different in this case. Monoclinic $KClO_3$ (Fisher; quoted by Friedel, 1926, p. 500) shears and twins in the same direction. The distinctive feature of $KClO_3$ is that slip can occur in only one crystallographic direction, the slip elements being (001) and [100]. Seifert (1944) has discussed the cause of polarity in slip. [*]

In $BaBr_2 \cdot 2H_2O$ "shear by a lattice repeat distance [shear for slip] exceeds shear for twinning, whereas the shear for twinning is the largest in $KClO_3$" (Friedel, 1926). For this reason, slip occurs more readily in $KClO_3$ and twinning in $BaBr_2 \cdot 2H_2O$.

13. Effects of Dimensions on Resistance Twinning; Twinning in Crystals with Few Dislocations

Resistance to twinning or slip is dependent on the dimensions of the specimen; metal specimens less than 0.01 mm in diameter show a pronounced increase in resistance to plastic deformation (Yakovleva and Yakutovich, 1940). The critical stress for twinning in extension is increased by a factor of 6-9 when the diameter of a

---

[*] Asymmetry in the ClO_3 group and displacement of the center of O_3.

monocrystalline cadmium wire is reduced to one-fourth; the critical stress for slip is roughly doubled. Twinning stresses of nearly 600 g/mm² are needed at a diameter of 0.09 mm, as against 50-70 g/mm² for thick specimens.

This effect was explained by a change in stress distribution around inhomogeneities and by the lower probability of distorted (overstressed) parts in thin specimens. Experiments with whiskers have shown that these have rupture strengths 10 to 100 times larger than normal and very high yield points (80-100 times the usual values). Price (1961) made zinc whiskers of diameters 20μ and less, which transmit electron beams, and also platelets parallel to the basal plane about 0.5μ thick. Electron microscopy (Elmiscope-1, Seimens; magnification × 10 000 to × 50 000) showed that these crystals were entirely free from dislocations.

In the first series of tests, the platelets or whiskers were stretched in a March tester, which gave the stress-strain curve (Fig. 81c), from which the mean critical stress was determined (50 kg/mm²). The specimen was observed (at × 100) under the microscope in the process and was later examined at higher magnification with a metallurgical microscope.

In the second series, the specimens were stretched while under examination in the electron microscope.

No dislocations appeared in the elastic range; basal slip was impossible when the strain axis lay in the basal plane, in which case twinning occured on $(10\bar{1}2)$ along $[10\bar{1}1]$. The twin layers occurred only in regions of stress concentration, namely at the ends, at corrosion pits, and at growth steps. The critical stress for platelets was 10 kg/mm²; that for whiskers was 60 kg/mm². These are comparable with the theoretical values for homogeneous nucleation. Price calculated the critical stress with allowance for the stress-concentration factor α (Neuber, 1937):

$$\alpha = 1 + \sqrt{t/\rho}\,(\pi - 60)/\pi.$$

Usually there was only one twin layer, which spread through the crystal. Displacement of the boundaries required only 2-20 kg/mm². Dislocations were often seen in the layers, and these accumulated in the form of secondary slip on the basal planes (sometimes on the pyramid planes).

Thompson and Millard's theory (Ch. 8, § 21) indicates that twin layers grow by spiral motion of a single twinning dislocation about a fixed dislocation node (pole mechanism); this process does not demand the generation of fresh twinning dislocations. Electron microscopy showed that the boundaries moved although no such dislocation configurations were visible, while complete dislocations with Burgers vectors of 1/3 $[\bar{1}\bar{2}10]$ were clearly visible along the boundaries.

Price concluded that his results gave no support to the pole mechanism.

Orowan (1954) found that the activation energy A needed for the nucleation of a thin twin platelet is

$$A = 2\pi\,\mu^2 b^2 \gamma/\sigma,$$

in which σ is the critical shear stress (about 50 kg/mm²), b (= 0.35 A) is the Burgers vector of the twinning dislocation, and γ is the surface energy of a twin boundary (about 20 erg/cm²). The result is A ≤ 1 eV, which is relatively small, on account of the small Burgers vector for twinning dislocations. Price supposed that the enlargement of twin layers in thin zinc monocrystals (which required 2-20 kg/mm², which corresponds to an activation energy less than 1 eV) occurs by repeated nucleation and displacement of twin dislocation, not by the pole mechanism. He stressed that a crystal transparent to the beam differs from a massive specimen in that twins occur before the first dislocation lines appear (the latter correspond to the onset of slip). The twinning stress is only about a quarter of the theoretical value, but the discrepancy for massive specimens involves a factor of 12. Twins are generated explosively (in overstressed regions) in massive crystals and grow in response to stress fluctuations and high rates of strain. On the other hand, in a nearly perfect crystal they grow at nearly fixed stress and at rates governed by the deformation rate. These differences Price ascribed to differences in the internal defects; he pointed out that all previous studies of thick crystals give evidence on the effects of structural imperfection, not on the process of twinning itself. A point here is that Price agrees with Konstantinova and Startsev (see Appendix 1) in observing dislocation nucleation at moving twin boundaries. The dislocations migrate conformably along their slip planes in response to stress; they retain their structure. These rows of dislocations are typical of polygonization walls (Liu, 1959); Livingstone (1960) has called this glide polygonization.

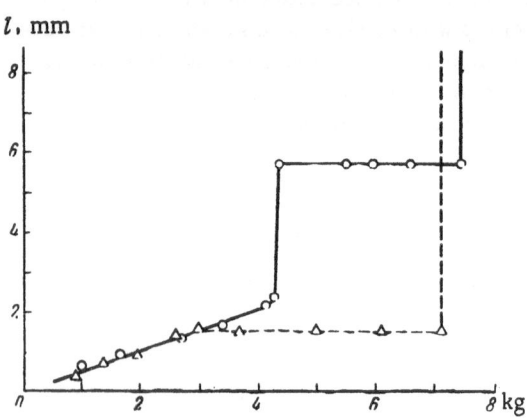

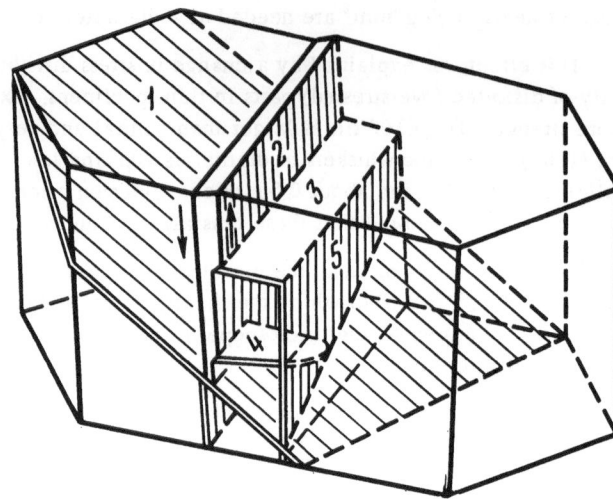

Fig. 82. Length l of an elastic twin as a function of load during intersection with a residual twin layer. Full line, residual layer far from loaded surface; broken line, residual layer near the surface (Startsev).

Fig. 83. Intersection of twin layers in calcite, with cracks on the cleavage planes: 1) residual twin layer; 2) elastic layer growing from the top downward; 3) crack with new twin layer 5 arising at its end; 4) second crack. The arrows indicate the direction of displacement of the atoms in twinning (Startsev).

13a. Effects of Real (Mosaic) Structure on Twinning; Effects of Impurities

Mechanical twinning is a gradual process of lattice reorientation. A real crystal has various types of defect and these, in general, interfere with the gradual reorientation and so increase the resistance to twinning (see Read, 1957, on defects described in dislocation terms and on the plasticity of crystals).

Our information on the effects of the real structure is rather scanty and relates mainly to sodium nitrate (Garber et al., 1947), quartz (Tsinzerling, 1958; § 4.6), and antimony monocrystals (Startsev, 1955).

Optically strain-free $NaNO_3$ crystals * vary in their capacity for twinning; some show very high resistance and break first (at 500-1000 g/mm^2). Only thin short elastic twins can be detected in these. Others break as soon as a twin layer has been formed, but the stress needed is much lower. The third type shows even less resistance, and all four stages (§ 2.6) can be observed. Thick plane-parallel layers can be produced and fixed before the crystal breaks.

X-ray studies (Laue patterns) have shown that the mosaic structure is responsible for the variation in resistance. Large mosaic blocks (up to 10 mm long), and blocks turned relatively through about 1° give the highest resistance (group 1). Twinning occurs most readily in crystals that give unsplit and sharp spots (group 3; ones in which the angles between blocks do not exceed 2-3'). Angles of 6-8' (group 2) in conjunction with large blocks do not interfere with twinning or twin growth. The principal obstacle to twinning appears to be microscopic distortion at the boundaries of the blocks.

Experience with quartz and calcite shows that reorientation occurs more readily the purer (more transparent) the crystal. No special evidence is available as yet for metals.

It is to be expected that the purer a crystal (metal or otherwise), the more mobile its twin boundaries (see § 13), but it should be considerably more difficult to produce twin nucleation in a perfect crystal, † on account of lack of overstress at lattice defects.

---

* Grown from the melt.

† Nucleation occurs at 50 kg/mm^2 in homogeneous whiskers free from growth steps (Price, 1961).

14. Interaction of Intersecting Twin Layers

Startsev et al. (1956) made a detailed study of the intersection of elastic twin layers with existing residual ones (thickness $1-2\mu$). The elastic twin was produced by applying a gradually increasing localized load to calcite (Fig. 62); it was observed by polarized light. Figure 82 shows the length of the twin as a function of load; the two are at first proportional, but then there is a stepwise increase (in this case, at 4.3 kg). The layer penetrates as far as the residual layer in its path; then it stops, but remains elastic. The dimensions then cease to vary with load, until (at 7.55 kg) the residual layer is penetrated. As it breaks through, it also becomes thicker; if the specimen is not then rapidly unloaded, it soon breaks.

The second (broken) line shows the behavior with a residual layer near the surface; no stepwise extension occurs. The elastic twin reaches the layer at 3 kg; penetration occurs at about the same load as before. The penetration occurs by the production of a secondary twin layer with the residual twin (Cahn's layer B; § 1.22). The direction of this secondary layer within B differs from that of the elastic layer; the latter continues parallel but displaced on the far side of B. This layer now persists when the load is removed, but the crack on the cleavage plane, which occurs where the two layers meet, vanishes. Figure 83 illustrates the intersection of twins in calcite.

Twin intersection in metals (antimony, bismuth, and zinc) and other opaque materials can be examined on cleaved surfaces in reflection (Figs. 84 and 85). Often a layer in zinc does not propagate through another; the two merely increase in width as the load is raised (Fig. 85). The wider the layer encountered, the more difficult it is to produce a secondary layer (Fig. 86). One layer sometimes absorbs another (Fig. 87).

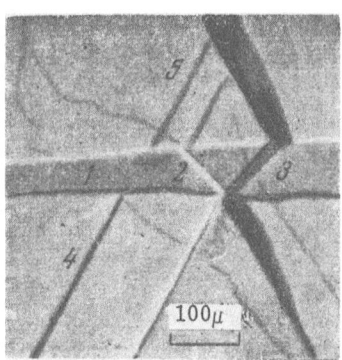 

Fig. 84. Interaction of twin layers in zinc: 1) residual layer; 2) and 3) secondary twin layers; 4) and 5) intersection of layers with discontinuity (Startsev).

Fig. 85. Interaction of intersecting twin layers in zinc, both growing simultaneously: a) initial stage; b) high load (Startsev).

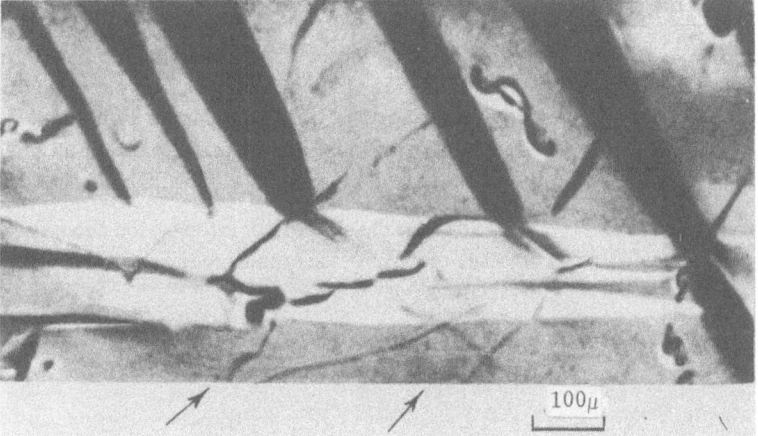

Fig. 86. Mutual interference of twin layers in bismuth (see Fig. 88, III, for scheme of process). On the left, the light layer terminates the dark ones; on the right, the dark one terminates the light one. The arrows indicate the secondary layers (Startsev).

Startsev has observed five forms of interaction (Fig. 88); the geometry of the intersection is governed by the state of stress (Fig. 88, II and III) and by the orientation relative to the force (Fig. 88, IV and V). If the orientation favors slip, a set of slip lines may occur on the far side (Fig. 89).

Figures 90-93 show slip and interaction in mechanical twinning; Fig. 93 shows that twin layers can pass through grain boundaries. The two grains may differ by 30° in orientation in the case of zinc.

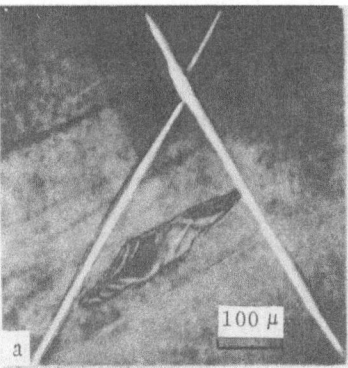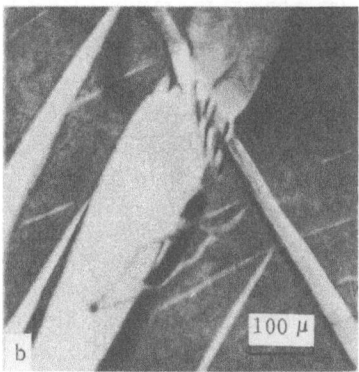

Fig. 87. Absorption of one twin layer by another in zinc: a) initial stage; b) load increased. The broad layer has partly absorbed the less favorably placed thin layer (Startsev).

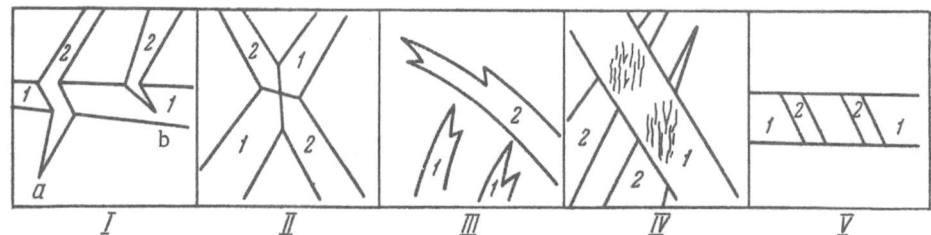

Fig. 88. Intersection of twin layers 1 and 2 (Startsev): I) intersection with formation of a) complete, b) incomplete secondary layer; II) cross structure produced by bending force normal to section (to cleavage plane); III) blocking of twin layers during bending in plane of paper (cleavage plane); IV) absorption of one layer by another, with loss of layer whose orientation is unfavorable to further growth; V) formation of secondary layers within a broad layer, which occurs for special orientations of the first layer and of the stresses localized within it (Startsev).

TABLE 6

Elastic Limit of Calcite as a Function of Temperature

Temperature, °C	Elastic limit, g/mm^2	No.
-142	546 [*]	1
-142	546 [*]	1
-149	695 [*]	2
-149	250	3
-119	335	4
20	244	5
400	233.5	5
400	323	5

[*] Yield point in calcite.

The rates of these various processes provide valuable information on the work-hardening of crystalline materials. The twin layers hinder propagation of slip (Fig. 92) and so increase the resistance to plastic deformation. Partial blocking of one layer by another also occurs at intersections, which should also increase the resistance; but Figs. 91 and 93 show that slip lines and grain boundaries are not insuperable obstacles to twin layers. [*]

The cracks at points of intersection can be points of onset of failure; they may reduce the strength of a crystalline body.

[*] See also Madsen (1956) on interaction between twin layers and with grain boundaries and kinks.

66

High temperatures favor lattice reconstruction in the twin position; twinning then requires less force, and larger residual twins may be produced. On the other hand, slip tends to become dominant, and this hinders twinning. The effect is thus dependent on the elastic limit with the higher rate of fall (slip or twinning). In general, high temperatures should favor twinning if slip does not occur or occurs only under special conditions.

One example of this is quartz, which shows no plastic deformation by slip. Twins (without change of form) are produced in low-temperature quartz by gradually increasing pressure from a steel sphere (Tsinzerling, 1940). The size of the twins increases with the temperature, other things being equal. At stress of about 1000 kg/cm² is needed near the $\alpha \rightleftharpoons \beta$ transition point, whereas one 15 times larger is needed at room temperature.

The shape of the twins is not substantially altered at high temperatures, though the outlines tend to become smooth and rounded rather than straight or zigzag (Tsinzerling, 1933).

Garber (1940) has given the resistance to twinning as a function of temperature for calcite. Twinning is the usual deformation process under ordinary conditions (slip does not occur up to the point of rupture). Table 6 gives the stress needed to convert an elastic twin to a residual one (the second elastic limit) as a function of temperature.

The elastic limit is roughly halved between -142 and +20°C, but there is then little change up to 400°C. Elastic twinning is the principal process between -142 and -119°C; the layers always arise at the point of application of the load. The specimen cracks or breaks up without giving residual twins in this range. The type of twinning does not alter between -142 and +440°C. The difference between the types of deformation at low and high temperatures is that cracking readily occurs at the site of loading at low temperatures, whereas an indent is formed at high temperatures. The indent, and the cracks prevent the elastic twins from vanishing on unloading, the result being reducing twins (§ 2.3). The crystal breaks before the elastic twins are converted to residual ones at low temperatures, so the second elastic limit lies at or above the yield point.

Elastic twinning in sodium nitrate at room temperature is analogous to that in calcite at high temperatures (the melting and decomposition points of calcite are 1339 and 825°C respectively, while the melting point of $NaNO_3$ is 627°C).

As regards metals, we know only that slip becomes dominant at high temperatures, but twinning is not entirely suppressed in zinc and cadmium (Boas and Schmidt, 1935; Ramsey, 1951), α-iron (Pfeil, 1926 and 1927), beryllium (Kaufman et al., 1950; Hausner and Pinto, 1951), uranium (Cahn, 1953a), and molybdenum (Cahn, 1954).

A similar effect occurs in synthetic corundum (leucosapphire, ruby), which deforms by twinning at 600-1700°C and by slip on $\{10\bar{1}0\}$ above 1700°C (Klassen-Neklyudova, 1942b and 1950). Creep occurs in bent or pulled corundum rods at 900-1400°C; here slip occurs on (0001) (Wachtman and Maxwell, 1957).

The evidence for magnesium is conflicting; Bakarian and Mathewson (1952) found that high temperatures favor twinning, whereas Barrett and Heller (1947) and also Ansel (1948) reported the reverse. Moreover, the twin law changes at high temperatures in the case of titanium (McHargue and Hammond, 1953).

Garber et al. (1955) measured the elastic limit as a function of temperature for twinning in beryllium monocrystals (Figs. 94 and 95). The only conclusion to be drawn for most metals is that the resistance to slip is more sensitive to temperature than is that to twinning.

This means that the slip resistance should increase more rapidly as the temperature is reduced, so twinning should be completely dominant at very low temperatures. In fact, copper crystals show slip on (111) at normal and high temperatures, whereas twinning on (111) occurs in response to compression at 4.2°K (Blewitt et al., 1957). It has often been observed (Basinski and Sleeswyk, 1957; Smith and Rutherford, 1957; Haasen, 1958) that jumps occur in deformation by compression in iron at temperatures between 298 and 4.2°K; these are caused by bursts of twinning.

Figure 96 shows the response of calcite crystals to hydrostatic pressure at various temperatures (Griggs and Miller, 1951); the specimens were cut to make the axis of compression perpendicular to the c axis, whereupon the crystal deformed by twinning. Curve 2 (for 150°C) lies above curve 1, while curve 3 (for 300°C) lies below curve 1. The reason for this is not clear.

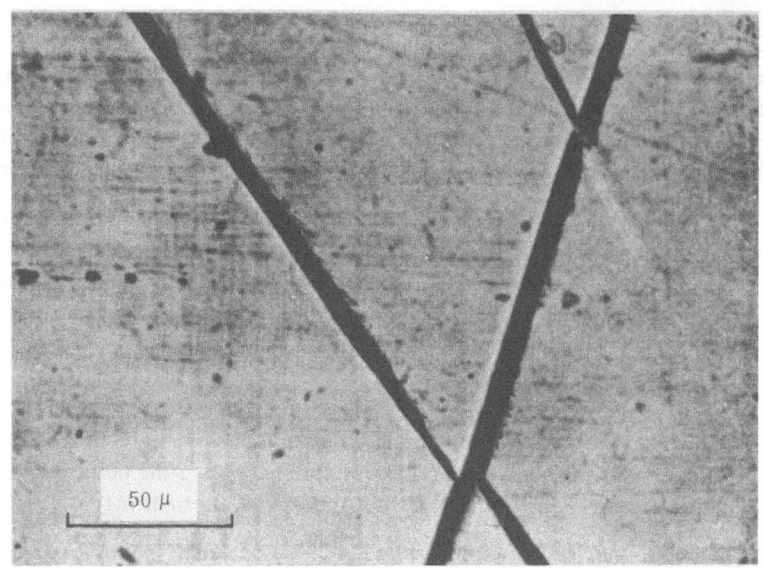

Fig. 89. Intersection of twin layers in silicon iron: top, formation of a secondary layer; bottom, production of slip (Hall).

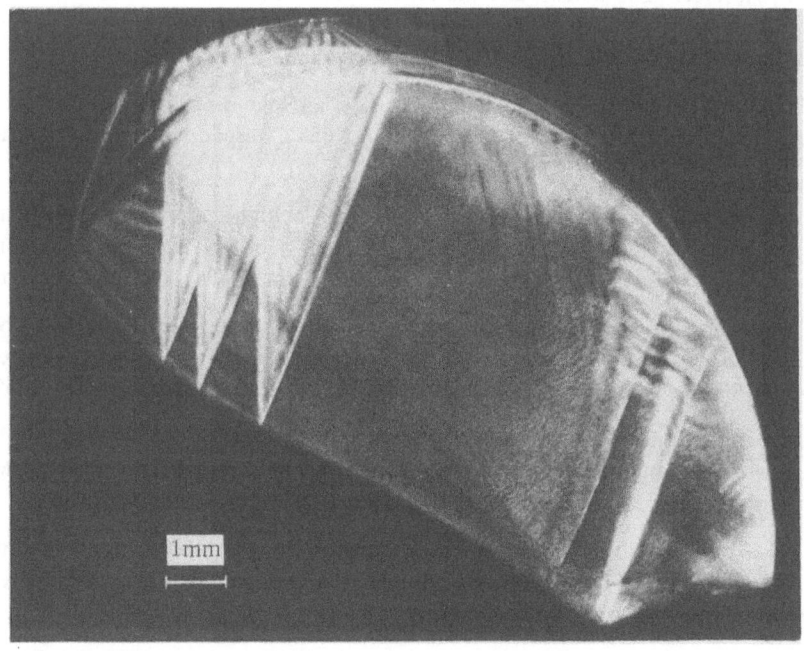

Fig. 90. Five twin layers in corundum; two traverse a region that had previously undergone slip. × 7 (Klassen-Neklyudova).

Curves 4-6 of Fig. 96 relate to calcite specimens having the compression axis parallel to the principal axis; slip is the only process here, when the crystals are subjected to hydrostatic pressure. Before Griggs' work it was assumed that slip does not occur in calcite. Garber et al. (1940) observed indentations in calcite subjected to localized stress at 400-440°C, but no signs of slip were seen (p. 67). Griggs' work demonstrates that slip can occur.

Hydrostatic Pressure. Slip occurs in calcite at high temperatures when other stresses are also present. The hydrostatic pressure increases the strength (delays breakage of the crystal).

Hydrostatic pressure enables one to produce twinning in crystals that do not show it under ordinary conditions; mechanical twins have been produced in several minerals. The crystals are first pressed into a powder (often sulfur or alum) in a thick-walled steel die and then are subjected to axial compression. Mügge (1886) produced twins in diopside and bismuth in this way. The former does not twin in the absence of hydrostatic pressure, while bismuth does twin but parts along the cleavage. In 1898 Veit (1922) produced twins in ruby by keeping the crystal under pressure (13,000 to 18,000 atm) for several hours. Johnsen (1914b) produced twins in lithium sulfate at 8000 atm in the same way.

Grühn and Johnsen (1917b) used axial compression in conjunction with hydrostatic pressure (50,000 atm) to produce a new species of twin having an unusually large s. Hydrostatic pressure can cause twinning on the unusual* $\{0\bar{2}21\}$ planes in calcite and dolomite (Turner and Ch'ih, 1951; Turner et al., 1954).

Bridgman (1955) found that material brittle under ordinary conditions became plastic under high hydrostatic pressure; the plastic deformation preceding fracture was increased by pressures of 20 to 75 katm in steel, molybdenum, tantalum, niobium, monocrystalline antimony, and γ-brass. The work-hardening was also increased. Plastic deformation by slip and twinning occurred in response to axial compression in synthetic corundum (Al_2O_3) exposed to 24,500 atm in a liquid. Corundum twins at 600-800°C and becomes plastic only at 1600-1800°C (Klassen-Neklyudova, 1953) under ordinary pressures. Slip occurs on the prism $\{11\bar{2}0\}$ planes usually, but hydrostatic pressure causes slip on unusual planes lying at 36° to the principal axis.

Twinning is facilitated by deformation under hydrostatic pressure if the temperature is also raised. For example, Heide (1931) was able to produce mechanical twinning in baryta ($BaSO_4$) only when the hydrostatic pressure (up to 16,000 atm) was accompanied by heating to 400°C.

Deformation Rate. Only qualitative studies have been made. The sound produced on deformation indicates that the lattice reorientation occurs at a high rate (§ 2.3). On the other hand, slip is gradual, and the resistance to shear has a negative temperature coefficient. This indicates that slip deformation is less likely to hinder twinning at high rates of loading; in fact, metallurgical experience shows that twins are most readily produced by hammer

* The usual twin planes in calcite are $\{10\bar{1}2\}$.

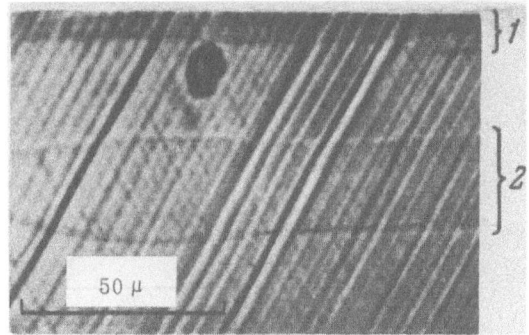

Fig. 91. Twin layers 1 and 2 traversing slip lines in crystals of bismuth. The breaks in the slip lines at the boundaries of the layers are caused by the production of relief on the cleavage plane as a result of twinning (Startsev).

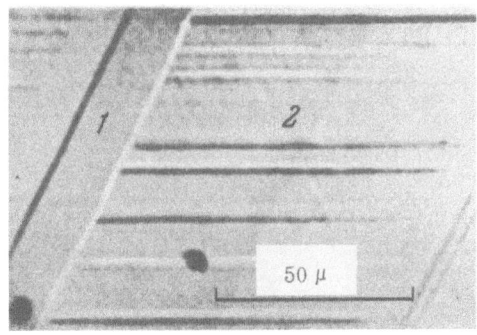

Fig. 92. Slip deformation after twinning in bismuth; twin layer 1 retards the slip line 2. × 540 (Startsev).

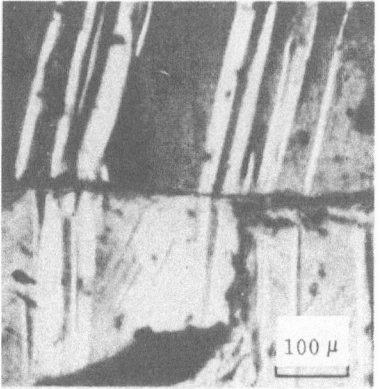

Fig. 93. Passage of twin layers across the grain boundary in a bicrystal of zinc (Startsev).

69

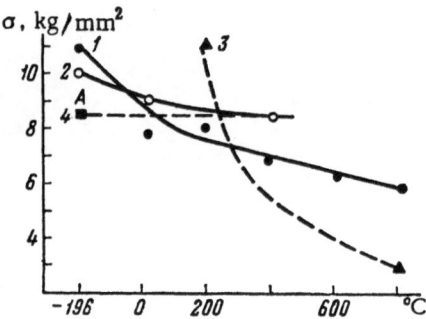

Fig. 94. Yield point σ as a function of temperature for twinning on ($10\bar{1}2$) in beryllium; forces applied: 1) normal to the faces of the first-order prism; 2) normal to the faces of the second-order prism. Curve 3 is for the yield point in slip on ($10\bar{1}0$) and ($10\bar{1}1$); curve 4 is for the stress producing the first cracks along the cleavage plane (Garber, Gindin, and others).

blows, and the probability of twinning increases at low temperatures. It was long assumed that twinning in α-iron occurred only in response to dynamic loads, but Garber et al. (1950) were able to produce twins in carbon-free steel (grain size 1.5-2 mm) by static bending. The twins arose at the points of greatest curvature.

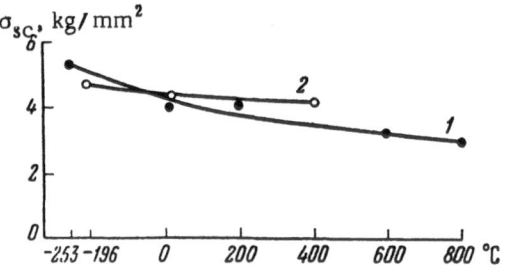

Fig. 95. Critical shearing stress σ_{sc} as a function of temperature for twinning on ($10\bar{1}2$) in beryllium; forces applied: 1) normal to the faces of the first-order prism; 2) normal to the faces of the second-order prism (Garber, Gindin, and others).

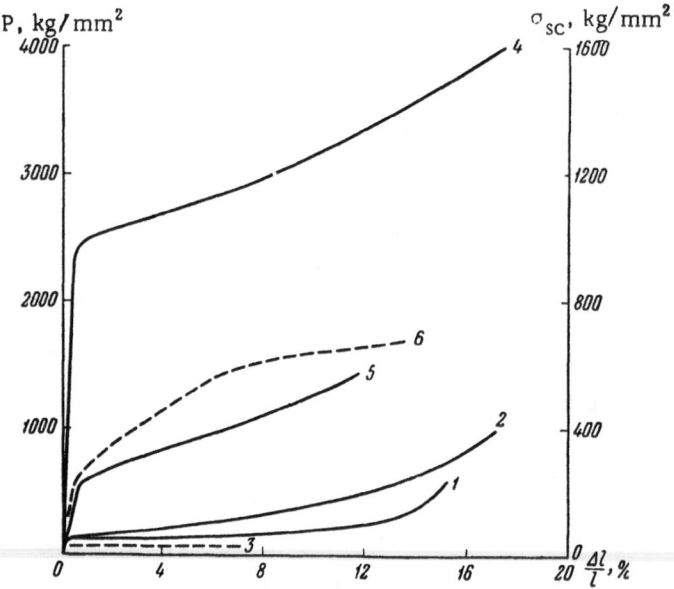

Fig. 96. Compression curves for calcite at several temperatures under a hydrostatic pressure of 10,000 atm; the abscissa is the strain (%); P is the axial pressure, and σ_{sc} is the critical shearing stress (Griggs). Curves 1-3 are for compression normal to the c axis, when twinning occurs; curves 4-6 are for compression parallel to this axis, when only slip occurs: 1) room temperature; 2) 150°C; 3) 300°C; 4) room temperature; 5) 150°C; 6) 300°C.

The important factors are thus the temperature, the loading rate, and the hydrostatic pressure (if any). High temperatures reduce twinning resistance, and high pressures tend to inhibit fracture; both facilitate twinning. Impact loads can eliminate slip, which tends to interfere with twinning.

16. Hardening of Calcite During Annealing; Elimination of Twin Layers by Annealing and Recrystallization of Layers

The work-hardening is reduced by annealing, as is the yield point (Klassen-Neklyudova, 1930; Blank, 1930); the internal stresses are relieved (Obreimov and Shubnikov, 1926). If the hardening from twinning were caused solely by internal stresses, then annealing should reduce the yield point, e.g., for deformed calcite.

This is not the case; calcite becomes stronger (Garber, 1946). Measurements were made of the stress needed to thicken a twin layer (i.e., of the yield point) for specimens twinned and then annealed at 350-510°C. The yield point lay between 185 and 262 g/mm^2, whereas twinning needed only 40-50 g/mm^2.

Calcite starts to decompose at 450°C, so the deformed and annealed specimens were carefully detwinned; the stress needed to produce fresh twins was then measured. This was only 90-100 g/mm^2, which was much less than that needed to displace boundaries strengthened by twinning, so the structure had not been disrupted by the annealing.

The structure along the boundaries is altered by the annealing; it may be that submicroscopic recrystallization or polygonization occurs in these areas at high temperatures.

Kuznetsov and Zolotov (1935) observed vigorous recrystallization of zinc monocrystals along twin layers. Garber considered this as evidence for distortion and stressed that he had never observed such recrystallization in calcite, sodium nitrate, and α-iron. Polygonization was also not revealed by x-ray examination.

Garber explained the hardening by annealing in terms of migration of vacancies to twin boundaries in response to residual stresses. However, it may be that redistribution of impurities may be responsible for the effect, for these tend to accumulate along the boundaries and retard further movement (segregation).*

The hardening may be related to local plastic deformation (e.g., by accommodation, § 1.16) if slip is also possible; residual stresses at the boundaries produce this.

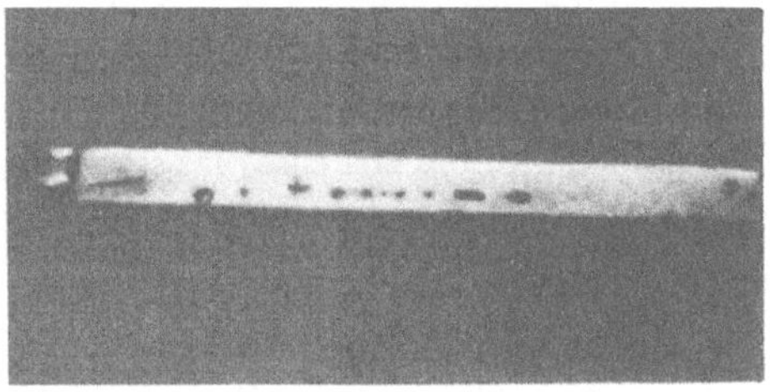

Interesting results have been reported for the effects of annealing on the shape and size of twin layers in calcite and sodium nitrate (Garber, 1947), bismuth (Gindin and Startsev, 1950), and coarse-grained carbon-free steel (Garber et al., 1950). Annealing at 100°C causes reducing elastic twins to shorten in sodium nitrate (melting point 308°C); at 306°C these twins were lost completely within 15 min. Prolonged annealing at this temperature removed the thin residual twins; the thinner they were, the sooner they were lost.

Thick residual twins were not affected, but thin layers of the original crystal enclosed between twins took up the orientation of the latter.

Fig. 97. Annealing of a twin layer in calcite. The dark background is the main crystal seen between crossed polarizers, the light band being the twin layer. The dark spots are regions having the orientation of the main crystal, which have been produced by annealing (Garber).

Bismuth gave similar effects; twin layers 50 μ wide were not lost in spite of many hours at 250°C, whereas ones 2 μ (or less) thick were removed completely. Wedge-shaped layers started to vanish at the thin end; lens-shaped ones gradually vanished from the edges.

Iron and calcite show a different behavior; a layer began to regress at several internal points simultaneously (Fig. 97), and the layer split up into parts, which then gradually grew smaller. This gradual change

---

* Impurities can also accumulate at slip lines and produce a similar effect.

shows that the annealing does not cause detwinning (in response to internal stresses during heating or cooling). The process resembles recrystallization, although new grains are not produced.

Startsev and Gindin observed twin layers of various shapes; those in bismuth were wedge-shaped, while those in iron were plane-parallel. The differences in the effects of annealing must be related to this difference in shape. Residual stresses accumulate at the end and along the side of the wedge, and these favor rapid reorientation, whereas both boundaries are coherent in a plane-parallel layer (§ 1.15), so the residual stresses are much smaller. Moreover, reorientation begins within the layer, because accumulations of vacancies and impurities may hinder the process at the boundaries (these cause hardening in twinning; § 2.10).

The lens-shaped regions in α-uranium (Cahn, 1954) are twins of the second kind with irrational K_1 planes; these straighten out upon annealing. This indicates that the boundaries tend to take up the orientation most favored by energy (§ 4.5).

17. Cracking Along Twin Boundaries, Secondary Cleavage, Fatigue Cracks, Cold-Shortness in Metals; Twinning and Deformation Textures

The energy of the atoms at the boundaries of a growing twin must be raised as a result of the loss of long-range order and also by the incoherence (loss of short-range order). The extent of the latter is dependent on the lattice structure (§ 1.16) and on any deviation of the boundary from the twin plane. If the boundary deviates from K_1, then "the energy increases rapidly with the angle of deviation" (Read, 1957, p. 223).

Large internal stresses may accumulate along these boundaries (§ 1.16); they often show more rapid recrystallization (§ 2.16), chemical etching (Mügge, 1932; Tsinzerling, 1933 and 1940; Zolotov, 1943; Reusch, 1867), and thermal etching (Cahn, 1953a). X-ray methods are sensitive to the lattice distortion for iron (Barrett, 1945) and zinc (Ancker, 1953); the stresses localize at one of the two boundaries in the latter case.

Internal stresses (mainly at the tip of a growing twin) and lattice distortion (along the boundaries) can act as causes of premature failure in crystalline solids. Cahn (1954) distinguishes two types of failure caused by twinning: 1) accommodation cracks; 2) parting at twin boundaries.

To the first type he assigns Rose channels (§ 1.17), cracks at the ends of twin layers, the cracks produced in the elastic twinning of barium chloride (Mügge, 1889a), cracks on ($10\bar{1}0$) in zinc compressed normal to the basal plane (glide plane), and cracks on (0001) within twin layers in zinc (Yakovleva and Yakutovich, 1950). The cracking in the last case was so extensive that the density was reduced appreciably. Iron and steel also show cracks at the ends of twin layers (Tipper and Sullivan, 1951); here they run at small angles to the boundaries.

To the second type he assigns the parting effect described in the mineralogical literature; this is a second cleavage (cracking on crystallographic planes to give mirror surfaces, although this is not directly a result of the lattice structure; Belov and Klassen-Neklyudova, 1948).

One cause of parting is stress localization at twin boundaries (§ 1.15) and slip sites.

Rubies often show signs of parting on rhombohedron planes; stones are often rejected as unsuitable for bearings for this reason (Klassen-Neklyudova, 1942b and 1953).

Zapffe (1953) examined parting in bismuth; Cahn (1954) observed it along the boundaries of twins on (112) in molybdenum crystals compressed at liquid-nitrogen temperature. The evidence on parting in iron crystals is not clear (Cahn, 1954).

Mechanical twinning is undoubtedly one cause of failure by fatigue in technical metals (sign-varying loads cause twinning and detwinning); it is also a cause of cold-shortness. Much work has been done on this (Davidenkov, 1938; Davidenkov and Chuchman, 1957; Shevandin, 1939a and 1939b). Cahn states that twinning can cause cracking in grains, but it is possible to have twinning without fracture and conversely (Geil et al., 1953; Rosenthal et al., 1952). Further, twins often arise at the ends of microcracks (Dereyttere and Greenough, 1954).

<u>Texture Formation</u>. Rolling, stamping, drawing, and so on produce a preferred grain orientation (deformation texture); the effect occurs also in rocks. Twinning may be responsible for some of the texture if the deformation rate is not very low or if the grains twin readily (e.g., calcite). Griggs (1951 and 1953) has examined the textures of marble from this point of view (marbles are produced under high hydrostatic pressure).

Calnan and Clews (1950, 1951, 1952) incorporated the lattice reorientation of twinning in their theory of the deformation texture in rolled α-uranium (orthorhombic); they used the latest information on the crystallography of twinning (§ 1.14) but "rather arbitrarily assumed that twinning occurs much more readily than slip" (Cahn, 1954). Toman and Simerska (1958) have done similar work on the texture of β-tin.

18. Change of Internal Friction on Twinning in a Simple Crystal

The internal friction is characterized by $\Delta = \Delta W/W$, in which ΔW is the energy lost in one cycle of oscillation and W is the total energy; Δ may be as high as 0.1, but low-loss materials have Δ of 10^{-7}.

The main causes of internal friction in crystalline materials are plastic deformation of the softest grains, flow at grain boundaries, and heat transfer between grains. The last occurs because the grains become differentially heated (thermoelastic effect). The two latter effects should be absent for monocrystals.

Structural defects in a real crystal can cause loss even in a monocrystal, e.g., mosaic blocks, slip lines, twin layers, and foreign atoms.

Dislocation theory relates internal friction to the displacement of dislocations (Zener, 1949). Cottrell (1958) considers that the boundaries of mosaic blocks (grain boundaries in a polycrystal) can retard the migration of dislocations. Similar effects can arise from oxide films on monocrystals, Kê (1950) explained the internal friction of aluminum containing a little copper in terms of the slow migration of dislocations surrounded by dissolved atoms. This effect could occur in a monocrystal.

The decrement may be affected by the actual process of twinning as well as by the presence of twin boundaries.

Worrell and Sievert (1951) found a peak in the internal friction for oscillating copper-magnesium alloys; this they ascribed to change in the width of the polysynthetic twins, whose existence was demonstrated on polished sections. Basinski and Christian (1951b) found that these "twins" are actually martensite platelets. The presence of these in the grains can increase the energy loss, as in the case of twin layers.

Cahn (1953a) states that twin layers produce similar effects in other metals, so the decrement can be used to examine twinning effects. Domain-boundary displacement in an antiferromagnetic (cobalt oxide; Fine, 1952) and in magnetite (Fine and Kenneey, 1954) causes a peak in the internal friction; the effect is analogous to that of twin layer displacement (§ 2.11). Jones and Munro (1950) subjected uranium to slow cycles of extension and compression; they found elastic hysteresis and energy loss, which they supposed to be caused by mechanical twinning and detwinning. Photographic measurement of the rate of expansion of twin layers in zinc (Siems and Haasen, 1958; 2000 frames per sec) showed that the speed was much below that of sound. This gave the stress τ^* corresponding to the frictional resistance from twinning; $\tau^* = 400$ kg/mm^2 for twin dislocations moving with the speed of sound. Similar calculations gave $\tau^* = 80$ kg/mm^2 for the movement of slip lines in the migration of edge dislocations in the glide plane.

In some cases at least, the displacement of twin boundaries must be responsible for the rise in the internal friction.

19. Twinning in Rochelle Salt

One special feature here is (§ 1.15) that mechanical twins enlarge by fusion of components of polysynthetic twins. The sense of migration of the boundaries is governed by the sense of the stress (Fig. 36c). Figure 98 shows the tangential stress τ_{yz} in an isotropic plate acted on by a localized load (the two colors correspond to stresses of opposed signs). Comparison with Fig. 36c shows that the direction of displacement follows the sense

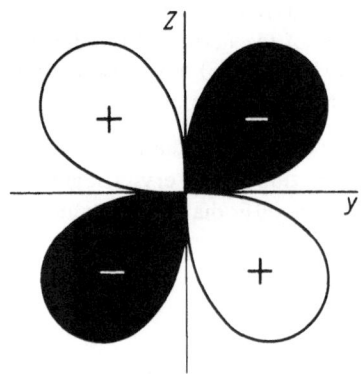

Fig. 98. Lines of equal tangential stress in an X-cut plate of Rochelle salt subjected to localized loading along the a axis (Chernysheva).

of τ_{yz}; in one quadrant the dark component absorbs the light one, while in the adjacent quadrant the converse occurs. The twin boundaries move in opposite senses in the two cases.

This indicates that the effect is analogous to mechanical twinning in calcite (§ 2.2) and essentially different from that in quartz (the sense of the motion is not dependent on the sign of the stress in quartz).

There are some special features for Rochelle salt, though.

Twinning and detwinning occur in calcite and the metals discussed above; the stage of mechanical twinning* is not observable in Rochelle salt (or in some other compounds, and in particular in alloys; see § 2.21), and all we see is detwinning (conversion of a polysynthetic twin to a monoclinic monocrystal).† An external force displaces the twin boundaries elastically until the whole specimen is converted to a monocrystal, after which we have elastic deformation of the monocrystal, which ends in brittle fracture. Chernysheva has observed this on Rochelle salt. The process reverses as the force is removed, but the rate of movement of the twin boundaries is dependent on the temperature and on the degree of perfection of the crystal.

If cracks occur during deformation, these inhibit the reconversion to the polysynthetic twin; enlarged components persist around the cracks. Chernysheva has observed similar reducing twins around natural inclusions.

One reason for the instability of the monoclinic monocrystal may be that the components of the polysynthetic twin are also regions of spontaneous polarization (Klassen-Neklyudova, Chernysheva, and Shternberg, 1948; Chernysheva, 1951). The electric axes in the components are antiparallel. The lattice is reoriented in response to the mechanical stress to bring the electric axes parallel over large areas, which gives rise to charges of opposite signs on opposing surfaces of the plate. These charges act to restore the original domain structure when the external force is removed, and the charges in adjacent domains (components of the twin) mutually balance. However, effects related to domain polarization cannot restore the original structure completely; the domains show a memory effect, presumably on account of some fixed defects in the structure.

A further distinction from calcite and all other crystals discussed above is that the angle of twinning is not constant (is a function of temperature). The angular displacement of the optical indicatrices in adjacent components as a function of temperature has a peak, as in Fig. 99 (Chernysheva, 1957; Indenbom and Chernysheva, 1957a and 1957b). The twin angle is proportional to the angle between the indicatrices, so this curve also indicates how the twin angle varies.

A temperature dependence for the twin angle should occur for all twins formed in second-order phase transitions (Landau and Lifshits, 1948).

The theory of ferroelectricity shows that the twin angle (angle of spontaneous deformation of the domains) must be a function of the field.

Rochelle salt shows brittle fracture without signs of residual deformation; the strain is a linear function of the stress up to the moment of fracture.

Optical studies of the domain structure indicate that the curve of elastic strain should have two parts. The first is related to the displacement of domain boundaries, while the second is related to elastic deformation of the lattice. Much work remains to be done on this.

---

* Twinning occurs during cooling as a result of passage through a point of polymorphic transformation.
† Between the Curie points Rochelle salt has monoclinic symmetry, outside this region orthorhombic symmetry.

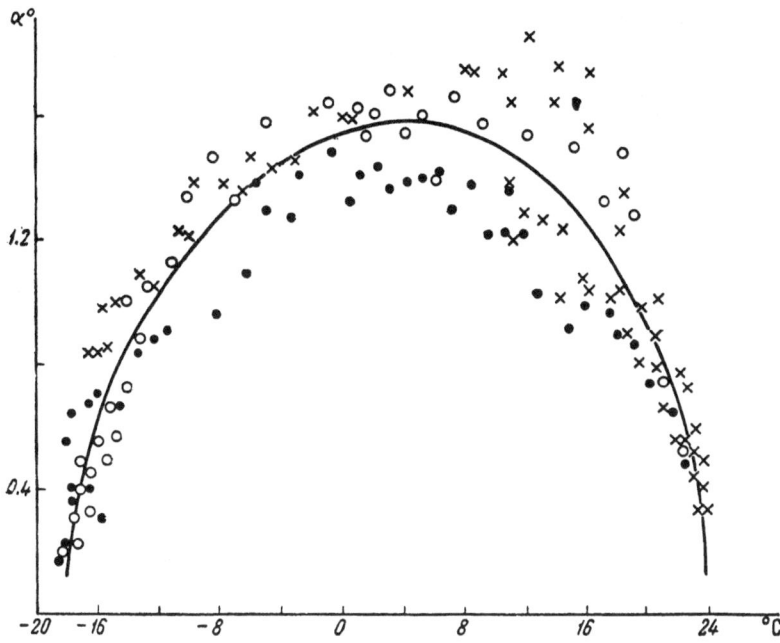

Fig. 99. Angle α between extinction positions of adjacent domains in
Rochelle salt (Chernysheva).

Finally, we may note that materials other than ferroelectrics show migration of boundaries between poly-synthetic twins in response to mechanical stress; it has been observed in alloys (§ 2.21). Before we turn to this, we must deal with some details of the relation between the components of the twin and the electric domains.

20. Twin Components as Regions of Spontaneous Polarization

The components of a polysynthetic twin in Rochelle salt (and in other ferroelectrics, e.g., barium titanate) are regions of spontaneous polarization (domains), as one may show by placing a plate (e.g., an X-cut plate of Rochelle salt) on the stage of a microscope between liquid electrodes (Chernysheva, 1951). A field parallel to the a axis produces an effect analogous to that of a mechanical stress; the boundaries are displaced and the pattern is altered. Reorientation occurs by expansion of one component at the expense of the other. The field strength can be made such that the conversion to the monocrystalline state is nearly instantaneous (occurs in less than 1/32 sec); the field required is dependent on the temperature and on the homogeneity of the plate (e.g., 10-60 V/mm at 20°). Figure 100 shows frames from a cine film (Chernysheva); part a is before the field was applied; b is with the field on; c to h represent gradual recovery with the field off. A reverse field produces a monocrystal by expanding the other component (the dark one), as in part j; parts k to p illustrate the recovery for this case. The mode of displacement of the boundaries when the field is removed is not the same in the two cases, for in the second case the lost component at first grows mainly in length and only later broadens.

The wedge-shaped twin layers temporarily (Fig. 100m) extend even beyond the initial boundary; then (Fig. 100n) they start to contract as they begin to broaden. This broadening is most rapid at the ends, which may divide. "Then the ends fuse together to give a sharp boundary at the place where it was before the field was applied. Restoration of this boundary is very rapid indeed, which may result in a ferroelectric Barkhausen effect" (Chernysheva, 1955).

Parallel displacement of the twin boundaries (horizontal) occurs during recovery in the first case (Fig. 100, a-h).

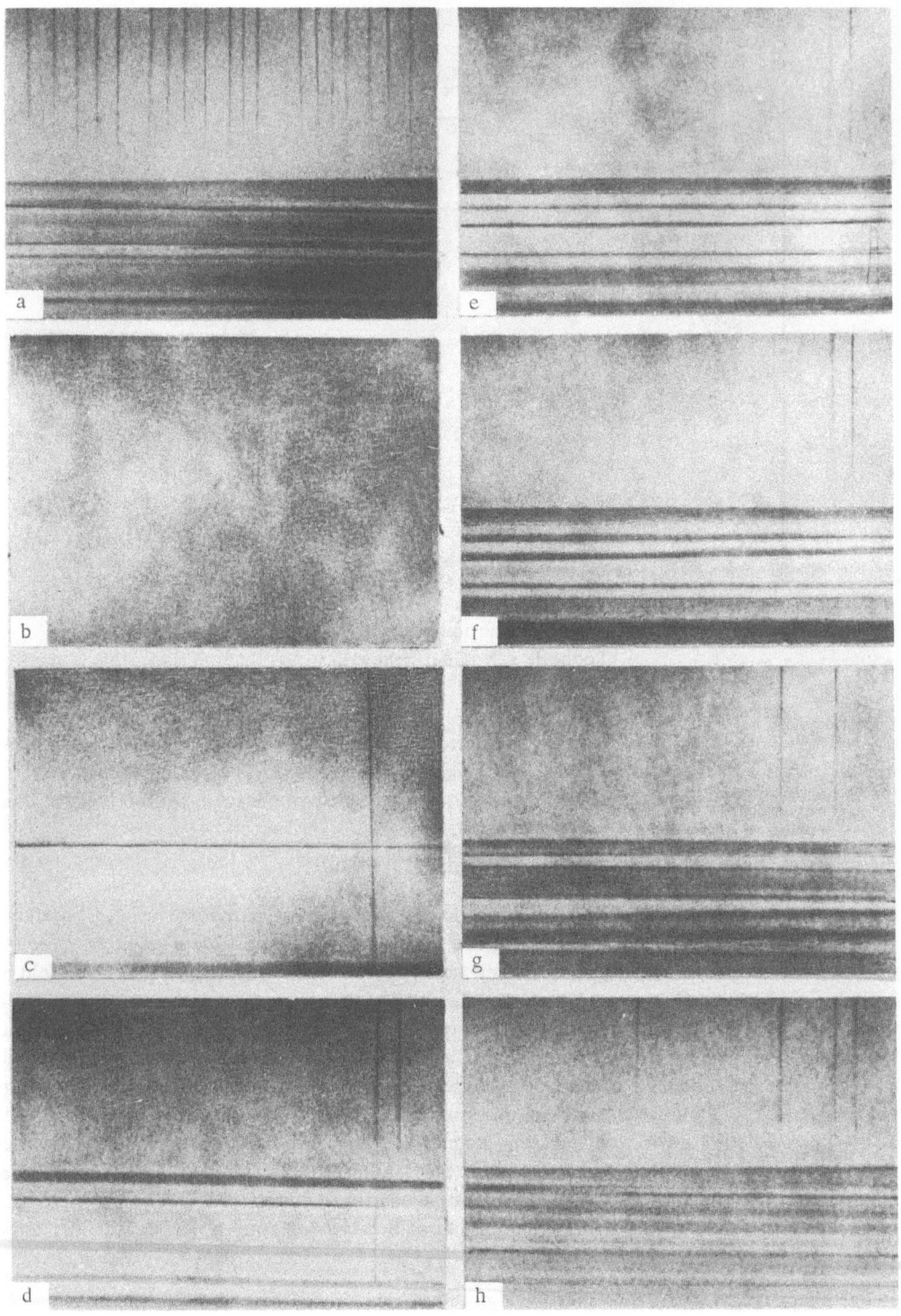

Fig. 100. Alteration of a polysynthetic twin in Rochelle salt by an electric field (crossed Nicols); frames from a cine film: a) and i) initial state; b) and j) monocrystal state

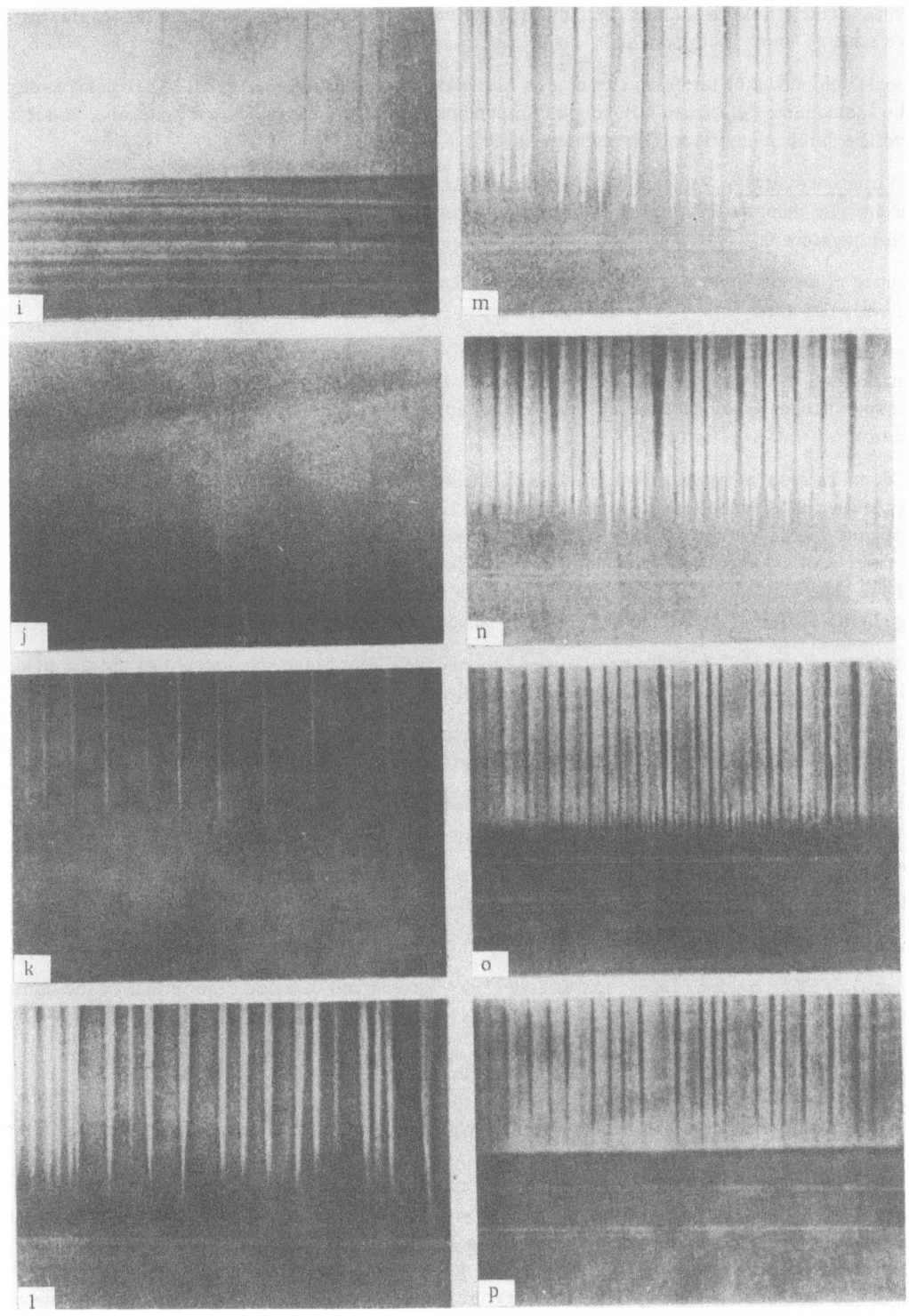

produced by fields of opposite direction; c) to h) and k) to p) recovery of twin structure
after removal of field (Chernysheva).

An external electric field thus alters the domain structure in much the same way as a mechanical stress; a transition to a state of lower energy occurs.

Many special properties of ferroelectrics can be elucidated from optical studies of twin-boundary displacement and of the spontaneous deformation in single components (domains); examples are hysteresis, polarization, unipolarity, and the Barkhausen effect (Chernysheva, 1955).

Domain Structure of Barium Titanate. $BaTiO_3$ resembles Rochelle salt in that the components of the polysynthetic twins are also domains. Twinning occurs by polymorphic transition; the material is cubic above 120°C, [*] the sequence (temperature falling) being cubic → tetragonal → orthorhombic → trigonal.

Ferroelectric properties appear at 120°C and below. The domain structure is complex, because there are several equivalent polar axes.

The tetragonal phase has 90° and 180° domains as its two basic types of domain structure. The polar axes of adjacent components meet at 90° in the first case and are antiparallel in the second. The 90° and 180° types may be present together, in which case the 90° ones are broad bands containing the 180° ones as a substructure (Fig. 101, a and b).

The 90° domains are readily seen in transmitted polarized light or from the surface relief in reflection. The boundaries (called 90° boundaries) are parallel to the (110) and (101) planes of the tetragonal phase. The 180° domains require special conditions for detection, because the optical indicatrices of adjacent 180° domains are identical. An electric field perpendicular to the [c] axis is required to make these domains visible in polarized light. The antiparallel [c] axes diverge slightly in response to the field, so the components are seen as having slightly differing extinction angles. Mechanical stress produces a similar effect.

The boundaries of the 180° domains can be seen by polarized light alone, on account of the distortion (§ 1.15) localized there; but they are best revealed by etching.

The 180° domains are inversion twins, while the 90° ones are twins by reflection in a plane normal to the twin direction (or twins by rotation through 180° about an axis normal to the twin plane).

The orthorhombic phase also has polysynthetic twins forming a domain structure; 90° and 60° (120°) domains occur. No search has been made for 180° domains.

The 90° boundaries separate optically distinct domains and run along the (001) planes; the 60° boundaries are parallel to the (011) planes and also separate optically distinct domains.

Only domain boundaries on the (100) pseudocubic planes have so far been observed in the trigonal phase.

Shirane et al. (1955), Sachse (1956), and Kenzig (1957) have given detailed descriptions of the domain structure of crystals of barium titanate.

21. Elasticity of an Alloy: Analogy with the Elasticity of Rubber

Elasticity resembling that of rubber occurs at low temperatures and near transition points in the following alloys: In − Cd (4.5 at.% Cd); Cu − Mn (88 at.% Mn); Cr − Mn (91.7 at.% Mn); Au − Cd (47 - 50 at.% Cd); In − Tl (18 at.% Tl); see Betteridge (1938), Worrell (1948), Carlile et al. (1949), Chang and Read (1951), Burkart and Read (1953), and Basinski and Christian (1954a and 1954b). Polysynthetic twins are produced in these by polymorphic transformation, as for Rochelle salt (§ 2.15), and mechanical stress displaces the twin boundaries. Figure 102 shows such a twin in an indium alloy containing 18 at.% Tl, which was produced by the conversion of the face-centered cubic (high-temperature) form to the face-centered tetragonal (low-temperature) form (transition point 100°C). A specimen bent in an ice-acetone bath shows one component (e.g., the light bands) expanding at the expense of the other; this can be seen by polarized light. The entire specimen can be made light by suitable bending; conversely, it can be made entirely dark by reverse bending.

---

[*] Perovskite lattice.

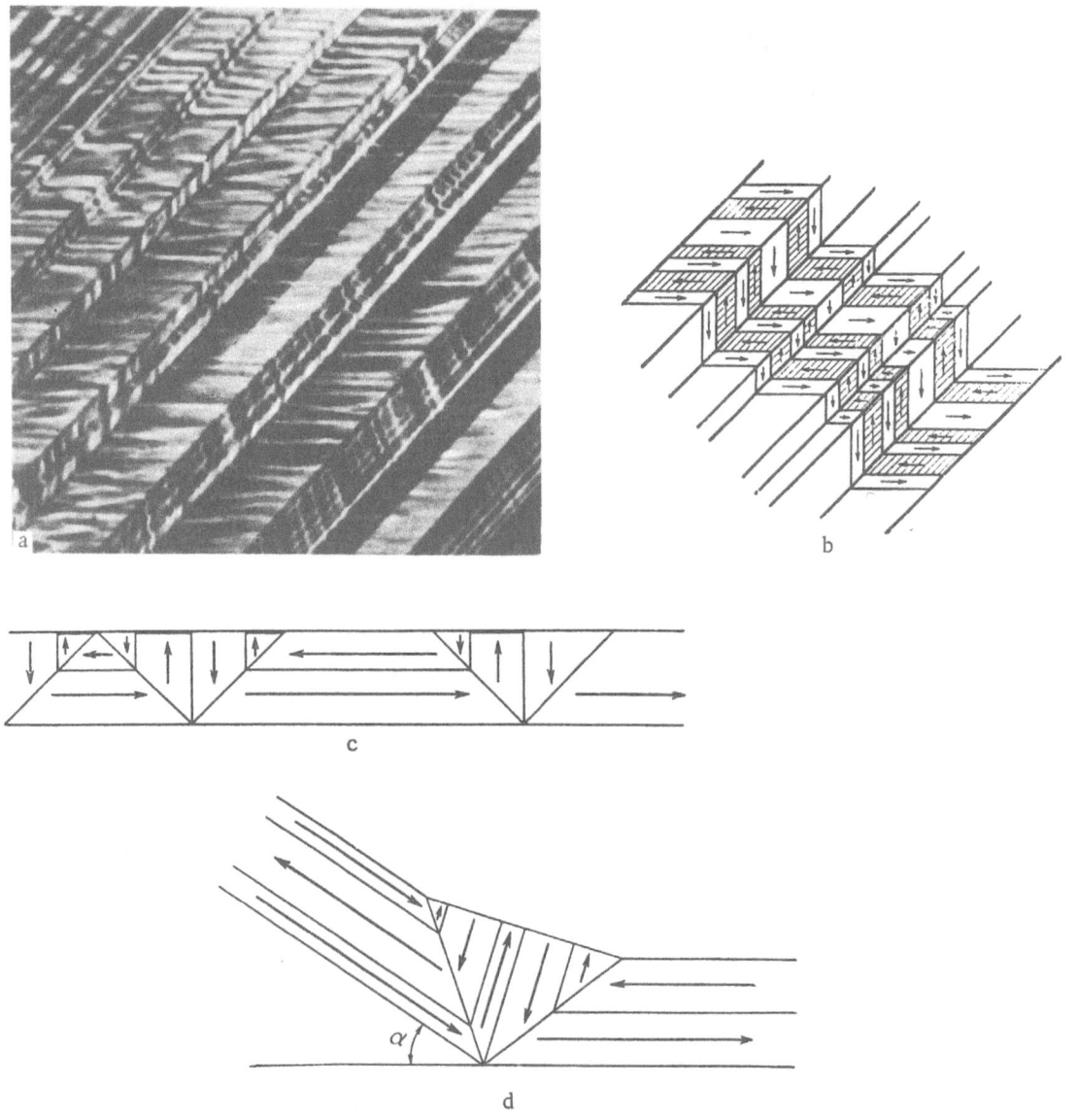

Fig. 101. Polysynthetic twin in barium titanate: a) in polarized light ($\times$ 50), with 90° and 180° domains visible; b) scheme of the domain structure (the thick straight lines represent 90° boundaries, and the zigzag ones 180° boundaries), with arrows representing the directions of the polar axes; c) scheme for a plate containing a polysynthetic twin, in which the 90° boundaries form wedge-shaped regions, the relief on the upper surface being different from that on the lower one; d) another instance of the effects of domain structure on surface relief in $BaTiO_3$; α = 1°15' (Merz).

Figure 103 shows Laue patterns from unloaded and loaded specimens; the first clearly shows double spots corresponding to the twin structure, while the second corresponds to a monocrystal. This alloy can be converted to the monocrystalline state by mechanical means, as for Rochelle salt.

The specimen straightens spontaneously if the bending forces are removed while it remains in the cooling bath. A specimen stretched and released under these conditions will contract as does rubber. Stretching is accompanied by a sound resembling the cry of tin under these conditions. This type of elasticity does not

Fig. 102. Twin structure of an indium-thallium alloy seen in polarized light; twinning produced by conversion of the cubic form to the tetragonal one (Burkart and Read).

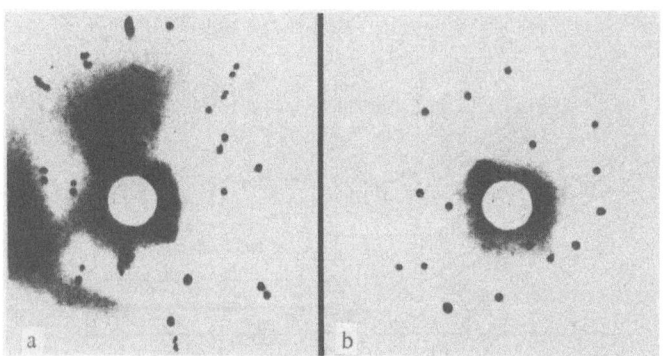

Fig. 103. Laue patterns of an indium-thallium alloy: a) unloaded; b) loaded (Burkart and Read).

occur in this alloy at room temperature. A specimen bent at a low temperature, heated to room temperature (still held in the bent form), and then recooled is no longer able to straighten out at a low temperature; in fact, it returns to the bent form if an attempt is made to straighten it.

This type of elasticity occurs in this alloy between -5 and -10°C (Burkart and Read, 1953). The alloy is plastic at room temperature, though work-hardening does occur. Basinski and Christian (1954b) have observed similar properties in this alloy, but over the range from -180 to 0°C; deformation at -180°C requires a rather larger force. The specimen rapidly straightens out after deformation at -180°C. The effect is dependent on the time of action of the load at temperatures not far below 0°C; the specimen straightens out rapidly if the load is soon removed, but the bent form persists if the load has acted for a long time. The time needed to produce residual deformation becomes shorter as the temperature rises. The plasticity is entirely regained at 20°C; the material can be bent into a ring. The plasticity increases with the temperature; spontaneous deflection under its own weight occurs near the transition point (about 100°C). A bent specimen heated to the transition point suddenly and spontaneously straightens out.

The elasticity above 100°C is analogous to that between -180 and 0°C, except that the forces are larger and that the time of action is unimportant. Elasticity in this range is a result of the martensite-type transformation; the mechanism is rather different from that for low-temperature elasticity. Basinski and Christian note that the stresses needed to deform the cubic modification are larger than those for the tetragonal one.

Metallographic and x-ray studies indicate that this elasticity is associated with displacement of twin boundaries. Figure 104 illustrates the effects of bending, while Fig. 105 shows interlocking twin components. Displacement of the boundaries requires very little displacement of the atoms in this alloy (Burkart and Read, 1953), so there is very little increase in the shearing stress along the boundary (Zener, 1949).

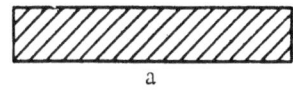
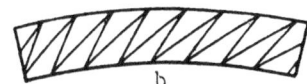

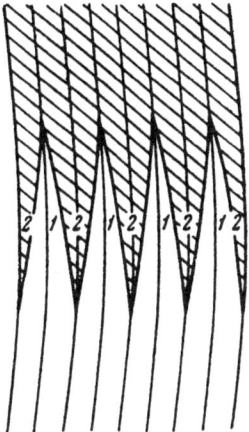

Fig. 104. Elastic bending of an indium-thallium alloy by displacement of twin boundaries: a) undeformed specimen [hatching parallel to the traces of the (101) planes, which correspond to the boundaries of polysynthetic twins]; b) positions of the traces of the (101) planes in a bent crystal whose upper and lower surfaces have become monocrystalline, the twin components in the body being wedge-shaped (Basinski and Christian).

Fig. 105. Bending of a crystal as a result of tapering twin layers (hatched); the plane of section is normal to [101]. The relief of the boundaries between the components is shown schematically (Basinski and Christian).

All the above observations are readily explained in terms of differences in the rates of the relaxation processes at the boundaries at high and low temperatures. The rates of relaxation at room temperature are so high that the displacements produced by a stress are instantly stabilized, so the new array persists when the force is removed (residual deformation). Relaxation is absent, or occurs very slowly, at low temperatures; the new position of the boundaries is not stabilized, and it is unstable with respect to the unstressed crystal, so the crystal reverts to its initial form when the load is removed.

Rubber-like behavior can also occur if the tetragonal lattice contains impurities distributed in a definite way. The moving twin boundaries encounter these and displace them to positions of high energy. These atoms are able to diffuse through the lattice to lower-energy positions at high temperatures, which stabilizes the new configuration. No such diffusion occurs at low temperatures, so the new configuration is unstable with respect to the unstressed crystal.

The instability of the monocrystalline state in Rochelle salt or $BaTiO_3$ is caused partly by the low relaxation rates and partly by the electric fields set up between the surfaces of the crystal (§ 2.19).

Table 7 lists the crystals at present known to show detwinning (monocrystallization) in response to mechanical stress. All these crystals give polysynthetic twins by polymorphic transition; the monocrystalline state is produced by mechanical stresses. The polysynthetic state is regained either rapidly or slowly when the stress is removed. The ferroelectrics (Nos. 4 to 8 in Table 7) can also be made monocrystalline by electric fields (mono-domain structure). Elastic monocrystallization should occur in any ferroelectric whose domains are components of twins that alter the form of the crystal. Triglycine sulfate and other such ferroelectrics show no change of form on twinning; the components have the same elastic constants, and monocrystallization is not elastic (§ 4.3 and 9).

TABLE 7

Detwinning of Polysynthetic Twins* (Elastic Monocrystallization)

No.	Crystal	Symmetry	Literature reference
1	$(NH_4)_3(HSO_4)_2$	Monoclinic	Fisher (Friedel, 1926, p. 493)
2	$Al_2O_3 \cdot 3CaO \cdot 3CaCl_2 \cdot H_2O$	Monoclinic	Friedel, 1926, p. 493
3	Rochelle salt**	Monoclinic	Klassen-Neklyudova, Chernysheva, and Shternberg, 1948; Chernysheva, 1950; Fisher, 1954
4	WO_3	Monoclinic	Rhodes, 1952; Tamsaki, 1958; Nakamura, 1956
5	$NaNbO_3$	Orthorhombic	Cross and Nicholson, 1955
6	$KNbO_3$	Orthorhombic	
7	Au−Cd	Orthorhombic	Chang and Read, 1951
8	$BaTiO_3$	Tetragonal (below 120°C) Orthorhombic (0 to -70°C) Trigonal (below -70°C)	Kay, 1948
9	In−Tl	Tetragonal	Burkart and Read, 1953; Basinski and Christian, 1954b; Betteridge, 1938; Worrell, 1948
10	In−Cd	Tetragonal	Carlile et al., 1949
11	Cu−Mn	Tetragonal	
12	Cr−Mn	Tetragonal	

* Polysynthetic twins are produced in polymorphic transitions.
** Elastic monocrystallization should occur in any ferroelectric whose domains are the components of twins causing change in the form of the crystal.

22. Distribution of Residual Stresses in Displacement for the Boundaries of Polysynthetic Twins and of Domains

Figure 36, parts b and c, shows the displacement of twin (domain) boundaries by the stresses set up in a plate of Rochelle salt by a localized load.

The twin (domain) structure also responds sharply to any residual stresses. For example, Fig. 106 shows the effects for Rochelle salt (stresses concentrated at the ends of cracks), and Fig. 107 does the same for the effects of inclusions (Chernysheva, 1955). This property can be used to examine the stress distribution in metals (monocrystalline or otherwise), in ferroelectric crystals (Rochelle salt, barium titanate, and so on), and in components made of these.

This new optical method differs from the earlier polarization method (which uses changes in the birefringence) because the effect is caused by displacement of the boundaries between regions whose optical indicatrices have different orientations (Indenbom and Chernysheva, 1956), instead of by distortion of the indicatrix

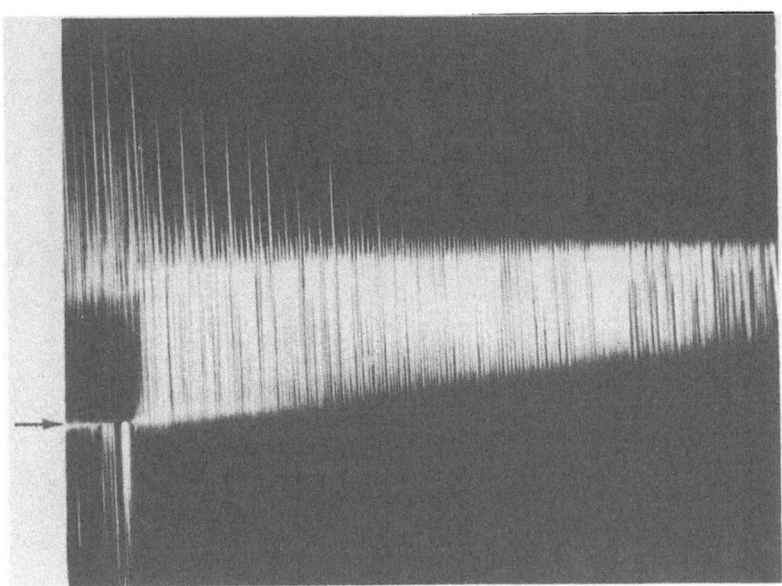

Fig. 106. Partial monocrystallization (dark field) of a polysynthetic twin in Rochelle salt in response to stresses localized at the end of a crack; crossed Nicols, X cut (Chernysheva).

(photoelasticity). This at once enables one to deduce the sign of the stress, which requires the use of sensitive-tint plates in the polarization method; the new method can also be applied to opaque materials. The effect has been used in this laboratory to check stress distributions calculated from the theory of elasticity for edge dislocations (Indenbom and Chernysheva, 1956). Figure 108 shows the lines of equal tangential stress $\tau_{xy} = \pm \tau_0$, which form a characteristic pattern around the dislocation. The stresses exceed τ_0 (in absolute magnitude) within the lobes. The black regions correspond to stresses of one sign, and the white ones to stresses of the opposite sign. Figure 107b shows a plate of Rochelle salt cut normal to the twofold axis, in which the light and dark parts form a pattern similar to that of Fig. 108; the light parts are produced by local broadening of the light components, and conversely. This broadening is associated with localized tangential stresses; in the equilibrium state (no mechanical or electrical stress), the components (domains) are plane-parallel plates of equal thickness, whose broad faces are perpendicular to the Y axis or the X axis. Mechanical or electrical stress causes one component to grow at the expense of the other; mechanical stress causes broadening in the component whose spontaneous* deformation ε_{yz}^0 coincides in sign with that of the tangential stress τ_{yz} in the twin plane along the twin direction. The components broaden as though the boundaries were subjected to a pressure $P = 4\varepsilon_{yz}\tau_{yz}$. The ratio of thickness at any given point is governed by the local τ_{yz} if this varies little within the thickness of a domain. One component is ejected completely above a certain τ_{yz}, so the specimen is locally a monocrystal. The two parts of Fig. 107 show that the light and dark regions agree well with the pattern given by theory (Fig. 108), although the anisotropy in the elasticity was neglected.

Figure 107a shows a lighter region on the right and a darker one on the left in the upper part of the plate; the two are reversed in the lower part, which corresponds to the presence of two dislocations differing in sign. The two are not independent but are related by an extra plane. The picture becomes diffuse if the focus of the microscope is altered; this means that the extra plane is bounded on all sides. The defect shown in Fig. 107a is a closed-ring edge dislocation lying in the XZ plane; rough calculations give the shear vector as a few hundred angstrom units, whereas the lattice parameter along this axis is about 14 A. This is therefore a macroscopic dislocation, and the extra plane is not simply atomic. Careful examination of Fig. 107a reveals a minute inclusion joining the the centers of the two dislocations. Figure 107b shows clearly a longish inclusion parallel to Z, which alters the domain structure as shown in Fig. 108, except that the patterns corresponding to the ends of the inclusion are displaced and partly overlap. This example illustrates the advantages of this new optical method; the displacement of the boundaries enables one to check the calculations on the stress distributions around dislocations, and it also enables one to detect macroscopic edge dislocations.

---

*That is, the deformation of the unit cell in the transition from orthorhombic to monoclinic.

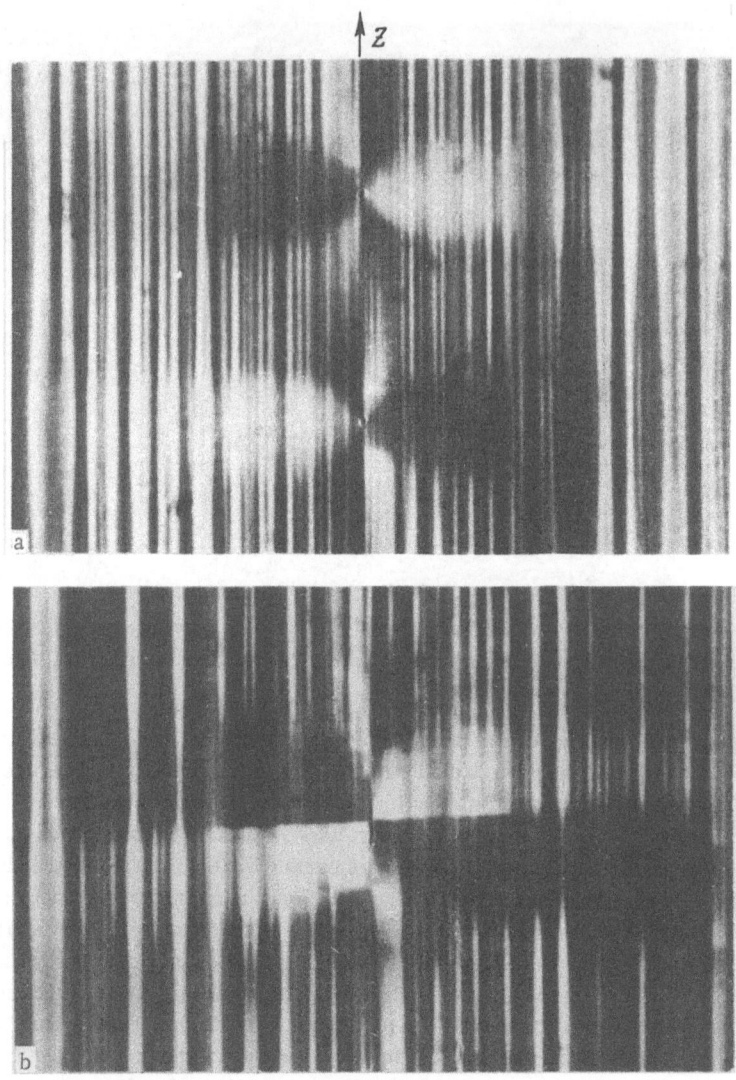

Fig. 107. Domain distributions: a) around a pair of edge dislocations differing in sign, extra plane along the Z axis; b) around an inclusion. Crossed Nicols (Chernysheva).

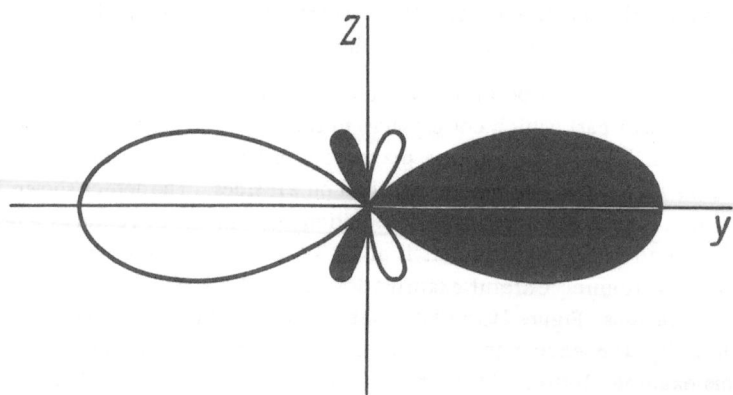

Fig. 108. Calculated lines of equal tangential stress $\tau_{xy} = \tau_0$ for an edge dislocation whose extra plane lies along the Z axis. The black regions correspond to $\tau_{xy} > \tau_0$ and the white ones to $\tau_{xy} < \tau_0$ (Indenbom and Chernysheva).

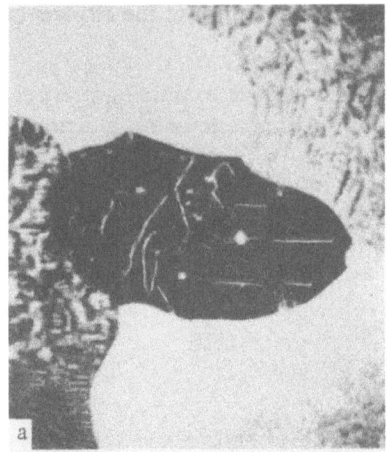

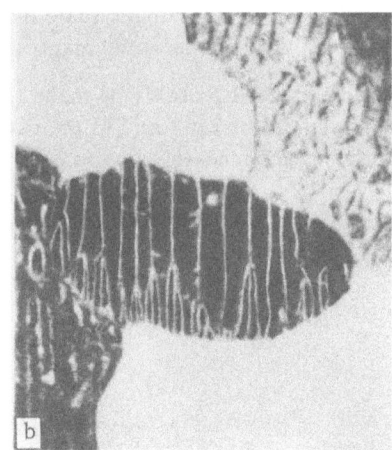

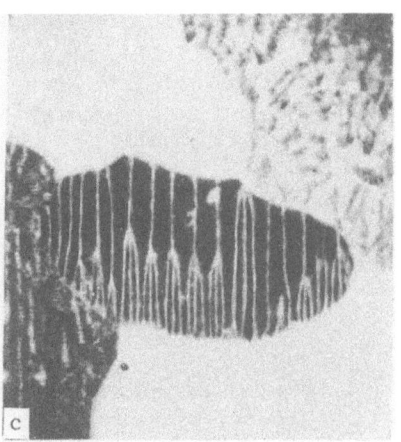

Fig. 109. Reorientation of magnetic domains in grains of silicon iron in response to tensile stress; the domains are revealed with magnetic powder. a) Specimen unloaded, only residual stresses present, domains horizontal; b) specimen loaded, some vertical domains; c) load increased. × 40. (Dijkstra and Martins).

The magnetic domains in a ferromagnetic (Néel, 1944) are sensitive to residual stresses of scale comparable with or larger than the domains (Dijkstra et al., 1954). It is rather more difficult to observe changes in these domains, but the boundaries can be revealed by the application of a colloidal magnetic suspension to the carefully polished surface. The particles are drawn to the boundaries on account of the fields between domains and so give rise to Bitter-Akulov figures (1903 and 1939). These patterns are now used in studies on the macroscopic stresses in steels (Fig. 109); the stresses set up by deformation and rupture can be examined, as well as residual stresses (Dijkstra et al., 1954). Kaczer (1955) has described a method of revealing the domains by means of a thin probe scanned over the surfaces. Changes in the direction of magnetization are recorded by an oscilloscope (Fig. 110). Shur and others (1958-1960) have published several papers on the effects of elastic stresses on the magnetic structures of crystals.

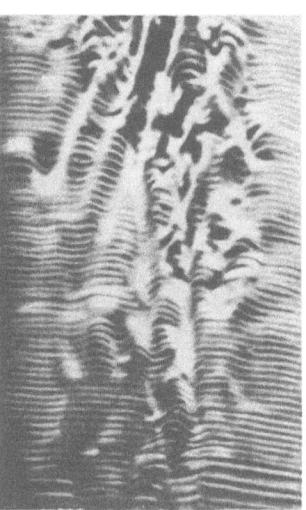

Fig. 110. Domain structure of a ferromagnetic (polished section of silicon steel). The domains are revealed by a permalloy probe moving over the surface, the change in induction current being recorded by an oscilloscope (Kaczer).

Stress distributions on the atomic scale may be examined by reference to the displacement of domain boundaries by mechanical stresses or by magnetic fields.

Considerable use will probably be made in the future of methods of examining stress distributions (and also dislocations and defects) in terms of the patterns given by polysynthetic twins, electric or magnetic domains, displacement of boundaries, and so on.

CHAPTER 2

TWINNING WITHOUT CHANGE OF FORM

§ 3. Geometry, Crystallography, and Relation to Atomic Structure

1. Laws of Twinning for Natural Quartz Crystals

Quartz is hexagonal and belongs to the trigonal-trapezohedral class (right-handed $\overset{\frown}{L}{}^3 3L^2$, left-handed $\overset{\frown}{L}{}^3 3L^2$).[*] A lattice of this class lacks a center or plane of symmetry; two varieties of quartz crystal (right- and left-handed) occur in nature (Fig. 111, a and b). Quartz forms twins in accordance with the Dauphiné, Brazilian, and Japanese laws, and also Leydolt twins.

Figure 112 shows three cases of twinning in the low-temperature form of quartz (Shubnikov, 1940). Figure 112a represents a Dauphiné twin; here two left-handed or two right-handed crystals interpenetrate. A Brazilian twin (Fig. 112b) has a right-handed crystal with its $(11\overline{2}0)$ prism plane in common with a left-handed crystal, the principal axes ($\overset{\frown}{L}{}^3$ and $\overset{\frown}{L}{}^3$ screw axes) being parallel. Nine types of Japanese twin (Fig. 112c) are found in nature (Kozu, 1952), which include intergrowths of:

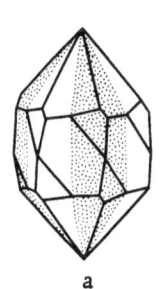
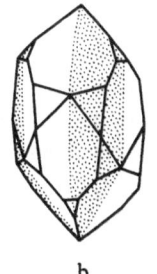
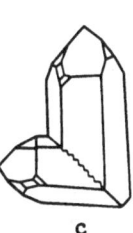

Fig. 111. Shapes of quartz crystals: a) right-handed; b) left-handed.

Fig. 112. Quartz twins: a) Dauphiné; b) Brazilian; c) Japanese.

---

[*] The lattice has a threefold screw axis ($\overline{L}{}^3$), the sense of the screw providing the distinction. Macroscopically, a quartz crystal has a simple threefold axis (L^3).

a) right with left, b) left with left, and c) right with right. The principal axes always form an angle of 84°33'. Leydolt twins arise by combination of Dauphiné and Brazilian twins.

Dauphiné twins are rotational (axial) twins; they may be described geometrically by rotation about an L^2 axis or about an L^3 axis. We shall see that they can be produced mechanically.

Brazilian twins and Japanese twins of type a are reflection twins. One component is right-handed and the other is left-handed, so the two lattices cannot be brought into coincidence by rotation; such twins cannot be produced mechanically. They are produced by intergrowth of two individuals, as may be seen from the inclusions at the twin plane.

Japanese twins of types b and c may be described as reflection twins or as twins obtained by rotation through 180° about an axis perpendicular to the plane of intergrowth. Stepanov considered that it should be possible to produce Japanese twins by the action of forces in the $(11\bar{2}2)$ plane, but experiment (Tsinzerling et al., 1960) shows that localized loading along this plane produces Dauphiné twins, not Japanese ones.

These examples show that mechanical twins are characteristically rotation twins; only for crystals of high symmetry are they simultaneously reflection twins.

2. Residual Mechanical Twins in Quartz Produced by Localized Loading; Methods of Detecting Twins, Shapes of Twins and Twin Axes

A quartz crystal always breaks without change of shape; prior to Shubnikov and Tsinzerling's work of 1932, it was considered that the fracture was absolutely brittle, but they showed that a localized load can cause twinning. Residual twins were produced by slowly forcing a steel ball (diameter 4-6 mm) into a quartz plate. Conical or pyramidal cracks (pressure or impact figures) extending to depths of 1-3 mm were produced, with their vertices at the point of contact. This point was also the source of residual twins, which extended for several millimeters. Figure 113 shows drawings made from photographs; Fig. 114 shows the traces of twin figures on the principal faces of quartz.

The trigonal symmetry enabled Shubnikov to give the spatial form (Fig. 115) for a Dauphiné twin for the ideal case of a force acting uniformly in all directions from the center of the crystal.

Good and regular figures are produced only if the loaded surface is flat, carefully polished, and lightly etched with hydrochloric acid. A second grinding (to a depth of 1-2 mm) and further etching are required to reveal the residual twins, which are seen because the etch figures in these are turned with respect to those in the rest of the crystal. These twins are not seen when the crystal is examined in polarized light, which shows that the L^3 axes are parallel.

Figure 116 shows two Laue patterns taken along L^3; the first is for undeformed quartz, and the second for mechanically twinned material. The first corresponds to a threefold symmetry axis; the second, to a sixfold one. The symmetry of the pattern for each component corresponds to L^3, but the transmission through the two components gives the appearance of an L^6 axis by superimposition.

The optical and x-ray studies show that mechanical twins are of Dauphiné type (§ 3.1; axial or rotation twins). Quartz is polar (lacks a center of symmetry), so the mechanical twins are purely twins of the second kind. The axes for Dauphiné twinning can be $L^3 = [0001]$, $L^2 = [10\bar{1}0]$, or $[11\bar{2}0]$ if the coordinate axes are chosen in another way.

3. Redistribution in the Formation of Dauphiné Twins; Ideal Spatial Form of a Mechanical Twin and Atomic Structure of Twin Sutures

The contours of Dauphiné twins can be wavy, curvilinear, rectilinear, and so on, but the figures produced at room temperature (Fig. 113) usually have their contours straight and parallel to the $\{10\bar{1}0\}$ prism planes. If we try to describe the mechanical twins of quartz as reflection ones, we must take $\{10\bar{1}0\}$ as the S plane, with $\eta_1 = [1\bar{2}10]$. However, there are several objections to any attempt to consider such twins as formed by a set of simple shear operations (as in the case of calcite), because the twinning-ellipsoid concept is not applicable to quartz. In the case of materials such as calcite (§ 1.2), there is either macroscopic displacement of parts of the crystal or division into layers having alternately the new and old orientations (polysynthetic twins). This effect

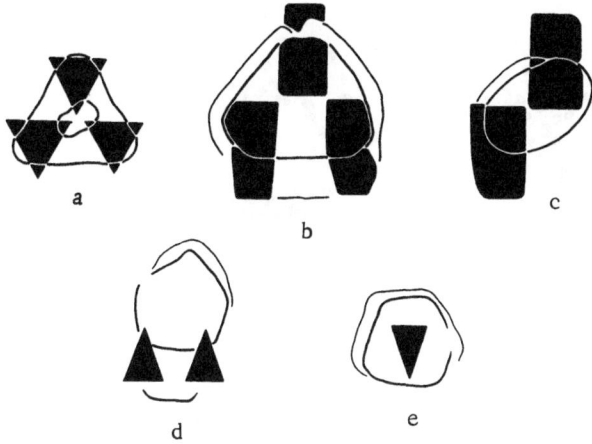

Fig. 113. Figures produced on faces of quartz crystals
by a steel sphere at room temperature: a) triangular
figure on a (0001) face, the three dark regions being
mechanical Dauphiné twins; b) two triangular figures
on a (10$\bar{1}$0) face, the three dark regions being the
twins; c) oval figure on a (11$\bar{2}$0) face, the dark regions
being the twins; d) figure on a (10$\bar{1}$1) face of the
positive rhombohedron, the dark triangles being the
twins; e) pressure figure and twin formed in it on a
(1$\bar{1}$01) face of the negative rhombohedron. All specimens were etched in hydrofluoric acid (Tsinzerling.)

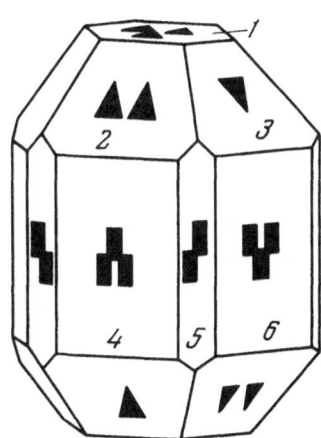

Fig. 114. Traces of mechanical
twins on the principal faces of
quartz: 1) faces (0001); 2) (10$\bar{1}$1);
3) (01$\bar{1}$1); 4) (10$\bar{1}$0); 5) (11$\bar{2}$0);
6) (01$\bar{1}$0) (Tsinzerling.)

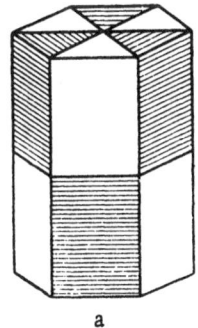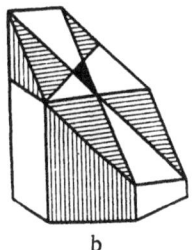

Fig. 115. Ideal form of a mechanical
twin in quartz for the case of a force
acting uniformly in all directions from
the center: a) section of the basal plane
and prism; b) section on an arbitrary
face. The black triangle corresponds to
the figure of Fig. 113d. (Shubnikov.)

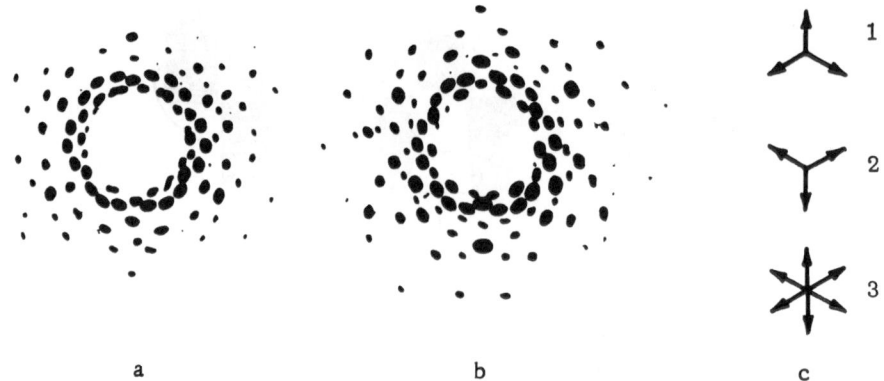

a   b   c

Fig. 116. Laue patterns produced along the L^3 axis in quartz: a) undeformed, pattern corresponding to a monocrystal; b) deformed region, rays passing through both components of the twin, so symmetry increased; c) scheme explaining the increased symmetry of the deformed region: 1) and 2) symmetry of each component separately; 3) symmetry from simultaneous examination of the two components. (Shubnikov and Tsinzerling.)

does not occur in quartz. Shubnikov and Tsinzerling supposed that twinning occurs in quartz without macroscopic deformation (residual deformation); only molecular rearrangement occurs.

Figure 117 shows that the simple shear of b is needed first, which is followed by rotation through 60° or 180° about the screw axes, in order to convert the initial array of a to the twin form of c. It also shows that rotation alone can produce the result; shear is not essential. Shubnikov concluded from this that the twinning of quartz cannot be reduced to deformation in simple shear; he doubted whether molecular groups could be turned through such large angles without breaking bonds.

He supposed that the atoms undergo small displacements without breaking bonds; the twinning involves no final change in the shape and size of the specimen. He used Bragg and Gibbs' (1925) model of the structure in this analysis.

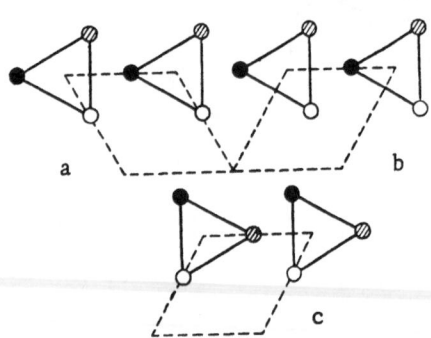

Fig. 117. Rearrangement of atoms in quartz in mechanical twinning: a) before; b) after shear; c) after shear and rotation through 60° around screw axes (Shubnikov).

Figure 118 shows the structure of β-quartz projected on a plane normal to the threefold screw axis; Fig. 119a shows the same for a simplified model, which contains only the silicon atoms. Figure 119b shows the simplified scheme for a Dauphiné twin, whose components meet along the line CD. The region with the initial structure lies on the right. Figure 119c shows the formation of a Dauphiné twin by the displacement of silicon atoms in the directions shown by the arrows. The atoms in each of the three planes at different levels are displaced by a fraction of a translation vector; the displacements are such as to leave the center of mass unchanged, so the shape of the crystal is unaltered, the array of atoms in one component being obtained by rotating the projection for the other component through 180° about the L^3 axis, which is normal to the plane of the projection.

No macroscopic displacement occurs, so Shubnikov described this as the third main type of residual deformation (the others are deformation by slip and twinning, as in calcite).

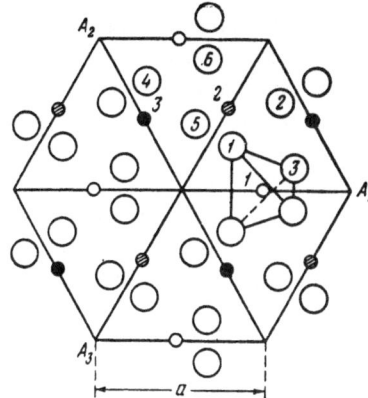

Fig. 118. Structure of left-handed
β-quartz (Bragg and others). The
large open circles are oxygen atoms;
the small open ones, silicon atoms
in the plane of section; the filled
ones, silicon atoms lying 2c above
the plane; and the cross-hatched
ones, silicon atoms lying c below
the plane; a = 4.90 A.

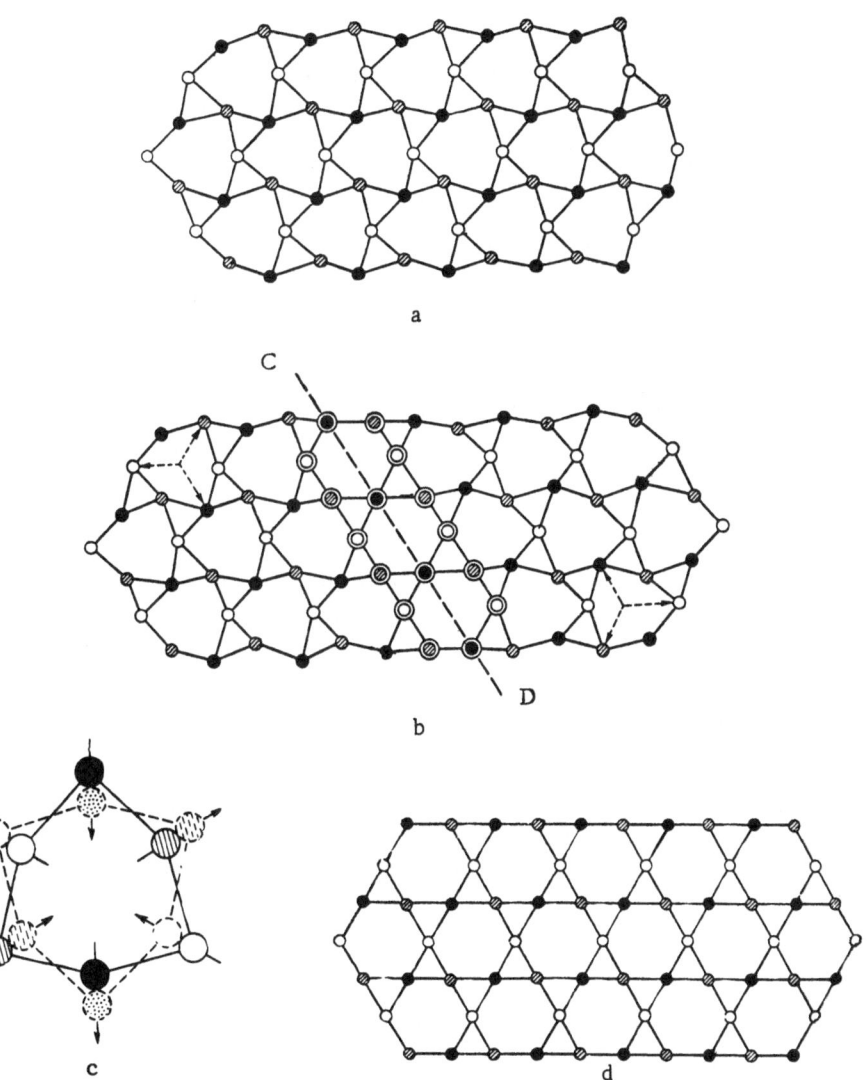

Fig. 119. Simplified scheme for the structure of quartz; the circles denote silicon atoms
in twinning (full lines denote the initial position, broken ones the new position); d) structure of the high temperature form, α-quartz.

Figure 119b shows that the array of atoms in the contact region differs somewhat from normal. The grouping along the line CD in β-quartz corresponds to sixfold symmetry (that of α-quartz). Figure 119d shows for comparison the structure of α-quartz projected on a plane normal to a sixfold axis. Figure 119c also shows that the silicon atoms pass through positions corresponding to the high-temperature form during the twinning operation. Figure 119b shows that a Dauphiné twin gives only slight elastic distortion immediately around the contact plane; the atoms are slightly displaced from their equilibrium positions. X-ray studies (Bond and Andrus, 1952) confirm that there are no macroscopic stresses. The surface energy of the contact must be slight. The properties of the twin boundary must be rather different from those for calcite; they may be curved or zigzag, and they should not give rise to cracks or transition zones (§ 1.19); they should also be fairly mobile (§ 4.5).

Figure 119b (arrows) shows the directions of the electric axes (Megaw, 1952), which are antiparallel.

§ 4. Production and Development of Twins

1. Strength of Quartz:[*] Role of Localized Stresses, Anisotropy of Twinning Forces; Flow at Room Temperature, Roles of Temperature and Time

The strength of quartz varies with orientation from 320 to 1400 kg/cm^2 in bending, from 850 to 1160 kg/cm^2 in torsion, and from 22,780 to 34,470 kg/cm^2 in compression (Berndt, 1917 and 1919). Mechanical twinning was first observed for uneven loading (localized loading), and also in torsion and bending at high temperatures (Tsinzerling, 1933, 1940, 1954b; Wooster, 1947). Loads of 50-1000 kg on the steel ball are needed to produce mechanical twins in quartz plates. Use may be made of Amsler test machines or of Brinell hydraulic presses; figures only 1-2 mm across may be produced with a small lever press.

These localized loads produce very high local stresses.

Twins are produced (Tsinzerling, 1933) when such loads act on the following faces: basal, (0001); right trigonal prism, ($\bar{2}110$); left trigonal prism ($2\bar{1}\bar{1}0$); hexagonal prism, ($10\bar{1}0$); principal (positive) rhombohedron, ($10\bar{1}1$); secondary (negative) rhombohedron, ($1\bar{1}01$).

Twin figures appear on ($11\bar{2}0$) and ($10\bar{1}0$) faces at room temperature more readily than on (0001)(Fig. 120).

Wooster's model for twinning (§ 4.2) can be used to estimate the resistance to twinning as a function of orientation.

If the force is inadequate to produce twin figures, the face may still show characteristic indentation pits.

It is usually necessary to leave the quartz under load for some time in order to produce twin figures at room temperature. The prism faces require only 15-60 sec, whereas the basal face requires 20-60 min at the same load.

The figures enlarge as the time is increased; flow (displacement of twin boundaries) thus occurs even at room temperature, as Fig. 121 shows.

Twinning occurs more readily as the temperature is raised; Tsinzerling (1940, 1943) estimated the yield point of quartz (the stress needed to produce an indent visible to the unaided eye) for room temperature and for temperatures near the α⇌β transition point (573°C) as about 1000 kg/mm^2 for the basal plane (the actual stress initially was much higher, of course), the value at the higher temperature being about 15 times less.

Impact loads do not cause twinning at room temperature, but a hammer blow on a steel ball readily causes twinning at 400-450°C.

The boundaries of a mechanical twin formed at room temperature are usually straight and parallel to the twofold axes (the electric axes) and also to [0001]; those produced in a heated specimen tend to be curved (compare Figs. 113 and 120 with Fig. 122), though most of the distortion occurs near the surface (the lines become as for room temperature at a certain depth). Twin figures produced at high temperatures extend far beyond the indent; mechanical twinning becomes the only process (crushing ceases) near the inversion point, and the quartz becomes plastic (Tsinzerling, 1943).

---

[*] See Sosman (1927) on the properties of quartz, and Fairburn (1933) on the relation of properties to structure.

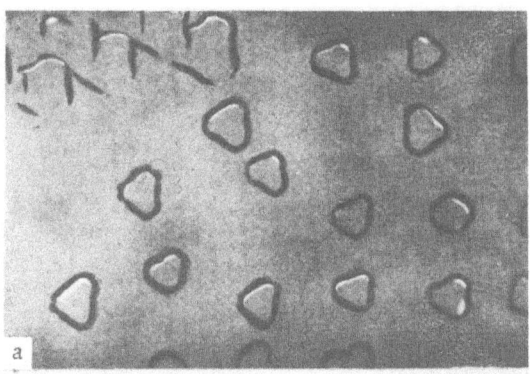

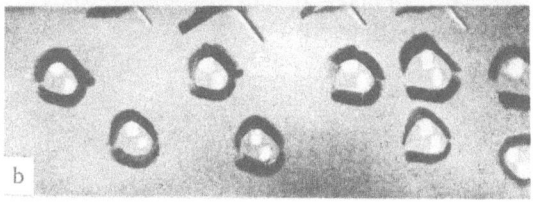

Fig. 120. Figures produced by indentation with a steel ball at room temperature on: a) (0001) at 200-300 kg; b) (1120) at 130-250 kg, with twins inside indent (Tsinzerling).

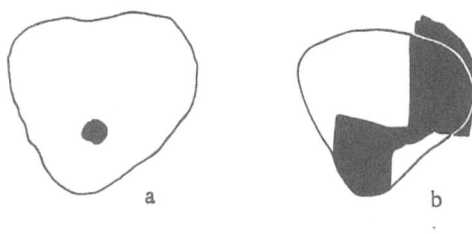

Fig. 121. Displacement of twin boundaries by a fixed force at room temperature; time of action: a) 20 min; b) 19 hr (Tsinzerling).

Of course, this is not strictly plasticity; the change of form characteristic of plastic deformation which occurs in calcite is absent from quartz (the twinning of calcite may therefore be considered as a particular form of plastic deformation).

2. Twinning of Quartz Plates Under Stress; Monocrystals Produced by Torsion and Vector Model of Twinning

Wooster et al. (1946a, 1946b, 1947, 1951) have made detailed studies of twinning in torsion, compression, and extension. Twins were produced in monocrystalline plates cut at various angles to the principal axes (Fig. 123; the angles shown are those defining the orientation). Residual twins were produced by all these types of deformation, but particularly in torsion of heated plates. Only a few random twin structures were produced in tension, because the plates broke at comparatively low stresses. A given stress distribution produced a characteristic distribution of twin regions (Fig. 124). Figure 125 shows that the pattern of twin boundaries is closely related to the state of stress.

Plates of various cuts with natural Dauphiné twins or with mechanical twins were tested in torsion at temperatures between 210 and 640°C (above and below the 573°C transition point). Measurements were made of the torque needed to turn the plate into a monocrystal.[*] Figure 126 shows the vector model of the capacity to detwin, derived from these measurements.

This vector model was subsequently given a theoretical basis from the surface of equal twin-energy density. Tsinzerling (1940) showed that a localized load caused twinning much more readily on the positive rhombohedron than on the negative one, twin figures being produced (Fig. 114). The vector model (Fig. 126) can be used to evaluate the figures occurring on the other main faces.

3. Stages of Twinning in Quartz; Sense of Displacement of Twin Boundaries and Relation to Sign of Stress

Calcite and metal crystals show four stages of twinning (§ 2.6), namely, elastic lattice deformation, elastic twinning (nonlinear elasticity), production of residual twins, and motion of twin boundaries.

Quartz shows the following stages: elastic lattice deformation, production of residual twins, and displacement of twin boundaries (flow).

---

[*]Wooster used plates 60 x 30 x 4 mm; torques of 55 to 470 g-cm were needed, in accordance with the temperature and orientation.

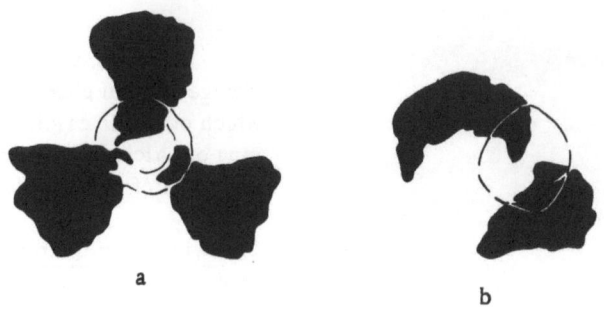

Fig. 122. Effects of elevated temperatures on the mechanical twinning of quartz: a) indent (circle at center) and twins (three lobes) from pressure on the (0001) face at 400-500°C; b) (11$\bar{2}$0) face at 550-560°C (Tsinzerling).

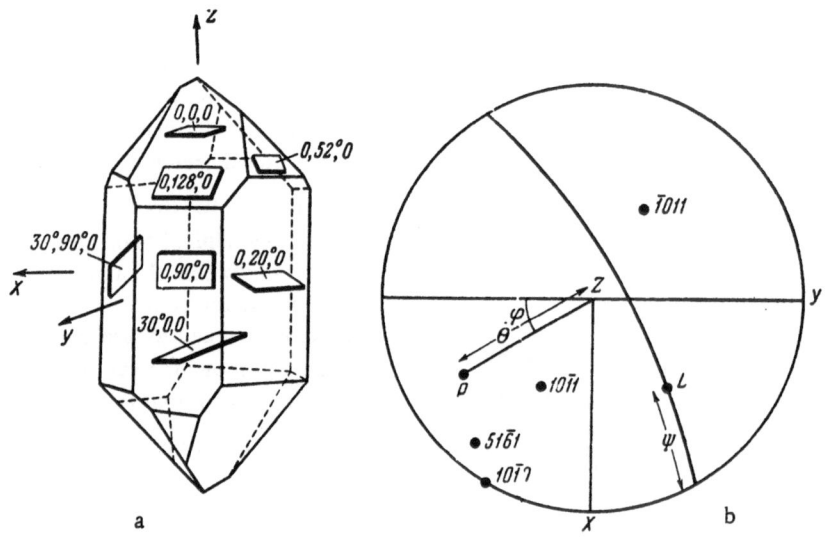

Fig. 123. Crystallographic orientations of the left-handed quartz plates: a) relative to the natural faces; b) in stereographic projection on a plane normal to the Z axis (Thomas and Wooster).

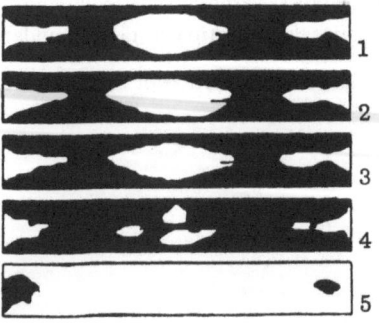

Fig. 124. Distribution of twin regions in an R-cut quartz plate deformed in bending above the inversion temperature; load applied as shown in Fig. 125, with the principal plane of the R-cut plate parallel to (10$\bar{1}$1), the long dimension being parallel to the L^2 axis. Level in plate: 1) 0.034"; 2) 0.030"; 3) 0.025"; 4) 0.020"; 5) 0.000" (Thomas and Wooster).

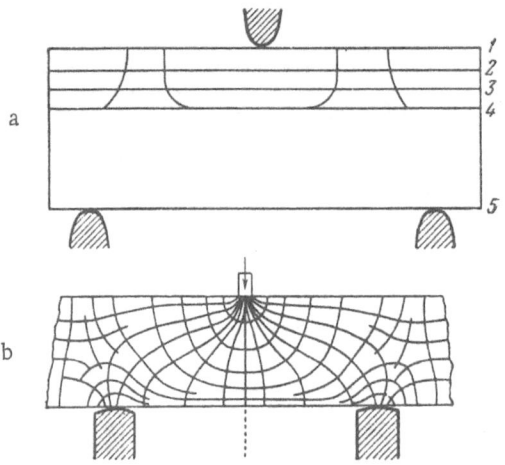

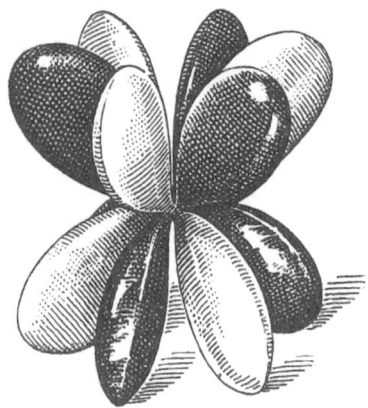

Fig. 125. Relation between the stress distribution and the boundaries of twin regions produced in bending: a) directions of twin boundaries in central part of plate; b) lines of equal principal stress for loading as shown. Level in plate: 1) 0.034"; 2) 0.030"; 3) 0.025"; 4) 0.020"; 5) 0.000" (Thomas and Wooster).

Fig. 126. Vector model for the capacity of quartz plates to detwin in torsion. The radius vector is the reciprocal of the minimum torque that eliminates twins. The dark regions correspond to orientations for which the twins are eliminated; the light ones, to orientations in which monocrystallization is not attained. The long edge of the plate was parallel to L^3, the torque being normal to the principal plane of the plate (Thomas and Wooster).

The elastic-twin stage should be absent, because this involves twinning with change of form, the twins being ejected by the elastic stresses in the matrix. The twinning of quartz involves no change of form, so reoriented regions should not be ejected.

To this we must add that ejection is possible as a result of the tendency to minimal surface energy, but this should not be an important effect in quartz.

Wooster et al. (1947) and Tsinzerling (1955) described a further very important feature of the mechanical twinning of quartz. The sense of displacement of the twin boundaries is independent of the sign of the stress (§ 19). For example, a plate of BT cut can be detwinned by clockwise or counterclockwise torsion; in either case, it is converted to a monocrystal bounded by a plane of the positive rhombohedron.

4. Patterns of Twin Boundaries and the Stored Elastic Energy

Dauphiné twins are revealed by etching quartz with hydrofluoric acid; lateral illumination then reveals pits of irregular shape. Thomas and Wooster (1951) used their twinning theory (§ 19) to elucidate the patterns produced by bending (a plate supported near the ends was loaded at the middle) and by rapid cooling of hot plates (Tsinzerling, 1941a). In the latter case, the twinning occurred as the crystal passed through the transition point (§ 4.6). Frocht's data (1941, Figs. 8 and 6) were used in the calculations (these were obtained from optical measurements on the bending of glass plates). Some additional and far from vigorously demonstrated arguments were used to deduce the state of stress and so to construct polar diagrams for the three principal stresses. The volume density of the elastic energy in the initial and twinned states was determined; the surface of the twin was taken as the surface having $\Delta W = 0$. Figure 127 shows that the observed patterns agree with the calculations, in spite of the lack of rigor. A similar study was made for twinning on passage through the $\alpha \rightleftharpoons \beta$ transition point.

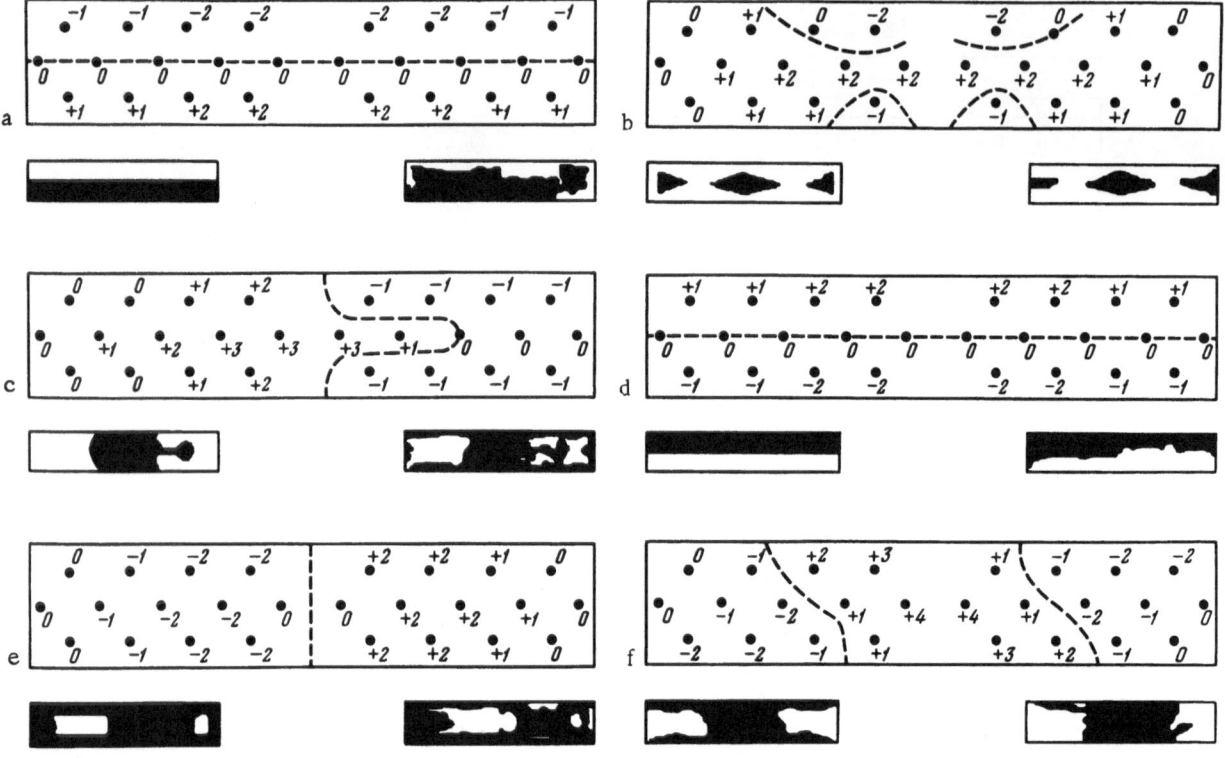

Fig. 127. Calculated and observed distributions of twin boundaries in a quartz plate loaded at the center. The small rectangles show the observed pattern; the large ones, the idealized array of boundaries (Thomas and Wooster).

5. Differences and Similarities in the Mechanical Twinning of Quartz and Calcite

The basic difference between quartz and calcite is that the latter twins with change of form. A sphere in the initial crystal becomes an ellipsoid in the twinned one, the twinning elements (K_1 and K_2) being the two circular (undeformed) sections common to the sphere and ellipsoid. Deformation propagates along the K_2 plane in a definite direction η_1; deformation along the other sense of η_1 is ruled out.

In quartz, the sphere becomes not an ellipsoid but a sphere turned through 60° (or 120°) about L^3, so the concept of twinning elements is not applicable.

The change in the potential energy is proportional to the stress (§ 19) when the twin boundaries in calcite are displaced, so the sense of the displacement is governed by the sign of the stress. * The change in the case of quartz is proportional to the square of the stress, so the direction of displacement is fixed (§ 4.3).

The absence of macroscopic deformation means that the distortion at a twin boundary of any orientation must concern only the short-range order; this is true for twinning with change of form only if the twin boundaries coincide with the twin plane. Any deviation from this produces discordant macroscopic displacements on the two sides of the boundary; macroscopic deformation and stresses occur.

Annealing in principle should displace the boundaries to reduce the surface energy, which should mean that they tend towards the twin plane. Complete coincidence is hardly to be expected, so macroscopic and short-range forms of distortion are present together. This explains the parting and cracking at twin boundaries

---

*We may assume that the pressure at the twin boundary is proportional to the tangential stress and acts in the twin plane along the twin direction if the crystal twins with change of form.

when there is a change of form. Such effects appear less prominent in quartz; cracks and parting are usually not associated with the boundaries of twins.

Wooster considered that mechanical twinning in quartz is a new effect not found in other crystals (calcite, metals), for which he proposed the not very precise name piezocrescence (growth in response to a force).

Twinning without change of form should occur in all axial crystals in which the rotation axis is a symmetry axis of the ellipsoid of homogeneous deformation (as for the thermal expansion or refractive index*), as well as for any reflection twin whose reflection plane is a symmetry plane of the ellipsoid (§ 19).

6. Morphology of Twinning in Quartz

Tsinzerling made a detailed study of thermally and mechanically twinned crystals which is incorporated in a dissertation (1958) and a monograph now in preparation. Tsinzerling established that natural quartz crystals may twin readily or with difficulty; inclusions, mosaic structure, and other defects hinder twinning. The capacity to twin is very sensitive to structure defects; twinning often follows the growth zones and sections, while the twin boundaries reveal the minutest details of the structure, especially defects that favor residual stress. Annealing a detwinned plate restores the twinned regions precisely as before; Tsinzerling called this the 'memory' of the crystal.

Moreover, quartz will twin in response to fixed or alternating electric fields when it is hot; this is a result of high mechanical stresses set up in the crystal. A monocrystal or twin can be preserved by applying a constant electric field to a heated plate; the field tends to remove impurities from the crystal, and this causes local irreversible defects (Chentsova, 1956). The crystal still gives the piezoelectric effect, though.

7. Practical Significance of Monocrystallization; Methods of Detwinning Quartz and Effects of Various Factors

Quartz crystals are used in optics, ultrasonics, and acoustics; optical components must not contain Brazilian, Japanese, or Leydolt (§ 3.1) twins. Quartz containing Dauphiné twins is acceptable for most optical components,[†] for the optic axes are parallel and the two components rotate the plane of polarization in the same sense.

Only monocrystals are acceptable for piezoelectric uses, for the L^2 (electric) axes are antiparallel in Dauphiné twins.

Natural quartz crystals free from twins are very rare (Shubnikov, 1940; Tyul'panov, 1955). Detwinning can be used in making piezoelectric plates (Tsinzerling and Shubnikov, 1943; Tsinzerling, 1941a, 1941b, 1941c, 1958; Wooster, 1947; Tsinzerling, 1948a, 1948b, 1954a, 1954b). Tsinzerling first used thermoelastic stresses to remove natural Dauphiné twins; these stresses were produced by cooling the plates rapidly from temperatures above the inversion point. During the war, Frondel (1945) in the USA and Wooster (1947) in Britain used Tsinzerling's method; subsequently, Wooster and Tsinzerling used torsion on plates heated to 400-550°C for the same purpose (Tsinzerling and Perekalina, 1955).

Monocrystallization by Thermal Stress. The plate (any cut) is heated to 600°C, which is above the inversion point; the resulting monocrystal of the high-temperature form is then cooled slowly to cause the low-temperature form to arise initially in some small area. The cooling is then continued in such a way that this center grows but other centers do not arise. About 80% of the area is thus made into a monocrystal without resort to special equipment.

---

*See Shubnikov, A. V., Flint, E. E., and Bokii, G. B. Principles of Crystallography [in Russian] , Moscow, 1940, p. 281 and 342.

† The differing atomic structures of the planes in the two rhombohedra cause the components of a Dauphiné twin to be revealed during the polishing of precision optical surfaces; this results in a loss of perfection.

Monocrystallization by Torsion. The plate (oblique or Y cuts) is twisted about specified crystallographic axes in a special machine while it is heated to 500-530°C (below the inversion point); it is then slowly cooled under load to room temperature. Stresses of the order of 200 kg/ cm^2 are needed. This gives 100% of the area in monocrystal form.

8. Twinning in Triglycine Sulfate *

This has the formula $(NH_2CH_2COOH)_3H_2SO_4$; it is monoclinic (Wood and Holden, 1957). The Curie point (transition from centrosymmetry to axial symmetry) is at 49°C. It is also stated that the material is triclinic (not monoclinic) below 49°C (Konstantinova, unpublished report).

Twins are always present in the crystals, which are grown from aqueous solution. Large twin regions usually occur in the growth sectors (Fig. 128), and sometimes in the growth zones.

  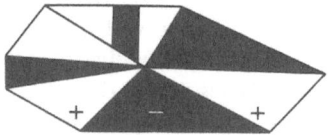

Fig. 128. Crystals of triglycine sulfate grown from aqueous solution: sector and zone distributions of the twins are seen. One component is colored black (Konstantinova).

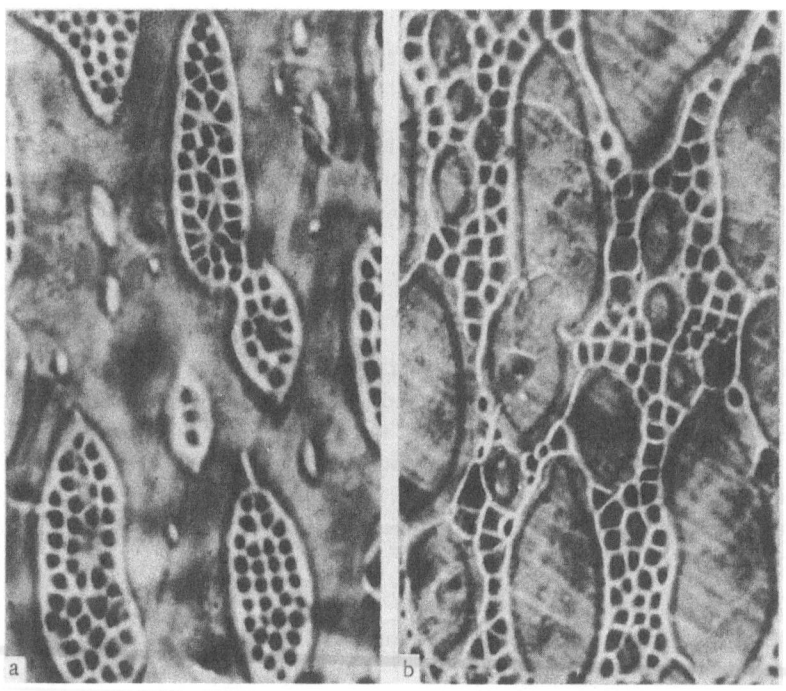

Fig. 129. Parallel cleaved surfaces of triglycine sulfate normal to the polar Y axis turned relatively through 180°; twin regions revealed by etching, the etch figures being formed where the negative end of the axis emerges (Konstantinova).

---

* § 4.8 and § 4.9 have been compiled from a report by V. P. Konstantinova in the Institute of Crystallography dated October 25, 1958.

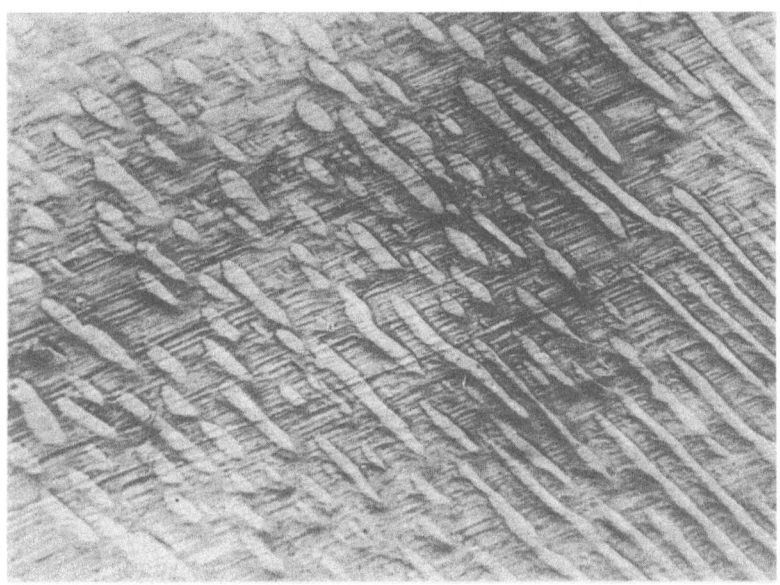

Fig. 130. Etched slice of triglycine sulfate cut normal to the Y axis; action of an alternating electric field. The lens-shaped components persist on the left, but a polysynthetic twin is starting to form on the right (Konstantinova).

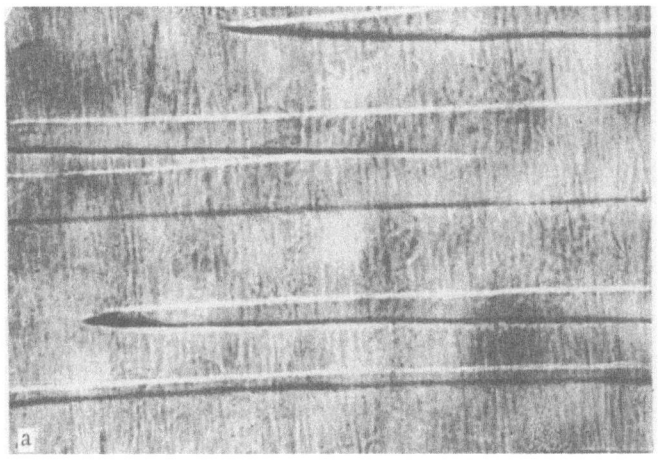

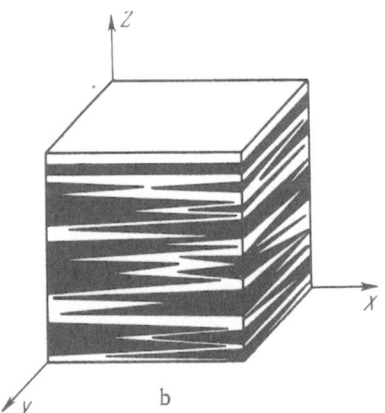

Fig. 131. a) Etched section of triglycine sulfate cut normal to the X axis; the boundaries of the polysynthetic twins form wedge-shaped regions; b) spatial model of a polysynthetic twin in triglycine sulfate (Konstantinova).

The twins are not visible by polarized light, so the optical axes of the components must be parallel. Etching reveals the twins; the boundaries stand out on sections normal to the a and b axes, but not on those normal to the c axis. There is often a notable distribution of the components in the form of globular inclusions (Fig. 129) in the growth pyramids. The globules tend to be elliptical and elongated in the direction of the a axis; the components are domains, as in Rochelle salt (§ 4.9).*

A specimen subjected to an electric field (§ 4.9) acquires the typical structure of a polysynthetic twin (Figs. 130 and 131), which resembles the polydomain structure of Rochelle salt (Fig. 35). The twin components here are related by inversion,† so twinning is not accompanied by change of form, and the elastic constants of the components are the same (§ 19). It is not surprising that the twin boundaries cannot be displaced mechanically.

9. Electrical Displacement of Twin Boundaries in Triglycine Sulfate

The material is a ferroelectric, the electric axes in adjacent components being antiparallel, as is readily demonstrated with a probe and electrometer. A constant field (1200-2400 V/cm) applied to faces normal to the b axis (Y-cut plate) or to the a axis (X cut) produces complete monocrystallization (Konstantinova et al., 1959).

Triglycine sulfate differs from Rochelle salt in that the monocrystals persist for several days at room temperature; reoriented regions start to appear on the second or third day, and the original structure is nearly regained by the fifth or sixth day. This recovery is much slower than that in quartz or Rochelle salt; complete recovery of the original pattern is not obtained in this case.

---

* The polar axis in Rochelle salt is a = X, whereas that in triglycine sulfate is b = Y.
† As for the 180° domains in barium titanate, see p. 79.

CHAPTER 3

TWINNING DURING PLASTIC DEFORMATION AND
FRACTURE OF CRYSTALS

§ 5.   Twinning in Polycrystals: Effects of Grain Size

Effects of Grain Size.   There are several papers on twinning mechanisms in crystal aggregates; the most systematic study is that of Garber et al. (1953) for technical iron (0.06% C).

Mechanical twinning may produce the following effects in an aggregate: residual change in grain shape, plastic change of shape consequent on secondary slip in regions reoriented by twinning, and cracking or rupture on twin planes (parting within grains).   Twin nuclei (elastic twins) arise at points of stress concentration in elastic twins.   The stresses concentrate primarily at grain boundaries, so these should be the sites of nucleation. Detwinning may also occur, of course.[*]

The grain size is found to have a pronounced effect on the nucleation and growth of twin layers (Davidenkov, 1938 and 1957).   Nucleation becomes more difficult as the grain size decreases (Garber, 1953). For example, plastic strain occurs (up to 20% of the initial length in the neck) when iron having a uniform grain size of 15-30 μ is pulled at liquid nitrogen temperature; slip occurs without appreciable signs of twinning.   Brittle fracture occurs under these conditions if the grains vary widely in size; twinning occurs in the grains of size 100-200 μ, but not in the small grains.   Coarse-grained (0.5-0.6 mm) specimens are also plastic at low temperatures, and this is largely the result of residual twin layers in the grains.   Twinning in a polycrystalline metal is favored by large grain size, by reduced temperatures, and by use of impact loading instead of slow loading.   For example, all grains in iron over 100 μ in size twin upon shock loading at -195°C (Garber et al., 1953).

Lance-type twin layers not extending to the far side of the grain often occur in large iron grains.   Very often the layers in adjacent grains have a common point at the boundary (Fig. 132).   This can arise from stress concentrations at the boundary, if the twinning directions of the adjacent grains are suitably disposed.   Further, a twin layer arising in one grain may extend to the boundary and produce there a stress concentration sufficient to initiate a layer in the adjacent grain (§ 2.20).

Metallography indicates that twin layers arise in the individual grains at 60° to the pulling direction; they lie at 60-70° to the neutral axis in impact bending.

---

[*]Grasc and Baudeau (1962) found that mechanical twins sometimes act as recrystallization nuclei in work-hardened and recrystallized beryllium.   The twin layers broaden at 620-650°C; their ends become rounded, and they break up into small regions.   Some of these vanish, while others become nuclei of fresh grains.

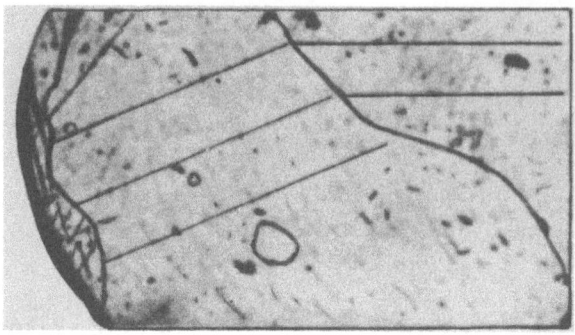

Fig. 132. Mechanical twinning in polycrystalline iron of
variable grain size; the twin layers have a common point
at the boundary between grains (Garber et al.).

§ 6.   Twinning and Fracture of Polycrystals

It is often found that the junction of two grains may be associated with a crack in one grain and a twin
layer arising from the same point in the other (Fig. 133a). This creates the impression (Garber et al., 1953)
that a crack in one grain can cause a twin layer in the next.*

It has often been supposed (Davidenkov, 1938; Shevandin, 1939b and 1940; Hall, 1960) that premature
brittle fracture in a technical metal is the result of the growth of cracks arising at the boundaries of twin layers.
This is supported by the well-known facts of second cleavage, parting, and production of perfect fractures along
the boundaries of twin layers (Klassen-Neklyudova, 1942a). Garber concludes from the brittle fracture of iron,
though, that the twinning accompanying brittle fracture is not the primary cause of failure.

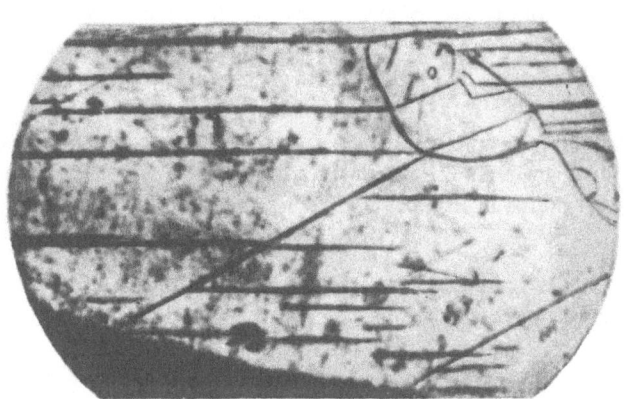

Fig. 133. a) Polished section of polycrystalline iron in region of brittle fracture; the horizontal lines are cracks, a
second grain with twins being seen in the top right-hand
corner, the twin layers having points in common with the
cracks at the boundary. × 100 (Garber et al.).

---

* This occurs also in chromium, which twins readily at low temperatures (Marcinkowski and Lippsitt, 1962;
Weaver, 1962).

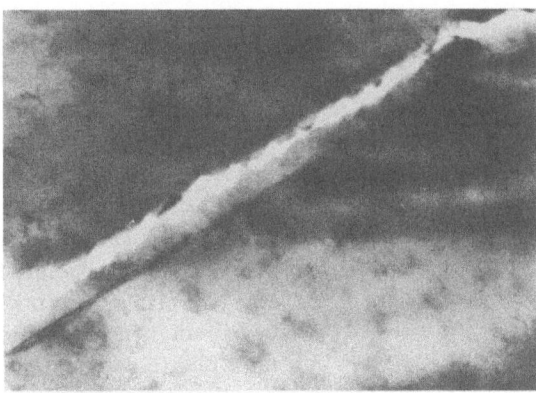

Fig. 133. b) Crack of sawtooth shape produced at one of the boundaries of a twin layer in a grain in a stretched film of iron. Electron micrograph (Orlov).

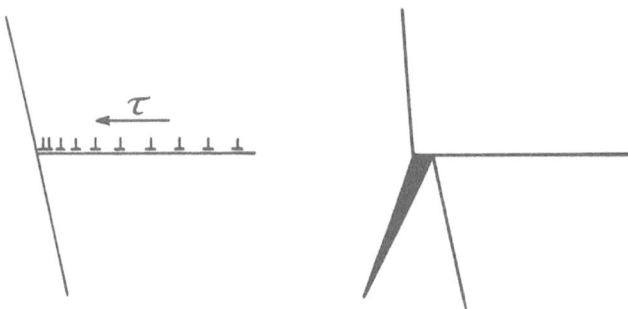

Fig. 133. c) Stroh's scheme for the formation of cracks.

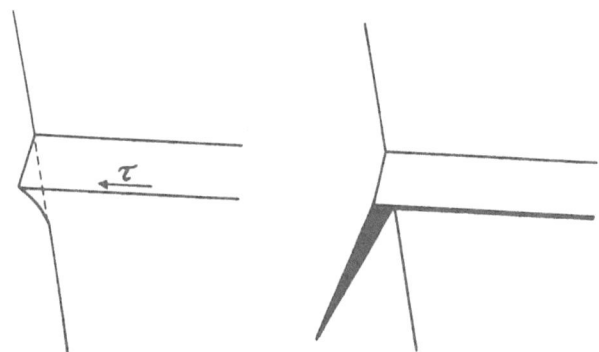

Fig. 133. d) Scheme for the formation of a crack along the boundary of a twin, with subsequent opening at the top of the twin layer (Orlov).

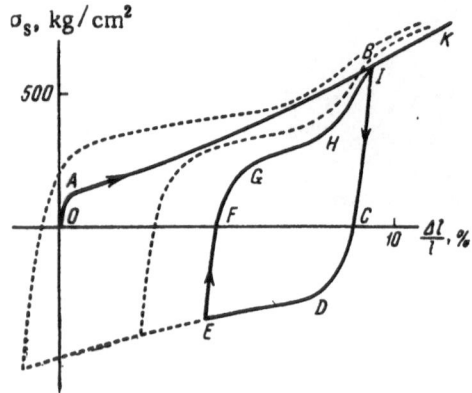

Fig. 134. Stress-strain curve for sign-varying torsion in a polycrystal of 99.98% magnesium. OAB represents loading, BC unloading, CDE loading in the reverse direction, EF unloading, and FGHIK loading in the initial direction. The broken lines represent the effects of loading to higher final stresses (same specimen). The shapes of CDE (D lower than B) and EGHIK (flat section GH) are caused by twinning in some grains and detwinning in others (Woolley).

Impact rupture at low temperatures produces virtually no twinning in the area where the cracks originate; the twin layers occur mainly in the area of plastic deformation around the site of impact. Twins occur in the fracture zone only in dynamic loading of heated specimens. Garber considers that brittle fracture of the individual grains and of the specimen as a whole is independent of the twinning and results from cracking of the grains along cleavage planes. He concluded from the plasticity of beryllium monocrystals that twinning actually facilitates plastic deformation in the brittle fracture of beryllium in compression; the reoriented twin regions retard the growth of cracks and favor secondary slip.

Twinning and slip in part relieve the internal stresses in crystals, but incoherent boundaries appear to act as points of stress concentration and so favor cracking.

Transmission electron micrographs for iron deformed at -196°C (Orlov, 1962) show many twin layers 0.1-0.3μ wide; very often there is a crack along one side of the twin layer, which usually has a characteristic sawtooth profile (Fig. 133b).

Electron micrographs of the dislocation structure of iron* indicate that planar accumulations of dislocations do not occur. This is important as regards the mode of initiation of cracks in brittle fracture. However, the conditions of Stroh's very general scheme for the formation of cracks (Fig. 133c) are similar to those that occur on the accommodation-zone side of a twin. Tangential stresses act along this boundary, so it may be considered as a planar accumulation of dislocations. Shear stresses can be relaxed by cracking along the junction of the twin with the matrix, the crack subsequently opening at the top of the twin layer along the cleavage planes (Fig. 133d).

This enables us to estimate the thickness of a twin capable of giving rise to a crack. Stroh's equation $nl\tau$ = 12γ (in which γ is surface energy, τ is shear stress, and nl is the shear from n dislocations of Burgers vector l) gives us that about 100 dislocations suffice to initiate a crack in iron, which corresponds to a shear step about 250 A high. The thickness such as to produce this displacement along the boundary (250 A) is found from the distance between twin planes (about 2.35 A for iron), which gives the shear of adjacent planes as $t/3 \approx 0.82$A (t is the translation vector). Then the thickness is

$$D = 250(2.35/0.82) \approx 720 \text{ A.}$$

Hornbogen (1961) and Sleeswyk (1962) also observed the formation of thin twin layers, which are the primary cause of rupture in α-iron.

Twinning can be the cause of premature failure in any crystal in which it gives rise readily to cleavage along the boundaries of twin layers. If this effect is not common (if the residual stresses at the boundary are not large), the twinning may actually increase the plasticity rather than favor rupture.

Grain detwinning can result from the capacity of a twin layer to vanish under reverse stress (Churchman, 1954). This detwinning in an aggregate affects the stress-strain relation; the effect is most pronounced (Fig. 134) for coarse-grained specimens (Woolley, 1954).

---

*L. G. Orlov, M. P. Usikov, and L. M. Utevskii. UFN, 1962, 76, No. 1, p. 109.

PART II

EFFECTS RELATED TO MECHANICAL TWINNING

CHAPTER 4

MARTENSITE PHASE TRANSITIONS

§ 7.   Cooperative and Other Phase Transitions

The phase transitions of solids are usually classified under the two headings of diffusion and diffusionless (martensite) phase transitions.

In the first type, the new phase is formed by nucleation and subsequent growth by the addition of single atoms or small groups (atom-by-atom mechanism); there is usually no clear correlation between the orientations of the old and new phases.

The relative displacements in this type of transition are of the order of interatomic distances or greater. The growth rate is low (only 10-100 cm/sec), and the rate as a function of temperature shows a pronounced peak. The activation energies for nucleation and growth are fairly large, being often comparable with those for diffusion and self-diffusion. The crystals of the new phase cannot grow at an appreciable rate at temperatures well below the recrystallization temperature; the high-temperature phase is readily supercooled, and it can exist in a metastable state for an indefinite period if the temperature is low enough.

This type covers all transitions that involve changes in concentration, such as decomposition of solid solutions, formation of solid solutions from components, and certain transitions in one-component systems (conversion of graphite to diamond, of white tin to gray tin, and so on).

The diffusionless type has some distinctive features that give it a resemblance to twinning; in particular, the rates are very high and the orientation of the new phase relative to the old one is strictly defined (§ 9). The first such transition to be studied in any detail was that giving martensite in the quenching of steel, so diffusionless transitions are also called martensite transitions.

The formation of martensite platelets has much in common with the mechanical twinning of metals, and so martensite transitions were long considered as purely mechanical effects resulting from quenching stresses in austenite (Scheil, 1929). Not till 1948 did Kurdyumov put forward the idea that the formation of martensite is a diffusionless phase transition and so one of the basic types of transition in solids.

The most important feature here is the regular and ordered (cooperative) displacement of large groups while the new phase is growing. Kurdyumov defined transitions of this type as follows: "Martensite transitions consist of the ordered reconstruction of the lattice, in which the atoms do not exchange places but are merely relatively displaced by distances not exceeding interatomic ones" (Kurdyumov, 1936).

The main feature differentiating martensite-type transitions from others is that the displacement is cooperative; it might be advantageous to call diffusionless transitions "cooperative" and the others "noncooperative" for this reason, the more so since the noncooperative type in a one-component system (tin or carbon) follows typical diffusion kinetics and the use of 'diffusion transition' might lead to unnecessary misunderstanding.

§ 8. Macroscopic Shear in Cooperative Transitions: Classification of Cooperative Transitions

Here the lattice alteration can be represented as the result of two successive shear operations. Figure 135 shows the lattice reconstruction in the conversion of austenite to martensite in carbon steels in accordance with the Kurdyumov-Sachse scheme. Greninger and Troiano (1949) proposed to consider the transition as a combination of macroscopic homogeneous deformation (primary or visible deformation giving rise to an intermediate lattice) with displacement within the intermediate lattice (no macroscopic deformation). This second component was called secondary or additional deformation. These methods of description are merely geometric schemes and do not define the actual paths taken by the atoms. The macroscopic deformation will in future be called simply change of form, which some consider as the most important feature of the transition. For example, Bilby and Christian (1956) base their definition on this feature: "Structural changes are classified as martensite ones if the atoms of the primitive lattice (which is defined by the cell chosen for the initial structure) are displaced to positions in the primitive lattice (as defined by some unit cell for the final structure) in such a way that the displacements constitute homogeneous deformation, which may vary as between adjacent small regions" (§ 4.2).

The formal analogy between twinning and cooperative transitions is so great that twinning can be treated as a special type of cooperative transition such that the structure is not altered. Twinning may involve change of form (as in calcite) or not (Dauphiné twins in quartz). Cooperative transitions may thus be divided into the following two groups, by analogy with twinning: 1) ones with change of form and 2) ones without change of form.

Transitions with change of form are ones in which we have primary (macroscopic) deformation and secondary deformation; the others are ones in which we have only secondary deformation.

Bilby and Christian's definition makes martensite transitions only a special type of cooperative transition, namely, the type involving change of form.

The type involving no change of form is more probable for crystals of complex structure, as in the case of twinning (molecular crystals and many-component alloys). The type retains the basic feature of cooperative transitions (cooperative displacement during lattice reconstruction) but may not have many of the features of martensite transitions (§ 9) resulting from the change of form (in particular, the additional deformation energy).

Abundant experimental evidence indicates that martensite transitions (with change of form) are the main cooperative ones in metals and alloys.

Martensite transitions have been reported for many alloys (iron-carbon, iron-nickel, cooper-zinc, and so on) and also for iron, cobalt, titanium, zirconium, uranium, lithium, sodium, cerium, thallium, and mercury.

Roitburd (1958, 1960, 1962) has considered some aspects of diffusionless transitions in hard metals and alloys.

§ 9. Main Features of Martensite Transitions

All martensite transitions have the following features to some extent; these features are the result of the cooperative displacement and of the anisotropy of the medium.

1. The crystals of the new phase grow at high rates even at temperatures near absolute zero. Forster and Scheil (1940) and Bunshah and Mehl (1953) have measured the rate of formation of martensite plates; in addition, theory indicates that the linear growth rate may approach the speed of elastic waves in the metal.

2. The initial and final phases have a strictly defined mutual orientation (as in twinning). Table III gives details of this orientation and also of the structures for the cooperative transitions in metals and alloys that have been examined in most detail. This table is from Bilby and Christian (1956). The evidence on habit planes is rather conflicting; moreover, some alloys show several such planes simultaneously (Liu, 1956; Margolin, 1958; Mehl and van Winkle, 1953).

3. A martensite transition may be structurally reversible; in that case, the initial lattice orientation is restored by the reverse transition, so the atoms follow the same paths but move in the reverse direction, as in the case of twinning.

4. Martensite transitions show pronounced hysteresis.

5. A martensite transition remains incomplete under isothermal conditions; it occurs over a considerable temperature range.

6. The metal is deformed by the transition; surface relief is produced.

The martensite phase starts to form at a certain temperature, which is dependent on the history of the specimen (heat treatment, cold working) and which is usually denoted by M_s or T_m. The transition remains incomplete under isothermal conditions, so the temperature must be reduced in order to produce more of the phase. This amount continues to increase down to some temperature M_f, but a certain proportion of the high-temperature phase usually persists below M_f.

The position and width of this temperature range are composition-dependent. Sometimes the range extends down to temperatures approaching absolute zero (Kulin and Cohen, 1950; Barrett, 1957). The crystals arise and grow very quickly, so the increase in amount occurs mainly by the production of fresh crystals, although some alloys show stepwise growth of existing platelets (Golovchiner and Kurdyumov, 1951). The martensite crystals often take the form of biconvex lenses, which is the result of elastic stresses in the matrix, as in the case of twinning (§ 1.16).

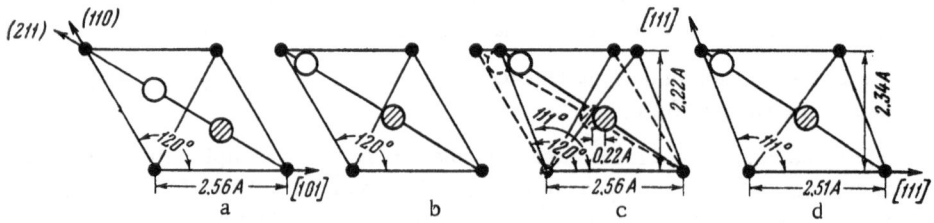

Fig. 135. Geometry of the transition from a face-centered cubic lattice to a body-centered tetragonal one in the martensite transition in carbon steel (Kurdyumov and Sachse): a) projection of the austenite lattice on (111); b) shear on the system (111), $[2\bar{1}\bar{1}]$ of the austenite lattice; c) shear on the $(\bar{2}1\bar{1})$, $[11\bar{1}]$ system of the martensite lattice; d) projection of the martensite lattice on (110) after alteration of the inter-atomic distances from 2.56 A to 2.51 A and from 2.22 A to 2.34 A. The filled circles denote atoms lying in the plane of section; the hatched circles, ones in the next plane; and the open circles, ones in the third plane.

Electron microscopy indicates that the martensite has a fairly complicated substructure.

Macroscopic shear accompanies the formation of martensite plates; this causes surface relief and so enables one to examine the formation of the phase directly. The change of form is homogeneous. Figure 135 shows typical martensite structures.

§ 10. Elastic Crystals of Martensite: Effects of Deformation on Martensite Transitions

The macroscopic shear accompanying a martensite transition can give rise to elastic martensite crystals in conjunction with the elastic stresses set up in the surrounding medium. This is analogous to elastic twinning

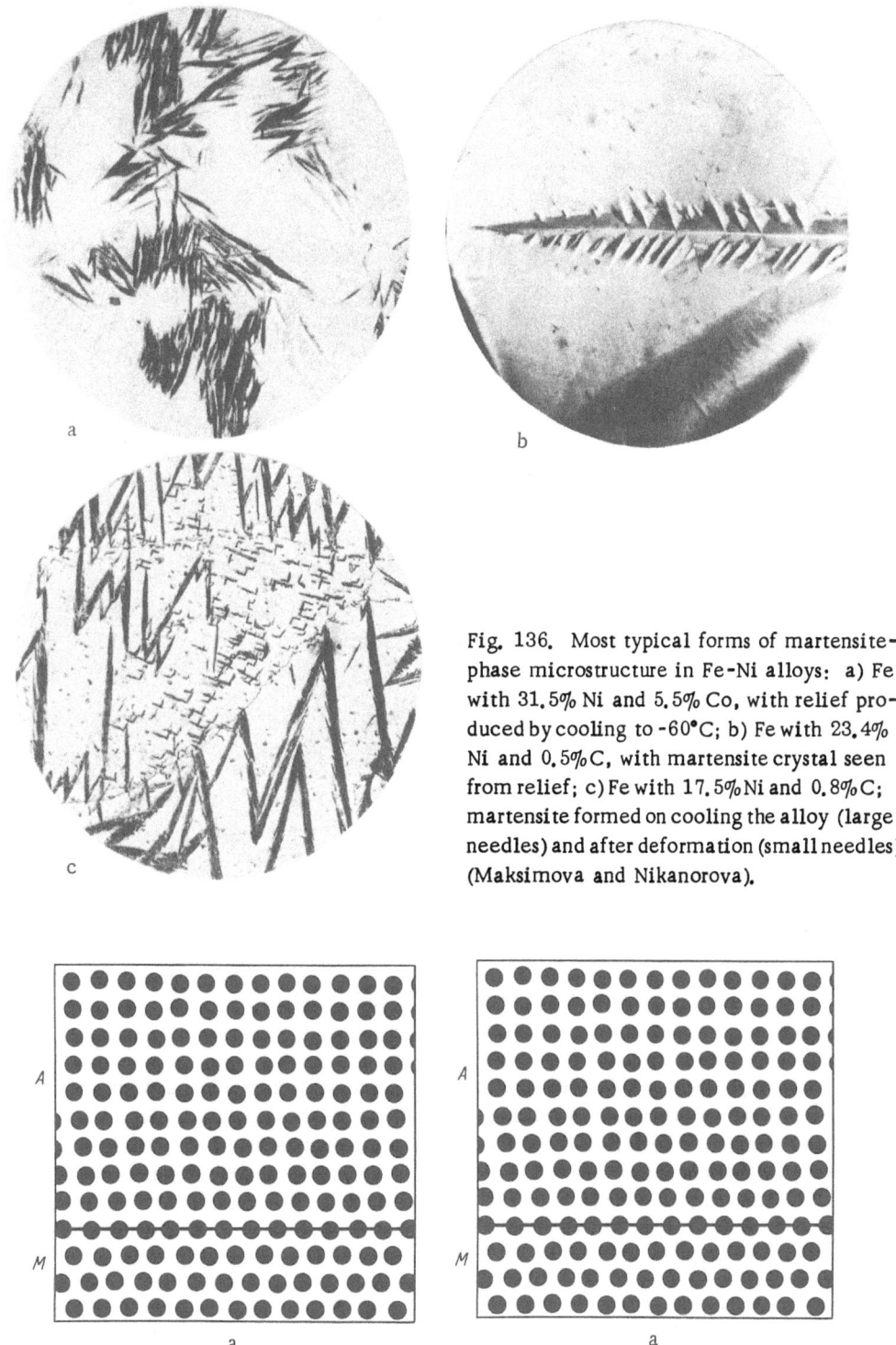

Fig. 136. Most typical forms of martensite-phase microstructure in Fe-Ni alloys: a) Fe with 31.5% Ni and 5.5% Co, with relief produced by cooling to -60°C; b) Fe with 23.4% Ni and 0.5% C, with martensite crystal seen from relief; c) Fe with 17.5% Ni and 0.8% C; martensite formed on cooling the alloy (large needles) and after deformation (small needles) (Maksimova and Nikanorova).

Fig. 137. Relation between the lattices of the initial and final phases in a martensite transition: a) coherence, with continuous transition from one lattice to the other; b) incoherence, with disorder at interface, which has resulted in interrupted growth in the martensite (Kurdyumov).

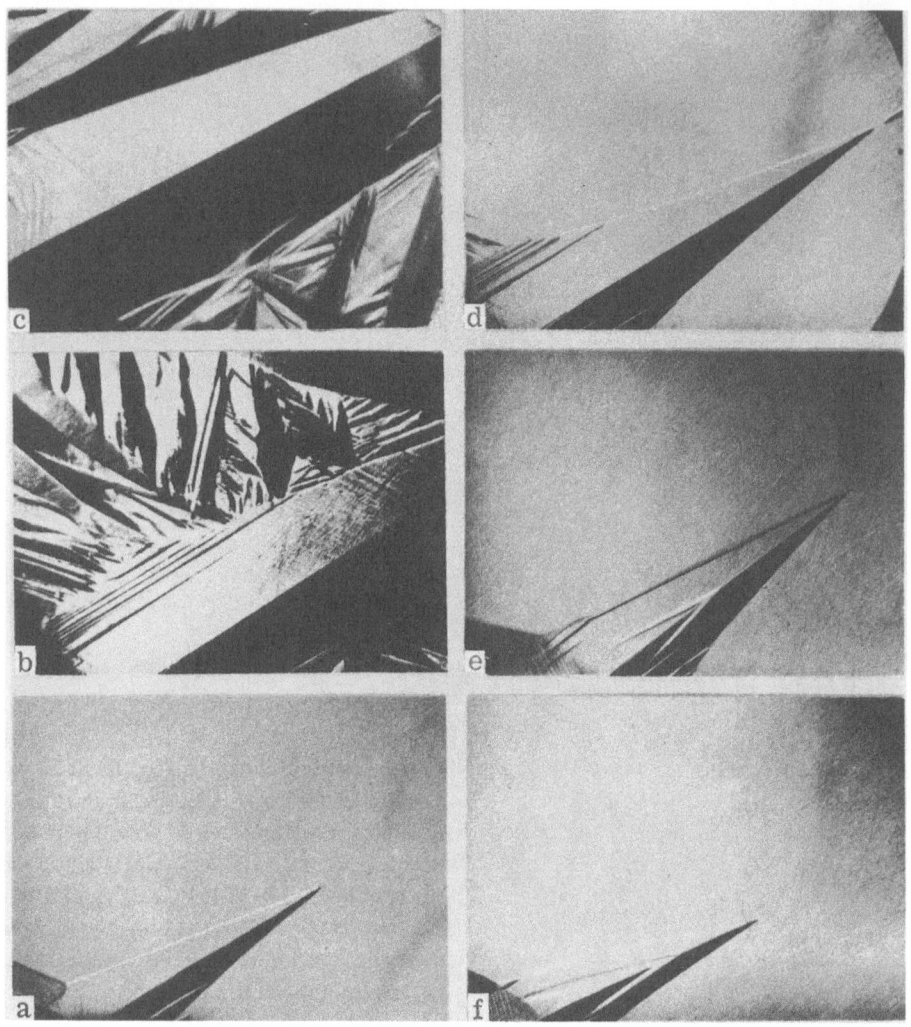

Fig. 138. Elastic needles of martensite phase in Cu with 14.7% Al and 1.5% Ni: relief produced on surface; a) to c) on cooling; d) to f) on heating, as observed by oblique illumination (Kurdyumov and Khandros).

(§ 2.3). The martensite grows by regular displacement of the atoms to their new positions; atoms that are neighbors in the initial lattice remain so in the new one. The interface is coherent (Fig. 137), but the elastic stresses at the interface increase with the size of the crystal, which can result in plastic deformation. This can cause lack of coherence and changes in the mode of growth. The thermodynamic potential is reduced by the growth of the martensite platelets, but the energy of elastic deformation increases. The martensite crystal will grow and shrink reversibly in response to temperature changes if thermodynamic equilibrium is reached before coherence at the interface is lost.

Elastic changes in size can be seen in response to tensile and compressive stresses. Kurdyumov predicted elastic martensite crystals, which were later observed by Kurdyumov and Khandros (1949) in aluminum bronzes. Figure 138 shows the response of elastic martensite crystals to temperature changes.

Mechanical stresses can produce martensite crystals within the range from M_s to M_f and also above the initial temperature of spontaneous formation.

Martensite cannot be produced by deformation above a certain temperature M_d, which is constant for any given metal. For example, cobalt monocrystals give no more than 50% of the low-temperature hexagonal modification even on prolonged cooling, but the conversion can be made virtually 100% by deformation. The highest temperature at which the $\beta \rightarrow \alpha$ transition can be produced by deformation is precisely the minimum temperature at which the reverse transition occurs spontaneously on heating.

In—Tl alloys (Fig. 102) give the low-temperature phase in response to stress even at temperatures above that corresponding to thermodynamic equilibrium (Basinski and Christian, 1954b). A monocrystal of In—Tl gives rise to the low-temperature phase with a plane interface when it is stretched at a temperature above that of phase equilibrium. This effect is accompanied by considerable stretching. The reverse transition occurs when the stress is removed; the specimen returns to its original size (§ 2.21).

Deformation of a metal sometimes gives rise to a phase that would not arise spontaneously; this is the case for Na, Cs, and Hg at low temperatures. The new modifications of Na and Cs have a hexagonal close-packed structure with many packing defects. The new modification of Hg is stable below 79°K (Swenson, 1958) and has a body-centered tetragonal structure.

CHAPTER 5

RECRYSTALLIZATION TWINS

§ 11. Processes in Deformed Crystals at High Temperatures

A gradual change in properties and structure occurs when a deformed material (monocrystalline or poly-crystalline) is heated; regions of altered crystallographic orientation are produced, the process being called recrystallization.* There are several distinct stages in the heat treatment of crystals:

1) The initial stage (stress relief or recovery); this gives rise to continuous changes in some properties (e.g., hardness) and to the start of structural change (polygonization);†

2) Formation of nuclei for the new orientation (primary recrystallization nuclei); and

3) Enlargement of these nuclei.

The latter two stages represent primary recrystallization; if this gives rise to small grains, it is often followed by secondary (selective) recrystallization, in which the larger grains absorb the smaller ones. The special features of this are that no previous deformation is needed; the surface tension of the grains provides the driving force. Selective recrystallization can give a monocrystal from a polycrystal, or a monocrystal of altered orientation from a monocrystal. It has often been remarked (see Burke and Turnbull, 1952, for example) that the recrystallization of a deformed crystal has much in common with phase transitions; primary or secondary recrystallization may be represented in terms of the rate N of generation of centers of the new phase (or orientation) and of the rate G of the linear growth of these centers. Burke and Turnbull (1952) deal with this; they also give data on the effects of various factors on N and G. Beck (1954) and Bürgers (1956) also deal with the recrystallization of metals.

§ 12. Nucleation of Recrystallization Centers

It used to be believed that recrystallization centers were produced mainly by deformation; Boas and Schmidt stated that these are probably deformation twins. For example, recrystallization in the twin layers always occurs within 1 min in a stretched cadmium crystal at 145°C or above, but new grains are formed in untwinned parts only above 240°C. Tin recrystallizes more extensively after compression than after pulling, because the first produces twinning and the second slip. Schmidt and Boas considered that distorted regions that grow at the expense of undistorted ones are the nuclei of recrystallization grains.

Several new ideas have recently been proposed; Becker and van Arkel consider that the nuclei are undistorted regions of a certain minimum size, which grow at the expense of the main distorted material.

---

*This sometimes occurs also at room temperature.

†Polygonization is the formation of substructure in deformed crystals by reconstruction along slip lines; this occurs by regrouping of dislocations in the slip planes into vertical rows, which form the boundaries between blocks.

Various views are held on the nature of these undistorted nuclei; some consider that they arise by fluctuation, distorted areas giving rise to larger fluctuations. Estimates have been made of the local elastic stresses needed for nucleation and of the sizes of the fluctuation volumes; on this basis, Burke and Turnbull considered that the fluctuation theory was hardly acceptable.

The block theory assumes that the nuclei are small blocks in the substructure. Two different hypotheses are used; in the first (theory of low-energy blocks) the nuclei are blocks that remain undistorted, while in the second (theory of high-energy blocks) some blocks become nuclei during annealing after the stresses have been relieved.

Cahn (1950a) proposed the following theory: The nuclei are formed during polygonization of highly distorted regions at the start of annealing. Polygonization gives small blocks differing slightly in orientation. Cottrell (1953, b and c) extended Cahn's hypothesis by assuming that initially the angles between the blocks are small, when they grow rapidly. Growth slows down when the angles become a few degrees. Eventually, one block becomes so large that the mobility of its boundaries increases with angle, in which case this block grows at a disproportionately high rate. This growth continues with an orientation steadily deviating from that of the rest of the material; ultimately, we have a new undeformed crystal separated from adjacent ones by boundaries at large angles. Beck and Hu (1952) compared the behavior of the subgrains in almost undeformed material; from this they concluded that Cahn's theory, in conjunction with Cottrell's mechanism, provides the best basis for describing the recrystallization centers in deformed monocrystals and polycrystals (Beck, 1954).

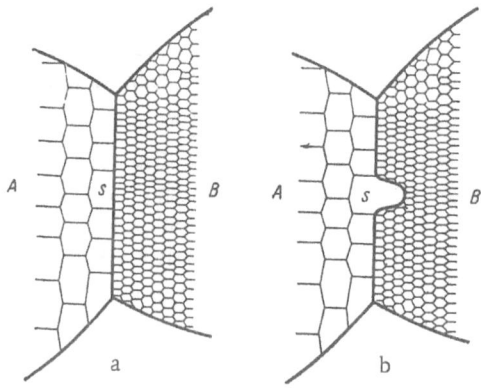

Fig. 139. Schematic representation of grains A and B both having substructure a) before and b) after the onset of boundary migration in response to the stored energy and to the annealing. Migration starts from grain A into subgrain s, which is the largest and which lies at the boundary between A and B (Beck).

Fig. 140. Part of a specimen of aluminum after rolling and annealing at 350°C for 150 min; the tongue-shaped projections arose by growth of blocks in the bottom grain, which absorbed material from the upper one (Beck and Sperry).

Beck's (1954) hypothesis (boundary migration in response to deformation) was very different; it applies solely to polycrystalline materials and is as follows. The recrystallization nuclei are substructure blocks adjacent to grains containing smaller subgrains. Let there be two grains A and B (Fig. 139); the subgrains at the boundary have an orientation with respect to grain B consequent on the high boundary stresses, so the boundaries of these subgrains can be highly mobile. The general boundary migrates into B, and annealing accelerates this. The boundaries between subgrains within A may also be mobile. In the end, A absorbs B (and perhaps other grains), which gives rise to a new grain with relatively few internal boundaries.

Beck's scheme derives from observations on high-purity aluminum deformed 40% by rolling and then annealed (Beck and Sperry, 1950); very few new grains were produced by recrystallization, though the boundaries of some grains were displaced by the absorption mechanism (Fig. 140). Other experiments (Anderson and Mehl, 1945; Tiedema, 1950) confirm this scheme.

§ 13. Development of Annealing (Recrystallization) Twins

Layers whose lattices have a twin orientation with respect to the initial lattice sometimes arise when a deformed material is heated. These are called annealing or recrystallization twins and are well seen on etched polished sections (Fig. 141).

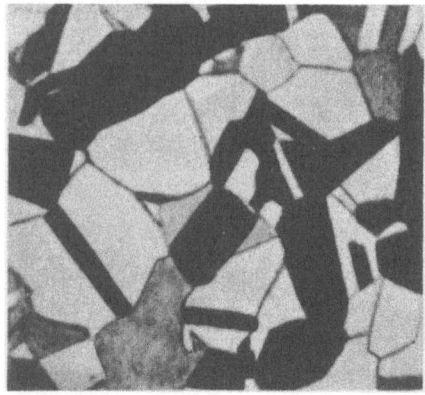

Fig. 141. Annealing twins in α-brass.

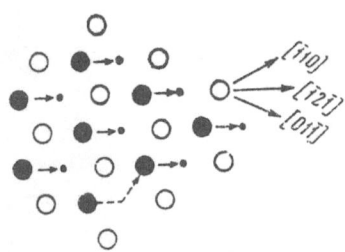

Fig. 142. Motion of atoms during twinning in adjacent (111) layers of a face-centered cubic lattice; the broken lines indicate the displacement in slip (Mathewson).

Fig. 143. Development of twin layers during grain growth or recrystallization (Hall).

Recrystallization twins are found in almost all face-centered cubic metals (copper, brass, zinc, iron, silver, and so on; see Table I).

Dunn et al. (1950) reported recrystallization twins in iron; these arise in the recrystallization grains in the case of deformed monocrystals (Klassen-Neklyudova, 1929). Such twins are very seldom seen in castings.*

Several hypotheses have been made as to the growth mechanism. Mathewson (1944) discussed a possible process for face-centered cubic metals; these give such twins more readily than others do, and the (111) planes in these can act as slip planes and as twin planes (slip usually predominates). He supposed that cold-working might cause one layer of atoms parallel to (111) to be displaced stepwise first along $[1\bar{2}1]$ and then along $[2\bar{1}\bar{1}]$, instead of along the usual slip direction $[1\bar{1}0]$ (Fig. 142). Such motion demands less energy.

Shear may occur only along $[1\bar{2}1]$, with the displaced atoms forming a layer corresponding to hexagonal packing. Such layers are called stacking faults (§ 23). The atoms in such a layer are in a twin position with respect to the initial crystal; a layer with irregular stacking is the nucleus of a twin. The twin plane is the (111) plane; the twinning may be treated as rotation through 180° (60°) on [111], and twins of this type are called spinel twins. Such a crystal, if annealed, shows the irregular stacking propagating to adjacent layers, which gives rise to a macroscopic twin. Mathewson's mechanism thus assigns a deformation origin to the nuclei.

Earlier workers (Boas and Schmidt, Barrett) had considered that annealing twins have a deformation origin. Blewitt et al. (§ 1.13) demonstrated that spinel-type mechanical twins can occur in copper.

Burke (1950) made a detailed study of the growth of twin layers in annealed crystals of copper and brass. A freshly recrystallized specimen was found to contain many annealing twins, but the width of the twin layers remained independent of grain size. The coherent boundaries did not move. The twins expanded only at the corners of grains in conjunction with the movement of

* Burke and Turnbull's theory (see above) is based on this. Whitwham and Lacombe (1961) give references to literature that indicates that numerous twins sometimes occur in solidifying aluminum, copper, germanium, and silicon under certain conditions.

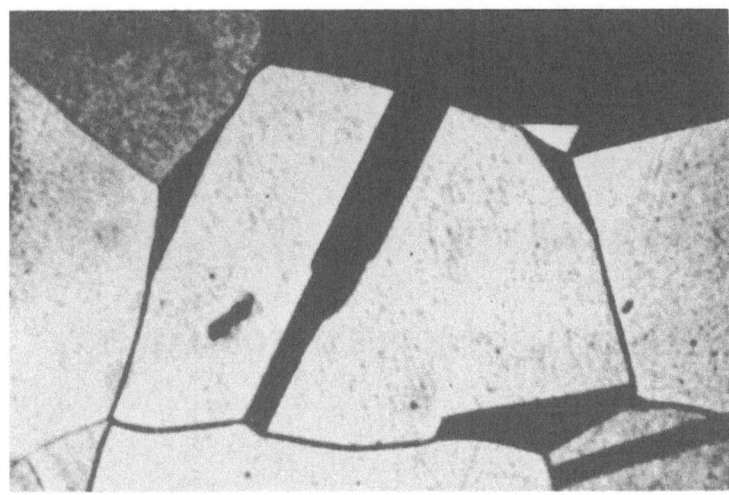

Fig. 144. Annealing twins at the corners of grains of α-brass.

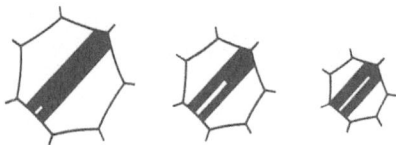

Fig. 145. Development of twin layers during grain growth or recrystallization in the presence of incoherent boundaries (Hall).

grain boundaries (Fig. 143). Figure 144 shows a grain with three twin layers. Burke also observed that parts with incoherent boundaries were displaced during grain growth (Fig. 145).

A new theory has recently been proposed; this was based on Fullman's (1951) and Burke and Turnbull's (1952) work. The decisive process for annealing twins is the migration of the twin boundaries during annealing. It is found (Harker and Parker, 1945; Smith, 1948) that the boundaries of a new grain may migrate until the energy corresponding to the junction of three

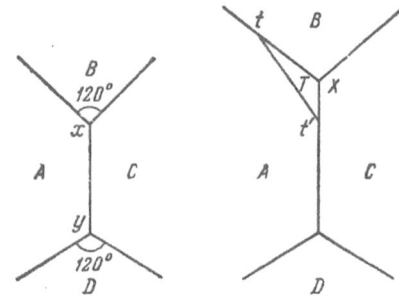

Fig. 146. Production of annealing-twin nuclei in the migration of the junction between three grains (Cahn).

grains is minimal. Equilibrium is attained when the three boundaries meet at 120° (if the surface tension is isotropic). Let A, B, C, and D (Fig. 146) be four grains meeting at two edges x and y. If the angle between two faces at x is less than 120°, then x will migrate into grain B; this is equivalent to the growth of grains A and C at the expense of grain B. During the growth of (say) A there may arise a layer with a packing defect (twin nucleus). This twin is unstable unless definite conditions are obeyed; stability requires that the total energy of the three faces of the twin T should be less than the total energy of boundaries xt and xt' for T with its previous orientation (that of A). This is possible, for the energy of the usual grain boundaries is dependent on the relative orientation of the grains, while the energy of a coherent twin boundary (such as tt') is very small. *

On this theory, the twins in a polycrystal are formed gradually during annealing by migration of boundaries. Twin nuclei are produced by recrystallization, not by deformation. Annealing twins are readily produced in polycrystalline copper, brass, iron, and silver, but they are very rarely found in aluminum. Fullman (1950) sees this last as a consequence of the relatively high energy of the twin boundaries in this case. If T, B, and C have nearly the same orientation, the energy of the boundaries between T and B and between T and C is very small; but then grain A lies roughly in a twin position with respect to B and C. Bürgers et al. (1953) have found that grains rich in twins actually take up twin positions with respect to the surrounding texture. Annealing twins can occur in a monocrystal by the movement of the junction between three subgrains.

---

*Friedel (1926) has discussed the crystallography of recrystallization twins; a brief but clear exposition is to be found in a paper by Whitwham and Lacombe (1961).

CHAPTER 6

LATTICE REORIENTATION IN INHOMOGENEOUS DEFORMATION

§ 14. Geometry, Crystallography, and Relation to Atomic Structure

1. Irregular Lattice Reorientation by Mechanical Stress

The lattices of the components of a classical twin are always inclined to one another at a definite angle and are rigidly joined at the surface of contact. The indices of the twin plane, the direction of twinning, and the twinning axis are small and are determined by the crystal structure. When crystals are mechanically stressed, reorientation of large regions of the lattice can occur as well as classical mechanical twinning and plastic deformation by slip; the orientation of these regions does not obey the rules of classical twinning. The angle through which the lattice is rotated depends both on the crystal structure and on the extent of the deformation. The plane of junction has a large (irrational, see § 1.3) index, and in general is not a plane of symmetry of the adjacent crystals. In individual cases the lattices are joined symmetrically along irrational planes of contact.

2. Historical

Lehman (1885) first reported the existence in minerals of regions joined at arbitrary and variable angles; these regions were considered as unusual twins. Friedel (1926) described the formation of imaginary twins. Mügge (1898) first distinguished between regularly oriented regions (twins) and irregularly oriented regions (kinks). He assumed that kinks are complex processes involving either classical mechanical twinning or plastic deformation by translation slip (§ 1.1). He also produced kinks by compressing crystals of various minerals. Brauns (1889) observed striations in natural rocksalt corresponding to the irrational plane (20.20.1) [close to (110)[*]] but could not produce similar striations by deforming rocksalt crystals.

Elam (1928) found that rotated regions are produced by inhomogeneous pulling of monocrystals of aluminum and silver; he called these regions mechanical twins. He found a rotation of 66° around [121] in one aluminum specimen.

Brilliantov and Obreimov (1935, 1937) found that rotated macroscopic regions are formed by compressing flat rocksalt specimens; the orientation of these regions is different from that of the flat twinned layer. They called this effect irrational twinning. Startsev (1940, 1941) and Klassen-Neklyudova (1942a, b) carried out similar experiments at higher temperatures. Stepanov (1937) and Stepanov and Donskoi (1954) found that unusual platelike regions are produced by pulling rocksalt crystals with a fixed orientation at high temperatures; Stepanov called this effect plating.

Orowan (1942) produced very unusual reoriented regions (kinks) by bending a cadmium crystal longitudinally.

---

[*]The (110) planes act as glide planes in rocksalt.

Since 1949 kinking has aroused increasing interest. Kinking has been produced by compressing monocrystals of cadmium (Orowan, 1942), zinc (Hess and Barrett, 1949; Gilman, 1954), and titanium (Anderson, Jillson and Dunbar, 1953), and by pulling monocrystals of iron (Holden and Kunz, 1953), aluminium (Calnan, 1952), zinc (Washburn and Parker, 1952; Gilman and Read, 1953), tin (Jackson and Chalmers, 1953), bismuth (Berg, 1934), magnesium (Chaudhury et al., 1953), titanium (Churchman, 1955), and copper (Dichl, 1956) (see review by Urusovskaya, 1960). Kinking in crystals of thallium and cesium halides has also been studied (Klassen-Neklyudova and Urusovskaya, 1953, 1955, 1956a, b, 1960).

Pfeil (1927) and Barrett (1939) found band-shaped reoriented regions (deformation bands) in metallic monocrystals (α-Fe, AlMg).

High-resolution microscopy has shown that substructures (microblocks) occur in metallic monocrystals and polycrystals (see reviews by Cahn, 1950b, and by Gifkins, 1955). As well as causing slip traces, deformation of metals causes these microblocks to break up into smaller (about 1000 A) disorientated blocks (Delise, 1953). [*]

The geometry of these effects indicates that most of them are connected with atomic processes.

3. Types of Lattice Reorientation: Irrational Twins, Kink Bands, Plates, Deformation Bands, Accommodation Bands, and Brilliantov-Obreimov Bands (Irrational Twins)

If a cleaved rocksalt plate (ratio of dimensions 1 : 10 : 10) is compressed along its short edge, ridges are formed on the sides (Fig. 147). Optical and x-ray measurements show that these ridges are rotated regions of the crystal (Fig. 148a, b) (Brilliantov and Obreimov, 1935). The angle between adjacent ridges varies from several minutes to several degrees, depending on the deformation. The boundary planes of adjacent rotated regions are not fixed in orientation; they are inclined to (110) by up to 1.5° if the deformation is carried out at room temperature (i.e., they have irrational indices) (Brilliantov and Obreimov, 1935). Figure 149a shows the side 1 (Fig. 147) of a compressed rocksalt crystal in reflected light. Figure 149b shows the surface 2 of this specimen in polarized light. Comparison of Fig. 149a and Fig. 149b shows that the ridge boundaries correspond to bright lines on the end face of the crystal; the lines are seen in polarized light on account of localized internal stresses at the boundaries of the rotated regions.

Kink Bands. Kink bands are produced by compressing crystals of kyanite, calcite, and mica (Mügge) (Fig. 150a-c), of cadmium (Orowan, 1942) and naphthalene (Perekalina, Regel', and Dubov, 1958) (Fig. 151a-c), and of CsI and a solid solution of TlBr and TlI (Klassen-Neklyudova and Urusovskaya, 1956a) (Fig. 152a-c). Kinks occur most frequently at the ends of specimens (Fig. 151c and 152b). Sometimes longitudinal compression produced one (Fig. 151a, 152a, c) or several kinked layers (Fig. 151b, c). The kinked layer in compressed monocrystals of naphthalene and of thallium and cesium halides forms an intermediate zone between two monocrystalline regions. This zone consists of many wedge-shaped regions rotated with respect to one another; the wedges are clearly seen in transmitted light.

Figure 152c shows a CsI crystal with one kinked layer and many slip traces both inside and outside the layer; the photograph was taken in polarized light. X-rays and polarized light were used to study the structure of the kink bands in thallium and cesium halides. The angle of rotation of the wedge-shaped regions with respect to the lattice of the original crystal changed stepwise from one wedge to the next; the wedge in the middle of the intermediate zone is turned through the greatest angle (up to 30-40°). Figure 153 shows a scheme for the formation of such an intermediate zone. The continuous lines are the boundaries of the wedge-shaped regions making up the kinked layer, while the dotted lines show the direction of slip traces in the deformed crystal. According to this scheme, the surfaces of the wedges during deformation tend to be perpendicular to the slip planes.[†]

Stepanov Bands (Plates). Plate-like regions (bounded by planes close to those of a rhombododecahedron) are formed by pulling rocksalt crystals in the [110] slip direction at 50-300°C (Fig. 154a) (Stepanov, 1937;

---

[*]Zankelies (1962) has observed crystallographic slip in nylon 66 and 610, and also the formation of kink bands.
[†]More precisely, to take the position of the bisector of the angle (close to 180°) formed by the lines of slip in adjacent regions of the kink (§ 14.4).

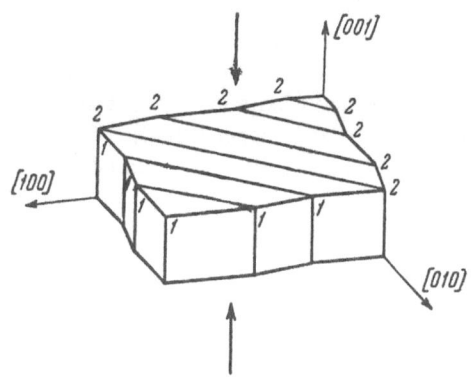

Fig. 147. Rocksalt plate deformed by compression and cleaved; residual deformation, with irrational twins. The rotated regions 1 form low ridges on the sides; the parallel faces of twins 2 appear also. The boundaries are seen only in polarized light.

Stepanov and Donskoi, 1954). The plates run the whole thickness of the crystal and lead to contours on the sides. X-ray studies of the region of contact show that the plates are rotated through various angles with respect to the main part of the specimen (Fig. 154b). The x-ray pattern shows clear resolution of the spots into two reflections with weaker lines between them; the angle of rotation of adjacent regions was 12° or more. Goniometric measurements showed that the plates were not twinned.

Deformation Bands. Deformation bands (dissimilar in appearance to slip traces) on the surface of the specimens can be produced by compression of monocrystals of α-iron (Pfeil, 1927; Barrett, 1939) and by rolling monocrystals of the alloy AlMg (Herenguel and Lelong, 1951). At first the bands are poorly defined (Fig. 155a) but as the compression is increased they become more contrasted (Fig. 155b). The photographs show that the bands have winding boundaries and that a second system of thinner bands occurs inside the bands. Barrett proposed that this second system should also

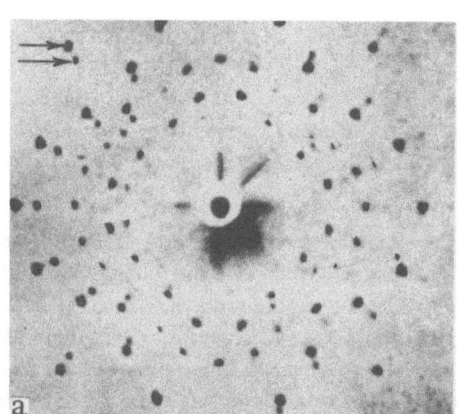
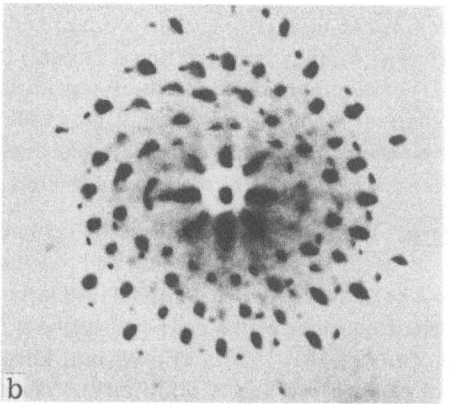

Fig. 148. Laue patterns along two <100> directions for the boundary of an irrational twin in rocksalt: a) beam normal to sides of plate in Fig. 147 (normal to edges of low ridges); the spots to right and left are doubled, but the ones on the vertical axis of the pattern are not split (the rotation axis of the lattice coincides with the edge of the ridge); b) beam normal to end faces (parallel to edges); spots split around a circle. Rotation axis of blocks as in Fig. 148a (Brilliantov and Obreimov).

be called deformation bands. X-ray studies showed that the lattice in the bands is rotated with respect to the lattice in the undeformed region of the crystal by an amount depending on the deformation. The relative angle of rotation of adjacent regions can be 15°, while the maximum rotation of the lattice in the band with respect to the original lattice can reach 90°.

Figure 156 shows a model (based on x-ray studies) of the structure of a crystal containing a deformation band. In this model the lattice orientation is represented by differently oriented cubes. In the middle of the deformation band the [111] direction (which tends to coincide with the compression axis) is perpendicular to the plane of the diagram. The crystal is compressed along [100] outside the deformation band. There is a gradual change of lattice orientation from [100] to [111] between these regions.

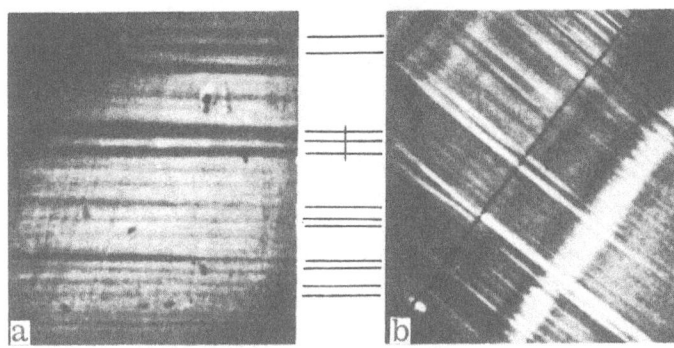

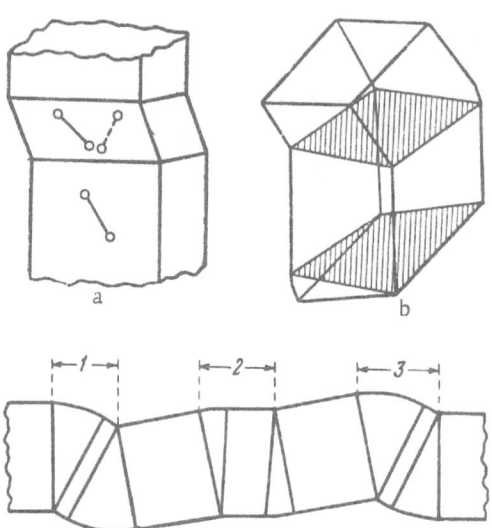

Fig. 149. Plates of rocksalt containing irrational twins produced by compression: a) side face (reflection, ordinary light), the light bands being the ridges 1 of Fig. 147; b) end face, polarized light; light bands are twin boundaries, which are seen on account of the birefringence at sites of residual-stress concentration (Brilliantov and Obreimov).

Fig. 150. Sketches of crystals containing kink bands produced by compression: a) kyanite, with an arbitrary direction in the initial crystal shown as altered by the distortion (full-line), the broken line showing the change for classical twinning; b) calcite, with kink bands on ends of a compressed crystal; c) mica, with kinks 1 to 3 (Mügge).

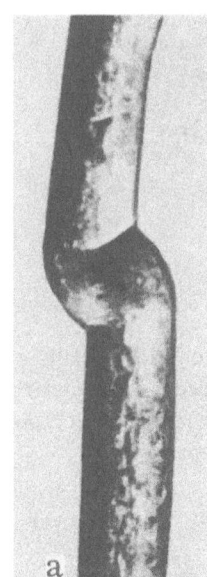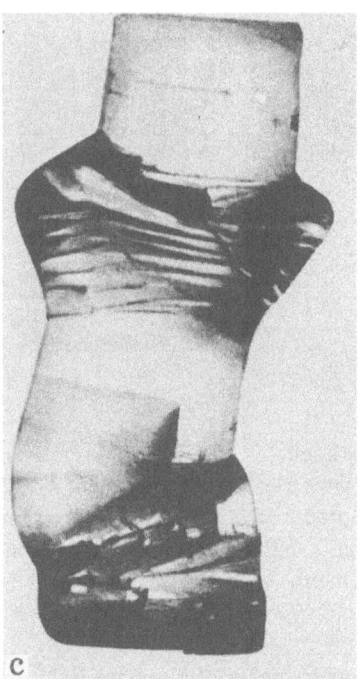

Fig. 151. View of crystals with kink bands: a) and b) monocrystals of cadmium (Orowan); c) naphthalene, the crystal generally appearing light on account of its inherent high birefringence, but the deformed regions appearing dark (Perekalina, Regel', and Dubov).

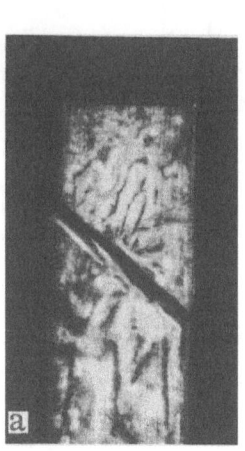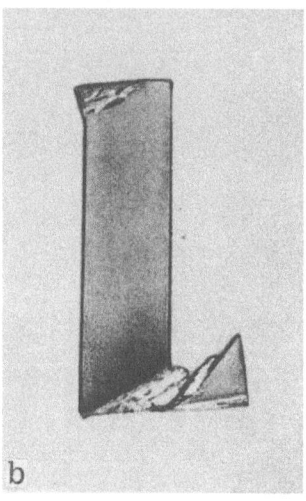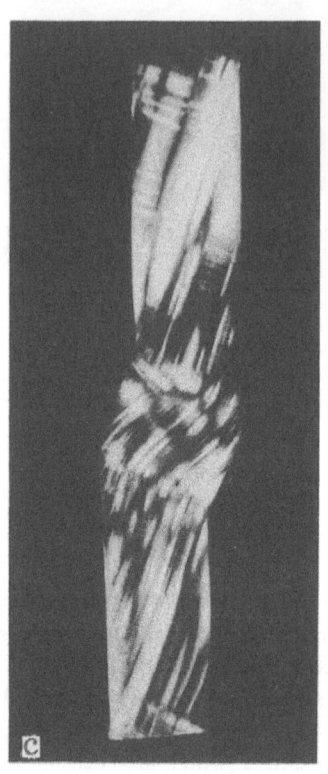

Fig. 152. Kinks in halide crystals: a) TlBr-TlI, crossed Nicols (details of the structure of the layer are given in Figs. 173-5; b) TlBr-TlI crystal compressed 13%, with kinks at ends; c) knee in CsI, the light bands being lines of slip, which change direction in the knee (crossed Nicols).

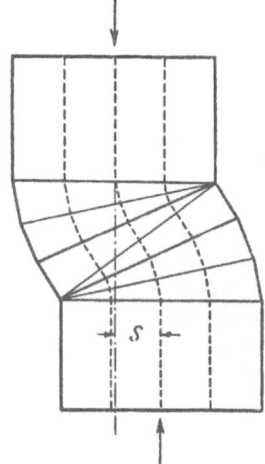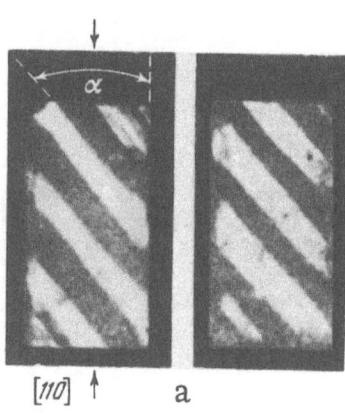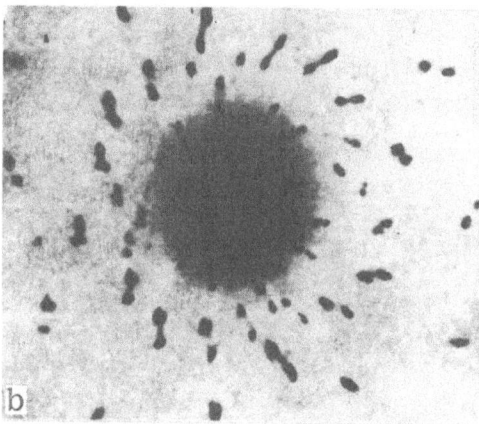

Fig. 153. Idealized scheme for symmetrical kink band produced by compression; s is the axial displacement.

Fig. 154. Division of rocksalt into plates by compression along [110]; a) two side faces in reflection, α being the variable angle between the boundaries and the compression axis; b) Laue pattern on [110] from boundary, which points to a transition zone between the light and dark bands (Stepanov).

Macroscopic groups of alternate light and dark bands are produced by rolling the alloy AlMg. Etch figures indicate that the lattice in the dark bands is twinned about [111] with respect to the lattice in the light bands, and that the lattice gradually rotates from one orientation to the other in the intermediate region between the bands. These bands are similar to the deformation bands in α-iron.

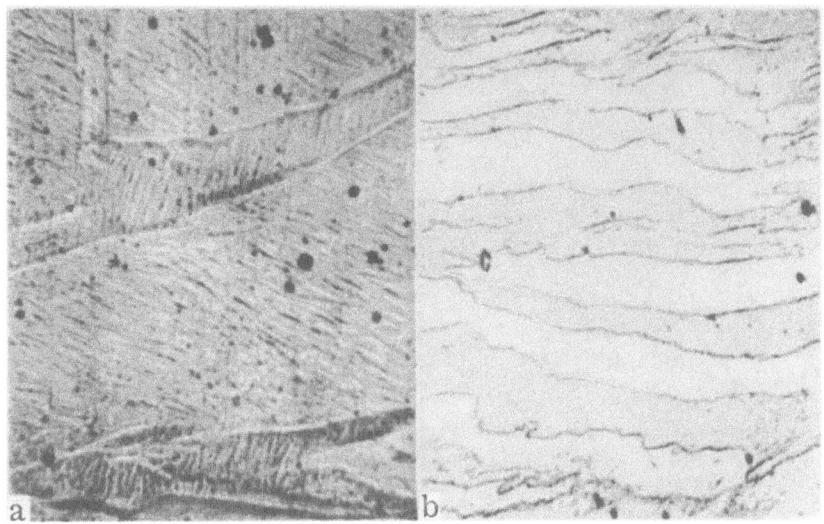

Fig. 155. Deformation bands in grains as seen on the surface of a compressed monocrystal of α-iron: a) initial stage of deformation; b) greater deformation; bands of higher contrast, and a second system of bands within them (Barrett).

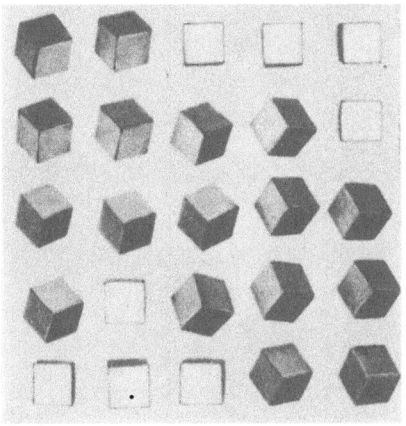

Fig. 156. Model of the gradual change in lattice orientation with the deformation bands of α-iron; the light cubes represent the initial orientation. Compression along [100] (normal to plane of figure). The reorientation is such that the [111] direction lies in the center along the line of action of the force. There is a gradual change of lattice orientation between [100] and [111] (Barrett and Levenson).

Fig. 157. Bands (shown by arrows) in deformed aluminum; reorientation does not occur within these. Plastic deformation causes transverse slip (with rotation), so the lattice orientation is altered outside these bands (Friedel).

Similar deformation bands have been produced in monocrystals of α-brass (Mathewson and Philips, 1916), copper and α-brass (Johnson, 1919, 1922), tungsten (Goucher, 1924a, b), silver and β-brass (Elam, 1922, 1928, 1936), and also in aluminum (Bürgers et al., 1931; Taylor, 1928). Similar bands are also found in polycrystalline metals. Barrett and Levenson (§ 2.17) used the formation of deformation bands to explain the occurrence of grains in rolled metals.

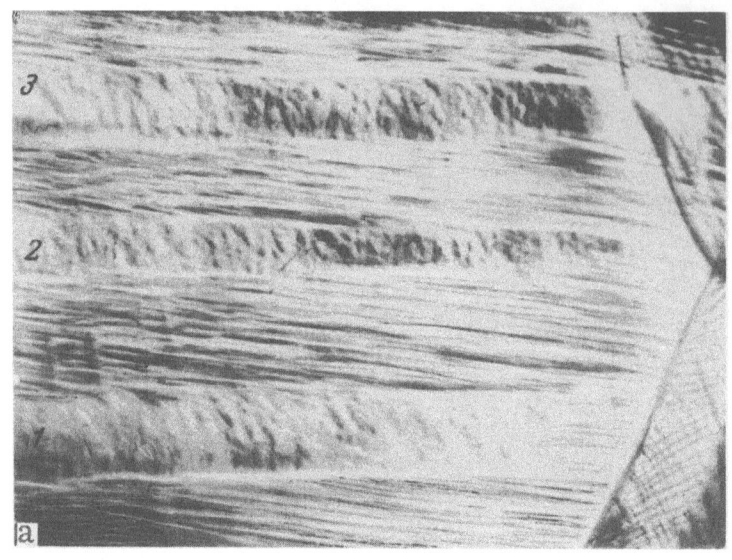

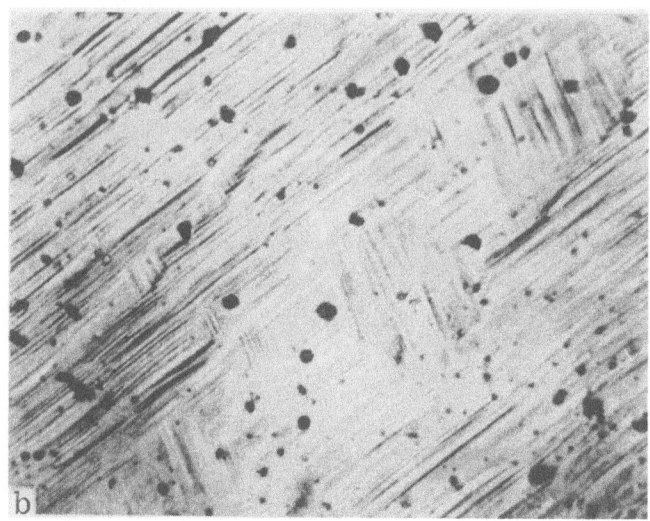

Fig. 158. Secondary slip bands in grains of deformed polycrystalline aluminum; the bands are formed by groups of secondary slip lines that took no part in the deformation of the main crystal: a) 1-3, wavy traces of secondary slip; b) rectilinear traces intersecting primary slip lines (Honeycombe).

The narrow bands that Cahn produced by pulling monocrystals of aluminum (Fig. 157) have also been called deformation bands (Cottrell, 1958). They can be explained if one system of slip directions occurs and if dislocations develop simultaneously towards one another, slow and finally stop one another; in this way the boundaries of the deformation bands are produced.

Another type of band was produced by pulling monocrystals of aluminum (Calnan, 1952). Honeycombe (1951) showed that this effect occurs in crystals with two systems of slip planes; the bands are formed by secondary slip in a system of slip planes that takes no part in the deformation of the main crystal (Fig. 158a, b).

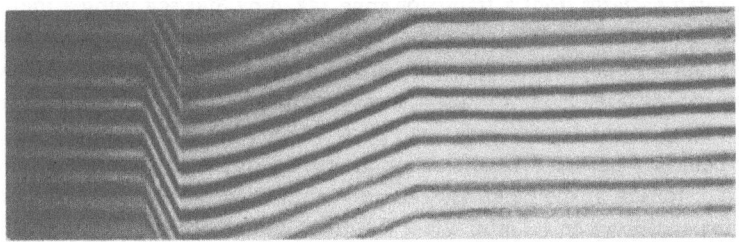

Fig. 159. Transition zone (accommodation region) adjacent to the lower boundary of a twin layer in zinc; interference microscope (Startsev and Lavrent'ev).

Accommodation Bands. In § 1.17 it was shown that transition zones (accommodation bands) are formed between the main crystal and the twin boundaries that are not parallel to the twin plane. These zones generally consist of several (3, 5, 7) sections (blocks) which are rotated with respect to the twin layer and to one another through fairly small angles. As yet such zones have been found only in twinned metals and in calcite. They resemble kink bands and Barrett bands in structure. Figure 159 is an interference microscope photograph of a cleaved zinc specimen that shows a transition zone adjacent to a twin layer; it does not show the blocks but the curvature of the interference lines in the accommodation region shows that changes in the positions of the cleavage planes have taken place (Startsev and Lavrent'ev, 1958).

Fig. 160. X-ray reflections from the transition zone (three lines on left) and from the main crystal (Startsev and Lavrent'ev).

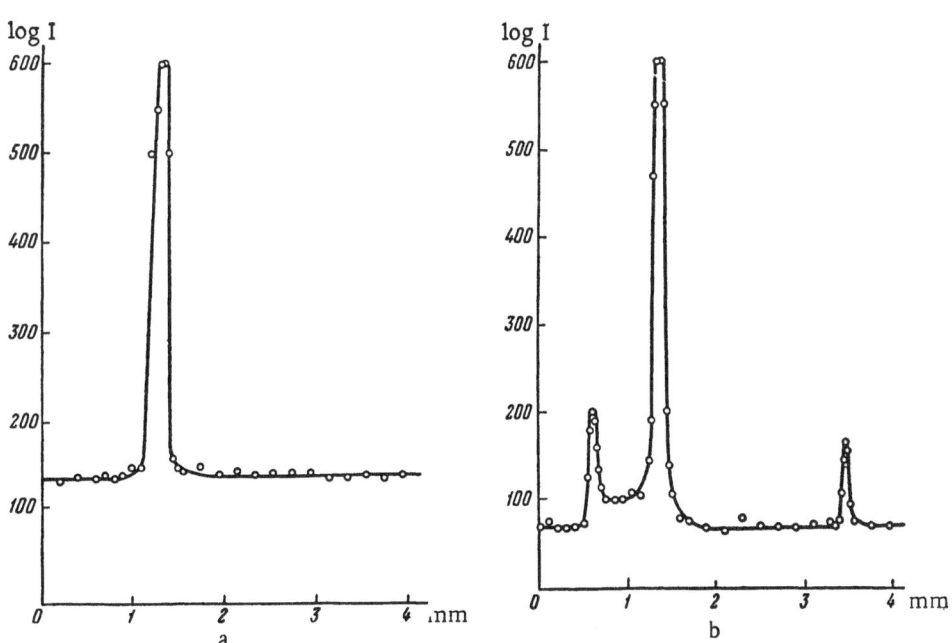

Fig. 161. Microphotometer traces of x-ray patterns from a zinc monocrystal: a) undeformed; b) deformed. The left-hand peak corresponds to the transition zone, the middle one to the main crystal, and the right-hand one to the twin layer. A diffuse background is seen between the left and middle peaks (Startsev and Lavrent'ev).

Figure 160 shows x-ray patterns from a transition zone, taken by Startsev using a method described by Brilliantov and Obreimov (1935, 1927). The number of reflections corresponds to the number of rotated blocks. The angles of rotation (about 10') depend on the width of the transition zone and are different for different twin layers. Microphotometer traces (Fig. 161a, b) show that a continuous background (diffuse scattering) occurs between the peaks corresponding to the accommodation region and the main crystal; this indicates that there is serious distortion of the boundary between the main crystal and the accommodation region (§ 1.17). Hirsch (1952) also found lattice distortion at the boundaries of adjacent blocks. The twin layers are about 0.05 mm wide, the transition region can be up to 0.1 mm wide, while the separate blocks are about 0.01 mm wide. X-ray patterns from a twinned and annealed zinc crystal (Fig. 162) show that the diffuse background and the line from the twin layer have been lost but the line from the transition zone remains; therefore the transition zone is more stable than the narrow twin layer.

4. Relation of Mechanical Lattice Reorientation to Slip and Twinning

Early workers (Mügge, Brilliantov and Obreimov, Elam) assumed that lattice reorientation can be brought about by a process similar to twinning and by inhomogeneous spreading of slip in the specimen. Brilliantov and Obreimov (1935, 1937) gave two possible schemes for the formation of reoriented lattice regions (Figs. 163 and 164a, b). The second scheme requires translational slip together with rotation of the slip plane; atomic layers must slip with respect to one another through a distance that is a multiple of the lattice parameter. The coherence of the lattice along the slip plane must not be broken, and distortions must accumulate at the boundary of the reoriented region. Brilliantov and Obreimov thought that the existence of translational slip required special experimental proof; they found reorientation by irrational twinning in rocksalt.

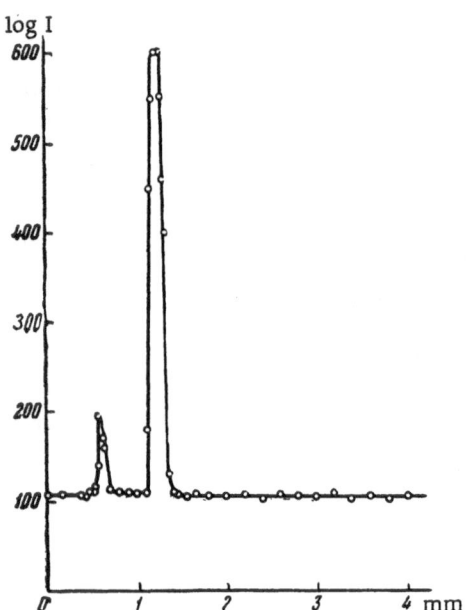

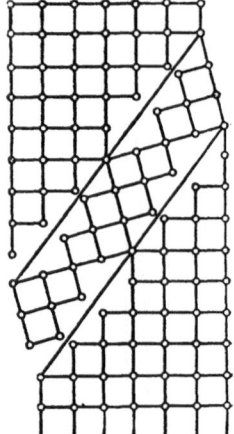

Fig. 162. Microphotometer trace from an x-ray pattern of a twinned and annealed zinc crystal; the line from the layer has been lost, as has the background (Startsev and Lavrent'ev).

Fig. 163. Lattice reorientation in rocksalt resulting in an irrational twin; the plane of contact has irrational indices, and the lattice is greatly distorted along this plane. The lattice is symmetrically disposed with respect to the boundary in the components (Brilliantov and Obreimov).

Orowan (1942) suggested that lattice reorientation with the formation of kinks and Brilliantov-Obreimov twins does not occur by twinning but by translational slip with rotation of the plane of slip (Figs. 165, 166 and

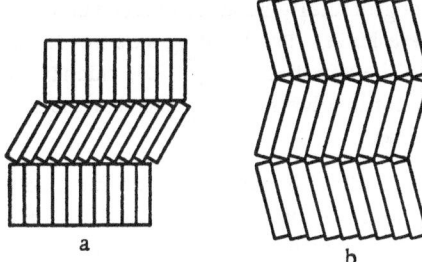

Fig. 164. Second scheme for the formation of reoriented regions (irrational twins); lattice reorientation by slip (with rotation) along a system of parallel crystallographic planes (atomic scale). Formation of: a) one block; b) three adjacent blocks (Brilliantov and Obreimov).

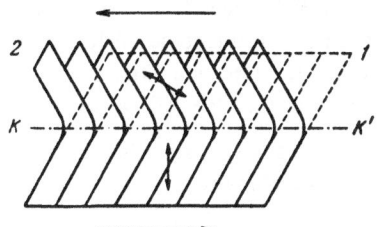

Fig. 166. Kinking mechanism; KK' is the plane of bending, whose position is governed by the state of stress. The long arrows indicate the directions of shearing stresses, while the short arrows illustrate the lattice orientation in the contacting parts.

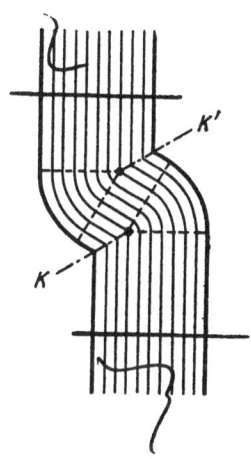

Fig. 165. Kink (knee) formation by slip on parallel crystallographic planes (thin full lines); the broken lines show the wedge-shaped region having slip with rotation. K and K' are the kink planes (Orowan).

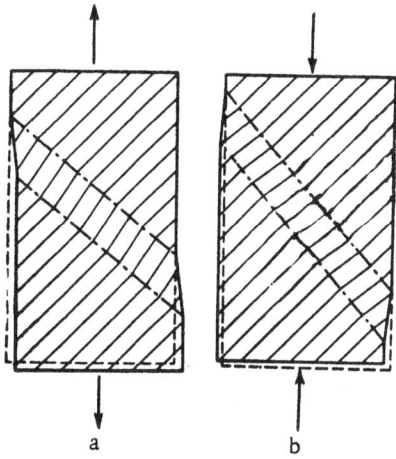

Fig. 167. Formation of a reoriented band (kink or irrational twin) in rocksalt: a) in pulling; b) in compression. The thin parallel lines are the slip planes; the broken ones are the boundaries of the irrational twin, which readily combine with a perpendicular system of slip planes (Orowan).

167a, b). Geometrical considerations require that the plane KK' must be a plane of symmetry with respect to the slip planes above and below KK', but KK' need not be a plane of symmetry for the lattices of the two parts of the crystal (compare the arrows in Fig. 166).

If the lattices are joined along slip planes, then $na = 2t \tan \varphi$, in which a is the lattice parameter in the slip direction, n is an integer, t is the thickness of the slip layer, and φ is the angle between the slip plane and the normal to KK' (Fig. 168). If φ is small, the boundaries of the reoriented bands in rocksalt can be confused with ordinary slip traces.

Hess and Barrett (1949) found that kinks are produced by longitudinally bending cleaved zinc monocrystals and that the (0001) plane is a smooth curve in the vicinity of these kinks. The bending axis lies in the (0001) plane and, in the initial stage of deformation, is perpendicular to the slip direction [$\bar{2}$110], which is close to the compression axis. Hess and Barrett assume that this process does not involve twinning and that the kinks are produced by slip together with bending of the slip planes (Fig. 169a).

Figure 169b-d shows a scheme for kinking as a result of longitudinal bending; this scheme involves motion and regrouping dislocations. In the initial stage the specimen is bent sinusoidally due to its loss of elasticity. Pairs of dislocations of opposite sign are formed in the bent slip planes when the stress reaches the elastic limit. As the stress increases the dislocations migrate and regroup in rows forming the kink boundaries (Fig. 169b, c). As the load increases the angle α increases (Fig. 169b). Then new dislocations are formed inside the kink bands at the new bending planes (Fig. 169c, d). Finally the slip planes inside the kink become completely bent. Gilman (1954) rejects the assumption that kinks originate during elastic bending of crystals. He also states that kinks resulting from loss of elasticity should be easily produced if zinc monocrystals are compressed in a rigid clamp with their plane of slip parallel to the axis of the specimen but should not be produced if crystals of any orientation are compressed in a hinged clamp; experimental results do not support this statement.

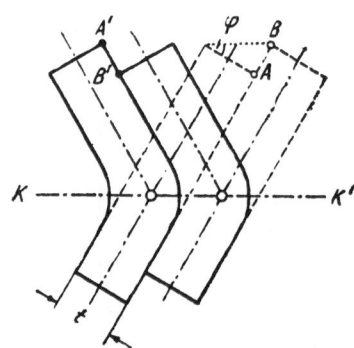

Fig. 168. Calculation of na, the relative shear of adjacent plates in deformation by slip with rotation (Orowan).

Gilman assumed correctly that kink formation is always preceded by plastic deformation with slip; kinks are formed because of retardation of slip at the ends of the specimen. Dislocations accumulate at the ends and either cause bending of the slip plane or, by regrouping in vertical rows, form boundaries of reoriented regions. The second possibility is more probable since it would lead to discontinuities on the slip planes and thus reduce the deformation of the crystal. Such discontinuities are clearly seen in kink bands in compressed crystals of thallium halides viewed in transmitted polarized light (Zemtsov, Klassen-Neklyudova, and Urusovskaya, 1953).

We have shown (using a rapid cinematographic method) that kink bands in the middle of transparent specimens are not produced by retardation of slip at the ends but by sudden redistribution of stress during plastic deformation. The tangential stress along the slip plane in the slip direction changes sign during the stress redistribution (§ 15.1).

It is clear that formation of deformation bands (Fig. 157) involves deformation by slip while the bands of Fig. 158 are a result of localized slip. Special study is required of the elementary processes involved in the formation of Brilliantov-Obreimov bands, Barrett bands, Stepanov plates, and various forms of kinks and accommodations.

Slip traces are always seen when the rotated regions of irrational twins are viewed in transmitted polarized light. This method also revealed slip traces when applied to the wedge-shaped regions of the kinks in crystals of CsI, CsBr and TlBr-TlI (Klassen-Neklyudova and Urusovskaya, 1955, 1956a, b, § 14). The appearance of slip traces in reoriented regions can be considered as supporting evidence for the translational scheme of lattice reorientation; however, it is possible that the traces are produced by secondary slip after reorientation (§ 1.13). Klassen-Neklyudova and Urusovskaya (1957) have done experiments designed to separate lattice rotation from deformation by slip.

The [100] slip directions are mutually perpendicular in crystals of CsBr, CsI, and TlBr-TlI (cubic system, CsCl lattice type); therefore, slip along (110) cannot occur if such monocrystals are compressed exactly along [100]. No slip traces can be detected in polarized light in kink bands of TlBr-TlI crystals compressed along [100]. Perekalina, Regel', and Dubov (1958) have observed kinks in naphthalene monocrystals that do not show slip traces in polarized light.

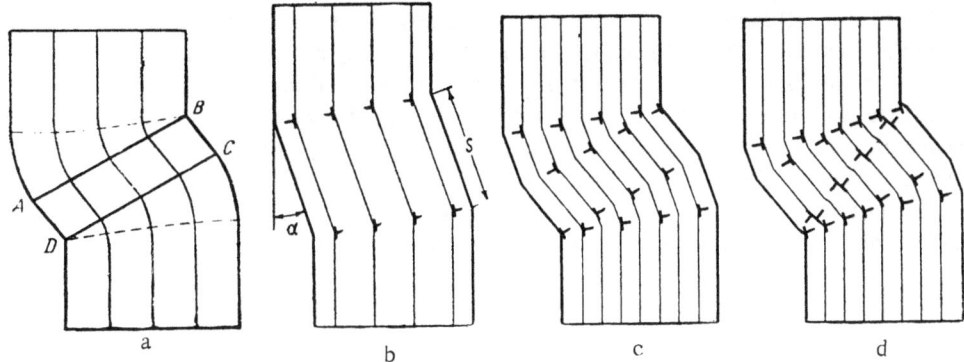

Fig. 169. Hess and Barrett's scheme for kinking: a) initial stage; b) formation of external boundaries of slip band by group of dislocations in rows by sign; c) production of new dislocations giving slip traces within kink planes; d) stage at which the rows of dislocations meet.

We have considered these results as evidence supporting the twinning process of lattice reorientation. However, it is possible that these napthalene and TlBr-TlI crystals contain very fine slip traces which cannot be detected in polarized light or by x-ray methods but which may be detectable by selective etching.[*] Slip traces are always clearly seen in CsI and CsBr crystals of any orientation when viewed in polarized light (Fig. 152c).

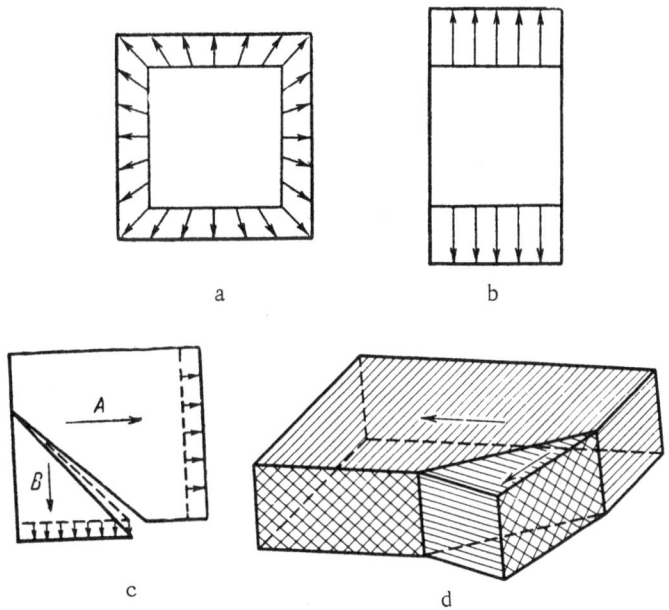

Fig. 170. Formation of irrational twins by inhomogeneous slip: a) deformation of cross section with four equivalent slip systems; b) the same, but two systems; c) different slip elements acting at different points, the slip directions meeting at 90°; d) irrational twin corresponding to the last; the side faces show the traces of the active slip planes (Indenbom and Urusovskaya).

---

[*] We have now found evidence for slip lines in naphthalene by studying the arrangement of cleavage cracks.

Indenbom and Urusovskaya (1959) have studied the formation of reoriented regions (irrational twins) in rocksalt; they showed that Brilliantov and Obreimov's scheme (Fig. 164a, b) is correct. If a rocksalt plate is compressed along the short edge (Fig. 147) and slip occurs equally along all four (110) slip planes (inclined at 45° to the compression direction) then a plate that is originally square will remain square (Fig. 170a). If slip occurs in two opposite directions, the plate is elongated (Fig. 170b). It is also possible for the deformation to be inhomogeneous with one slip system in region A (Fig. 170d) but two slip systems in region B; the elongation directions in A and B meet at 90°. If the specimen remains intact, then the regions A and B must rotate with respect to one another through an angle that depends on the amount of deformation by slip; the angle between the faces of the two regions is equal to the angle of rotation of the lattice. If the specimen is uniformly compressed, the boundary between A and B must lie along (110) and must experience similar distortions due to the elongation of the cross section in two mutually perpendicular directions (Fig. 170c); otherwise the boundary will be a source of macroscopic stresses. Indenbom has treated this effect quantitatively.

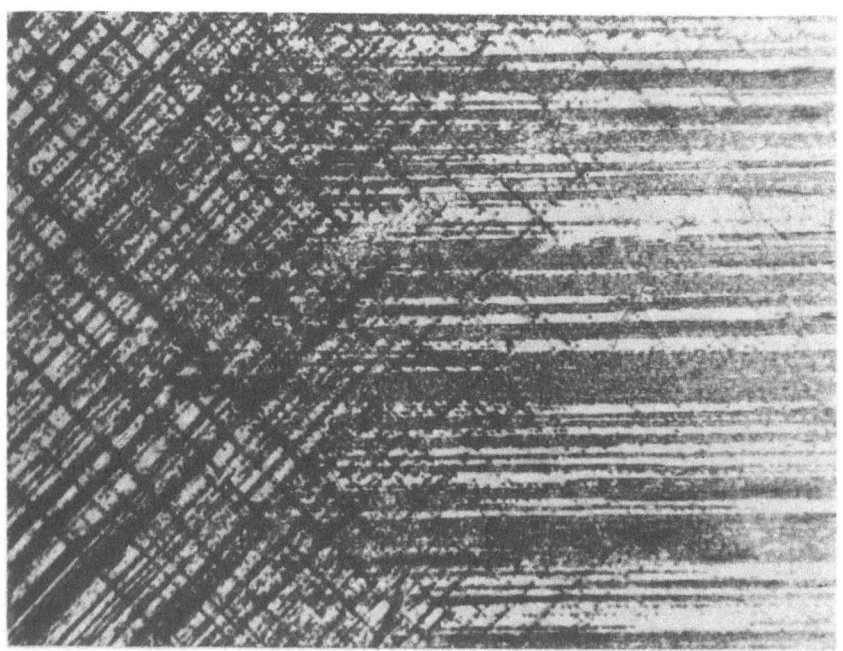

Fig. 171. Side face of a deformed LiF crystal etched with 3% hydrogen peroxide; the array of etch figures in the adjacent components indicates that different slip planes are absent. There is no boundary (Indenbom and Urusovskaya).

The role of deformation by slip in lattice reorientation during irrational twinning of NaCl is confirmed by selective etching, interference studies, and stress measurements by optical and x-ray methods.

The initial stages of reorientation (angle of rotation 1-2°) can be studied by selective etching; separate dislocations are then seen in the etch figures. A series of etch figures showed slip traces together with both edge dislocations and spiral dislocations (not visible in polarized light). The slip traces in Fig. 171 correspond exactly with the scheme of Fig. 170d. If the rotation is caused by twinning, Fig. 171 should show a sharp boundary.[*] Interference studies also showed that the boundary is not sharp but that the two regions merge smoothly into one another.

---

[*]Strongly deformed specimens (angle of rotation 3-4°) sometimes show boundaries due to the appearance of cracks.

128

Before deformation, a network of mutually perpendicular striations appeared at the end of the crystal. After deformation, the square meshes of the network became rhombs and were drawn out in mutually perpendicular directions in adjacent regions. Laue patterns, taken in the compression direction along the boundary, showed diffuse spots corresponding to two mutually perpendicular slip directions.

Very high internal stresses occur at the boundary between rotated blocks (Obreimov and Shubnikov, 1926). If the specimen is homogeneously compressed, the boundary between the rotated regions is a plane of symmetry of the two regions [(110) in rocksalt] and no internal stresses need occur. If the boundary is not a plane of symmetry, macroscopic stresses arise which try to rotate the boundary to a symmetrical position. The jump in stress at the boundary is proportional to the angle of rotation of the block and to the angle between the boundary and the plane of symmetry (Indenbom and Urusovskaya, 1959).

Doubly-refracting bands (visible in polarized light) at the ends of the crystals showed that stresses occur on both sides of the boundary. Measured stresses in NaCl and LiF agree in sign and magnitude with calculations based on a translational slip scheme (Fig. 172a-c). The scheme for the formation of kink bands (Fig. 172c) needs further study since it requires the appearance of slip lines inside the reoriented regions and these are not found experimentally. The rate of lattice reorientation in the initial stages of kink formation also needs to be quantitatively estimated (§ 15.2), while the formation of accommodation bands requires further study.

5.   Optical and X-Ray Evidence on the Structure of Kink Bands

The structure of reoriented regions in transparent crystals can be studied by transmission of ordinary or polarized light; comparison of optical and x-ray results can be very informative (Klassen-Neklyudova, 1953). Photomicrographs of kinked layers are shown in Fig. 173a, b (ordinary transmitted light) and Fig. 174 (polarized light). Figure 175a shows a scheme for the structure of a specimen in which a kinked layer occurs, while Fig. 175b shows the structure of the same layer according to optical studies. These figures show that kink bands consist of a large number of elongated wedge-shaped regions. The well-defined outer boundaries of the layer are not parallel and have irrational indices; one of the boundaries ends in a crack (Fig. 173a). A number of parallel slip traces exist inside the wedge-shaped regions; these

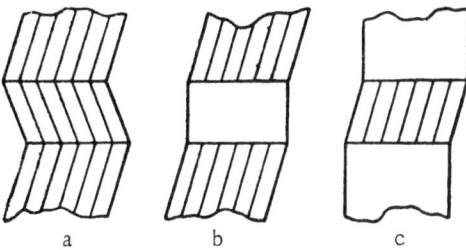

Fig. 172. Formation of reoriented regions by translation: a) one of Brilliantov and Obreimov's schemes; b) scheme for deformation bands corresponding to Fig. 157; c) kink bands (Indenbom and Urusovskaya).

slip traces continue outside the well-defined boundaries of the layer and form the steps A and B in Fig. 175a (these steps are not due to classical twinning). The slip traces are always perpendicular to one of the boundaries of the wedge-shaped regions (Fig. 175) and change direction abruptly at these boundaries. The orientation of the slip traces in adjacent regions shows that the lattices in these regions are oriented unsymmetrically; the angle of rotation of the lattices in these regions can be found from the angle between the segments of the slip traces. Two fan-shaped groups of wedges are formed if the kink band develops symmetrically (Fig. 153).

Figure 176 shows Laue patterns from a kink band using both a broad beam covering the whole layer (16 wedges), and narrow beams covering a smaller number (5, 2, or 1) of wedges. Each Laue spot is split into a number of reflections equal to the number of wedges seen under the microscope; also the angle between the outer reflections is equal to the maximum disorientation of the lattice measured in the photomicrographs, while the angle between adjacent reflections is equal to the angle between adjacent segments of the slip traces. The x-ray studies also show that the lattice rotates about an axis in the (110) plane (slip plane) and perpendicular to [100] (slip direction).

Compression of TlBr-TlI crystals along one of the possible <100> slip directions can lead to rotation of the lattice through very large angles (compression of up to 13-15% of the original height of the specimen) (Fig. 177a, b). Cracks occur at the wedge boundaries for such large deformations. The lattice rotation was estimated from the orientation (Fig. 177a) and from the symmetry of the impact figure (Fig. 177b).

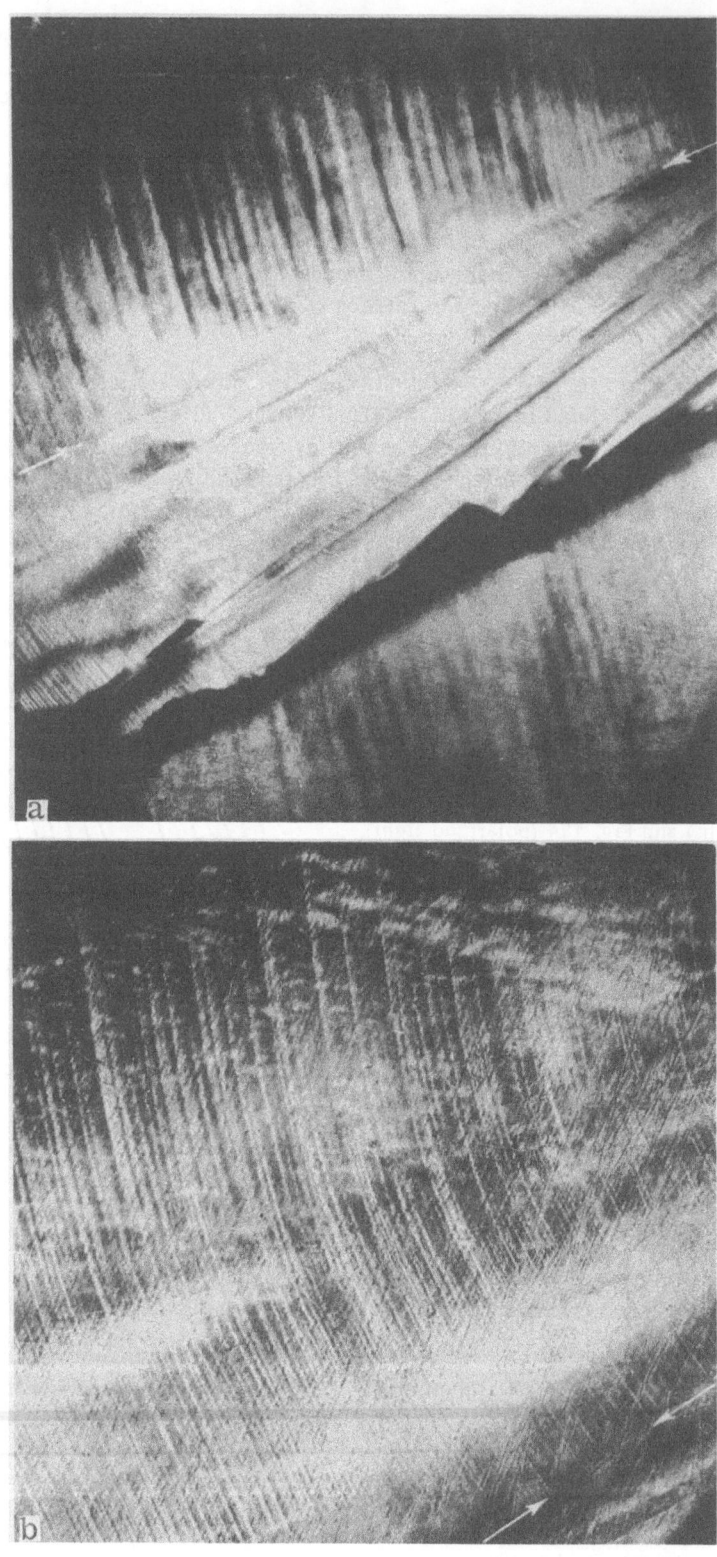

Fig. 173. Photomicrographs (ordinary light) of kink bands in a TlBr-TlI monocrystal (Fig. 152a): a) two external boundaries of band, and wedge-shaped regions from left to right, each wedge bearing several parallel slip lines; b) upper boundary of slip band (denoted by arrows) at high magnification, with bending of lines above the boundary caused by change in direction of the slip lines at barely detectable wedge-shaped regions.

130

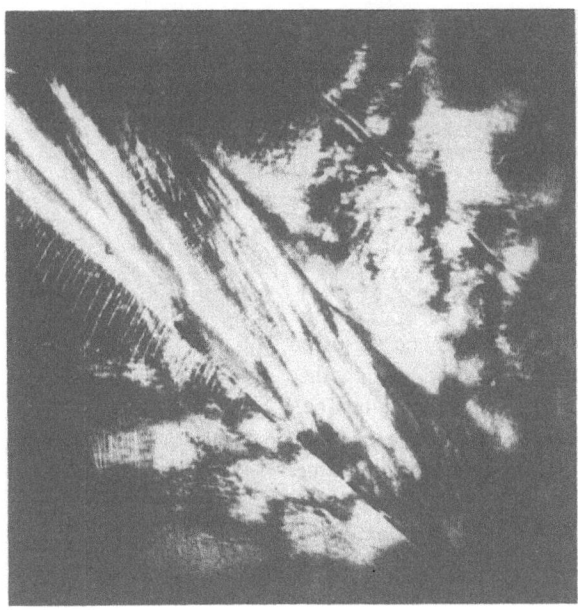

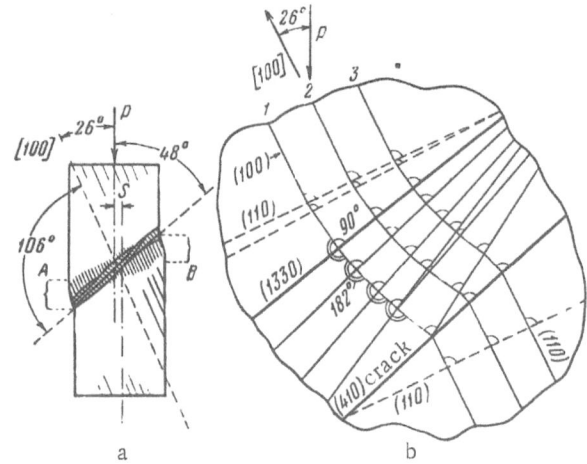

Fig. 174. Photomicrograph of another part of the same band (Fig. 152) in polarized light; the light areas are ones of stress concentration.

Fig. 175. Structure of kink bands in TlBr-TlI (Fig. 152a): a) array of bands relative to [100] and to the compression axis; the fine hatching represents the slip lines; b) refraction of slip lines 1-3 at the boundaries of wedge-shaped regions; the thick lines denote the external (prominent) boundaries. The sharp boundaries of the six wedge-shaped regions are denoted by thin full lines; indistinct boundaries outside the kink bands are shown broken.

Kinks are observed in naphthalene crystals of any orientation (Perekalina, Regel' and Dubov, 1958); the reoriented regions are easily distinguished by their extinction positions.

6. Effects of Orientation on the Production of Mechanically Reoriented Regions

The crystallographic orientation of the specimen with respect to the direction of the applied stress is important in the production of kinks, irrational twins, deformation bands, and plates in monocrystals by inhomogeneous stress distributions (§ 14.1).

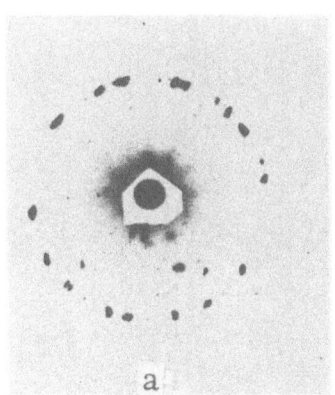

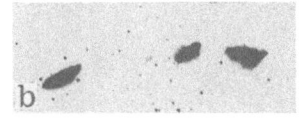

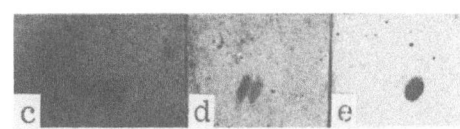

Fig. 176. Laue patterns from parts of a kink band (Fig. 152a) in TlBr-TlI: a) broad beam covering whole band (the diffuse spots correspond to a lattice rotation of 38°); b) beam covering 16 wedges within band; c) beam covering 5 wedges; d) beam covering 2 wedges; e) beam covering only one wedge.

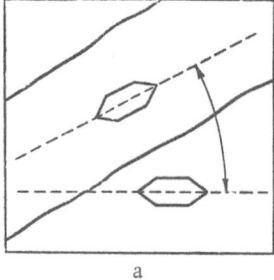 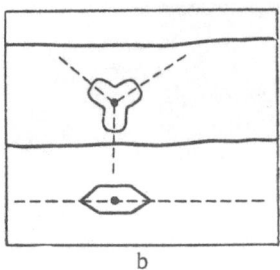

a   b

Fig. 177. Two parts of a kink band in TlBr-TlI, compression
axis [100]; an impact figure is shown within and outside the
band: a) section parallel to (110) and normal to the lattice
rotation axis in the band, with impact figure elongated along
[110], which clearly forms a substantial angle with the same
direction in the initial crystal; b) plane of section parallel
to another (110) plane and also parallel to rotation axis; the
symmetry of the figure outside the band corresponds to an L^2
axis, while that of the one inside corresponds to an L^3 axis.

Gilman (1954) studied the effect of orientation on the formation of kinks during the compression of very
high zinc crystals (height more than ten times the diameter) (Fig. 178). He found that the crystal bent smoothly
without kinking if the only plane of slip (0001) at 20°C was parallel to the principal axis of the specimen (the
compression axis). Kinks began to appear when the angle (X_0) between (0001) and the principal axis of the
specimen reached 2.5°. The width of the kink band increases with X_0, while the boundary of the band becomes
indistinct when X_0 reaches 20°.

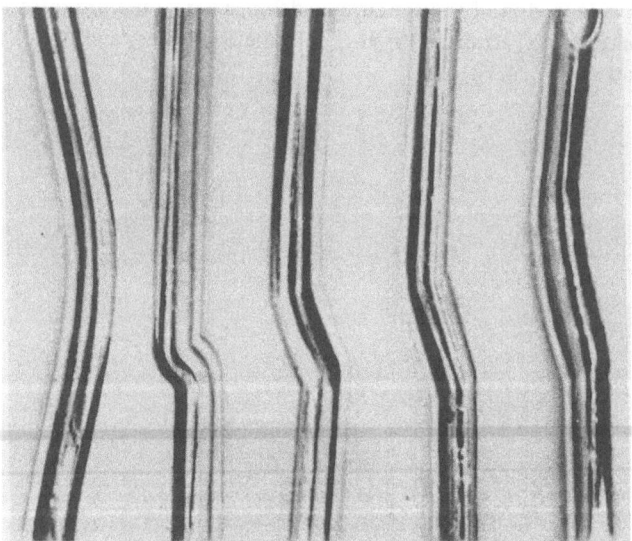

Fig. 178. Width of kink in relation to orientation of specimen
in compression for zinc monocrystals. The angle between the
axis and the slip plane is (from left to right) 0, 2.5, 7, 10, and
20° (Gilman).

Regel' and Govorkov (1958a, b) compressed short (height about twice the diameter) monocrystals of zinc and found kinks when the compression axis was parallel to the slip plane; they also found kinks in the initial stages of deformation when the compression axis was perpendicular to (0001). These different results indicate that the height-to-diameter ratio may affect the stress distribution.

Kinks appear in crystals of thallium and cesium halides (cubic system, CsCl lattice) if the angle (χ_0) between the compression axis and one of the <100> slip directions is less than 25°. If χ_0 is large the kink bands become very broad and indistinct (Klassen-Neklyudova and Urusovskaya, 1955, 1956a; Dubov and Regel', 1957; Regel' and Berezhkova, 1959). The range of orientations that lead to kinks can be considerably extended by changing the temperature and the rate of testing (§ 15.3). Kinks can be produced by compressing monoclinic naphthalene crystals (one slip plane, two slip directions in this plane) of any orientation (Kochendörfer, 1941). However, the external appearance of the kinked layer in naphthalene does depend on the orientation (Perekalina, Regel', and Dubov, 1958).

Kinked layers occur at different angles to the compression axis for different orientations. This effect is not easily seen with zinc crystals (Fig. 178) but compression of crystals of naphthalene and thallium and cesium halides shows that kink boundaries tend to be perpendicular to slip planes (see Fig. 153). The available evidence indicates that well-defined kink bands (with large angles of lattice rotation) are produced during compression if plastic deformation by slip is difficult due to the orientation of the specimen.

The orientation of the monocrystal must allow deformation by slip if kinks are to be produced by pulling. The production of kinks by pulling also depends on inhomogeneities in the structure of the monocrystals and inhomogeneous stress distributions in the specimens (e.g., additional bending of the specimen in the clamps). The boundaries of the kink bands in pulling, as in compression, are perpendicular to the active slip direction. Kinks were not produced by pulling (and bending) TlBr-TlI crystals at room temperature when the pulling axis was inclined at a small (or zero) angle to the slip direction; the critical stress for kink formation under these conditions lies above the yield strength of the crystals.

General discussion of this topic is difficult because the effect of orientation on kinking has only been studied in detail for the compression of zinc and naphthalene; also the processes leading to the formation of kink nuclei are not well understood.

§ 15. Production and Development of Reoriented Regions

1. Stress Distributions Producing Irrational Twins and Kink Bands

Irrational twins, kink bands, and deformation bands are produced in crystals if inhomogeneous stress distributions occur during deformation. Irrational twinning (not the usual plastic deformation by slip) occurs in rocksalt during the compression of very thin cleaved plates or high prismatic specimens.

Kinks are found more often at the ends of the specimen (where the stress distribution is most inhomogeneous) rather than in the middle (Fig. 152b). Kinking is also encouraged if the specimen is not squarely placed in the jaws of the press (Fig. 179).

No kinks were produced by pulling crystals of thallium and cesium halides of various orientations (Klassen-Neklyudova and Urusovskaya, 1956a). Specimens of one orientation (stress along [100]) cracked without noticeable plastic deformation; specimens of other orientations showed plastic deformation by slip. We only succeeded in producing kinks by pulling when the specimen was fixed at an angle in the machine (Figs. 180 and 181). The plates produced by pulling rocksalt (Stepanov) must be due to bending during fixing in the machine.

Kinks are produced in metallic monocrystals (cubic, hexagonal, and rhombic systems) by pulling as well as by compression and rolling.

Barrett-Levenson bands are produced by rolling, drawing, or pulling polycrystalline metals; this is because the homogeneity of the stress distribution is disturbed by the presence of grains of different orientations.

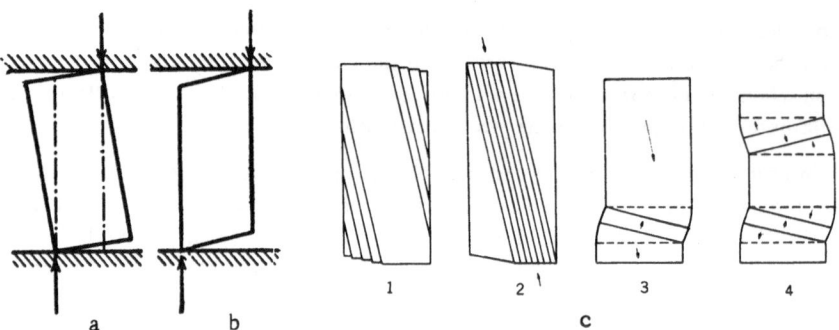

Fig. 179. a) A method of compressing monocrystals to produce kinking using an oblique setting. b) A method of compressing monocrystals to produce kinking using a specimen with
oblique ends. c) Formation of kinks by compression: 1) slip by compression of corners, compression axis deviating to the right; 2) axis deviating to the left, set of slip lines produced; 3) first
kink plate produced by local reversal of slip direction; 4) two wedges formed at plate by second reversal of slip; at the top a second kink, which deflects the direction of the compression axis.

Kinking in metallic monocrystals depends on the existence of defects in the structures of the crystals
(Klassen-Neklyudova and Urusovskaya, 1956a). Gilman and Read (1953) found that kink bands in zinc usually
occur near damaged regions (surface scratches or notches). Regel' and Govorkov (1958a) produced kinks by
pulling impure zinc monocrystals.

Local overstress produced classical twins or kinks (accommodation bands) in metals but in thallium and
cesium halides it produced groups of slip traces (Klassen-Neklyudova and Urusovskaya, 1955; Urusovskaya,
1956a, b).

Kinks produced during pulling of metals are due to surface defects (scratches or notches), to poor center-
ing of the specimen in the machine, or to inhomogeneities in the structure of the specimen (e.g., mosaics or
regions with different yield strengths due to inhomogeneously distributed impurities).

We have attempted to follow the process of kink formation using a rapid cinecamera (up to 5000 frames
per second); synchronization was achieved by using shock-type compression. Kinks were observed in transparent
monocrystals of CsI and naphthalene. The method was used with transmitted polarized light to follow the stress
distributions during deformation of CsI and the lattice reorientation in naphthalene; ordinary transmitted light
was used to study the details of kink formation. The specimens were cut so that only one slip system could be
involved in deformation by slip (one of the six {100} planes for CsI; (001) for naphthalene).

The results of this method show that kink formation is always preceded by extensive plastic deformation
for a given orientation of the specimen to the slip system (Fig. 179c). The basic process of kink formation also
involves the same slip system. A sudden stress redistribution immediately precedes the appearance of a kink
band; this redistribution is due to plastic distortion of the ends of the specimen, causing the compression axis to
deviate repeatedly from the axis of the specimen. Each time this happens the tangential stress, acting along the
slip plane in the slip direction, drops to zero and may even change its sign in some parts of the crystal; this
sign reversal can lead to local reversal of the slip direction. Thus slip in the initial small kink bands is always
in the opposite direction to the preceding plastic deformation. Further stress redistribution is due to localized
slip in kink bands that causes a relative shift of the regions separated by a kink, and to stress concentrations in
the kink boundaries caused by the fact that the rotation of these boundaries lags behind the plastic deformation
in the kink bands. These processes lead to the formation of reoriented wedge-shaped regions next to the initial
small kink band; the slip direction is reversed in each successive series of wedges.

Kink formation in pulling must occur rather differently. Stress redistribution cannot be connected with
changes in the deformation axis, but in the initial stages of pulling it is caused by oblique setting of the specimen in the clamp. Only one narrow kink band is produced.

134

Thus kink formation is caused by loss of plasticity (not loss of elasticity) in the specimen. Stress redistribution is important both before the initial kink formation and during lattice reorientation.

2. Rate of Kink-Band Formation, Displacement of Kink Boundaries by Mechanical Stresses, and Effects of Impurities

It is necessary to distinguish the rate of initial kink formation and the rate of movement of the kink boundary (compare § 2.9 on twinning); the second process will be much slower. These rates have not been studied. It is known that kinking is sometimes accompanied by sounds or crackles in the specimens (e.g., in metals (Gilman, 1954) and thallium halides). The very plastic crystals of CsBr, CsI, and naphthalene are deformed almost silently. We have shown (using a cine-

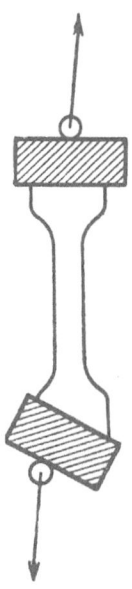

Fig. 180. Attachment of specimen in test machine for production of kinks by pulling.

Fig. 181. TlBr-TlI monocrystal of square cross section pulled in Fig. 180; the dark inclined band at one side is a narrow kink band.

camera at 4000 frames/sec) that broad kink bands can be produced in less than 250 μsec. Cracks can occur at the boundaries of reoriented regions if the angle of rotation of these regions is of the order of degrees (Fig. 190e); the cracks cause transparent crystals to become partly cloudy. It is not clear whether the sounds produced during deformation are caused by the rapid reorientation of the lattice in the kink or by the formation of cracks.

The deformation curves of CsI (§ 15.4) indicate that there are stages in kink formation. The rate of lattice reorientation is very high in the initial stages of deformation and the curves show a correspondingly sharp decrease in stress (Fig. 187). Further kink growth occurs relatively slowly under a nearly constant load (Regel' and Berezhkova, 1959).

If the load is gradually increased on a specimen in which the disorientation of the lattice is small (no cracks at kink boundaries), the kink boundaries gradually move parallel to themselves (Washburn and Parker, 1952; Gilman, 1954). These shifting boundaries leave no trace; breaks in slip traces at the kink boundaries move with the boundaries. This movement is thought to be evidence for the movement of dislocations under stress (Read, 1953). Lattice disorientation in kinks produced by compression can be reduced or eliminated by pulling. The boundaries move towards the middle of the kink but the number of slip traces is unchanged. Comparable loads are required for the formation, removal, and rotation of kinks (Gilman, 1954).

135

Kink formation is made easier by impurities (Forsyth, 1951); this may explain the kink formation and large-scale lattice reorientation in TlBr-TlI (Klassen-Neklyudova and Urusovskaya, 1957).

The rate of the initial stage of kink formation and the effect of impurities on initiation and growth of kinks require detailed study.

3. Effects of Temperature and Deformation Rate on the Formation of Kink Bands and Brilliantov-Obreimov Bands

Kinks can be formed at any temperature but most work has been done at room temperature. Gilman (1954) has studied kinking in compressed zinc monocrystals at -196°C. He found that kink layers are narrower at low temperatures and that the angles of rotation of the sections of the lattice inside the kinks are larger. There are more well-defined boundaries between rotated regions and thin cracks often appear at these boundaries. The stress required to cause kinking is increased (the maximum in the compression curve is 1.7 times larger at -196°C than at room temperature). Finally, kink boundaries can still move as the stress is increased.

Regel' and Govorkov (1958a) have studied pulling of zinc and AgCl at high temperatures; the rates of pulling were varied by a factor of 10^4. No kinks are produced by pulling annealed AgCl crystals at any temperature, rate of pulling, or orientation.[*]

Kinks are produced in zinc monocrystals by both pulling and compression between 20-300°C.[†] High temperatures and low deformation rates favor kink formation in specimens that permit slip along (0001) (Fig. 182a-c and 183). Localized slip is usually observed at high temperatures (up to 300°C) due to the high rate of softening especially in the most overstressed regions of the specimen. During deformation the slip plane tends to rotate and become parallel to the pulling axis (Boas and Schmidt, 1938). Local lattice rotation disturbs the stress distribution and aids kink formation. Kinks are produced near the necked regions in strongly deformed crystals.

At room temperature, medium and large loading rates lead to a uniform distribution of slip traces along the length of the specimen so that the specimen is stretched into a ribbon; kinks can be produced but only at the clamps.

The distribution of slip in specimens oriented so that plastic deformation by slip is possible is not uniform at low temperatures and low rates of pulling; stresses favorable to kinking can arise at the boundaries of lattice regions that have rotated due to rotation of the slip plane.

The effect of rate of deformation on kinking must have a different explanation in specimens oriented so that plastic deformation by slip is not easy. In fact, kinking is favored by high rates of deformation when simple translational slip cannot occur.

The most characteristic kinks in zinc crystals are produced by compression (with any orientation) at 200°C. Slip occurs along other planes at temperatures near the melting point and ordinary translational slip occurs instead of kinking (Fig. 183, 5); kinks cease to appear in compressed zinc monocrystals above 350°C. Kinks are produced in CsI monocrystals (oriented so that plastic deformation by slip is difficult) at all temperatures from 20-600°C. Thus, although kinking can occur at almost any temperature, there is usually a most favorable temperature range for each substance. Polygonization may accompany deformation at high temperatures; this causes the wedges (reoriented regions) to enlarge and reduces the dislocation density. This can also occur when deformation is complete, at room temperature (e.g., TlBr-TlI).

The maximum lattice rotation in Brilliantov-Obreimov bands increases as the temperature is increased (Startsev, 1941).

---

[*]We cannot give definite information about kinking in compressed AgCl crystals.
[†]Regel' and Govorkov did not study this effect at low temperatures.

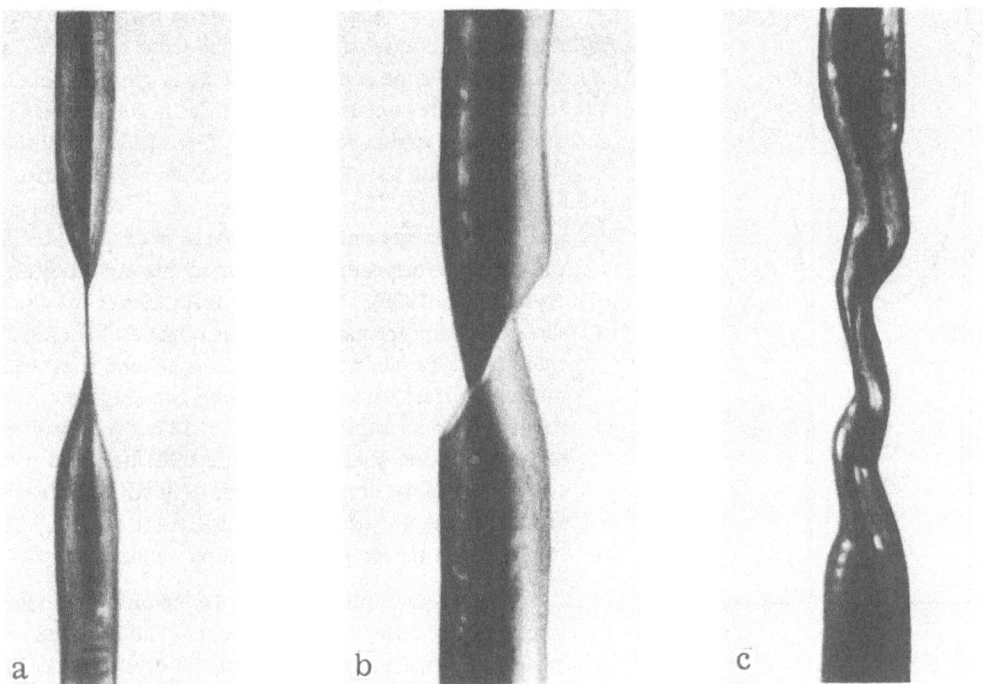

Fig. 182. Stretched monocrystals of zinc: a) 200°C, medium (0.08 mm/sec) and high loading rates, deformation by slip only, central part of specimen converted to a flat strip; b) 300°C and 0.00074 mm/sec, sharply localized slip; c) 200-300°C and 0.00074 mm/sec, kinks resulting from production of bending moments (Regel' and Govorkov).

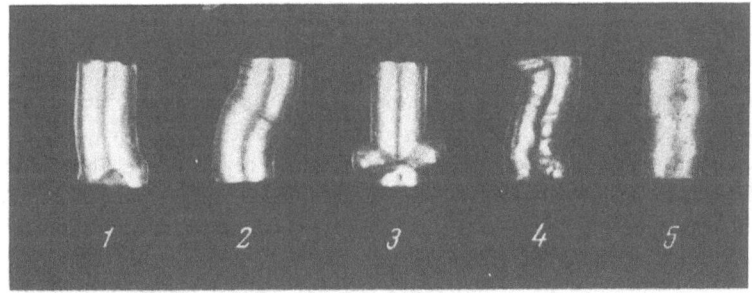

Fig. 183. Compressed zinc monocrystals containing 0.1% Cd; compression axis parallel to (0001) slip plane. Strain rate 0.1125 mm per min; temperatures, °C: 1) 24; 2) 100; 3) 200; 4) 300; 5) 400. The most characteristic kinks appear at 200°C (Regel' and Govorkov).

4. Shape of Stress-Strain Curves in Kink-Band Formation; Test of Law of Critical Shear Stress

The stress-strain curves for the compression of crystals of zinc, TlBr-TlI, naphthalene, and CsI at various orientations are shown in Figs. 184, 185a and b, and 186. Gilman (1954) found a drop in stress, as well as various critical stresses, in the curves for zinc specimens with $X_0 = 2°$; this drop is connected with contraction of the specimen when the first kink appears. The next specimen ($X_0 = 4°$) showed another sharp drop in stress due to the appearance of a second kink. The height of the maximum in the stress-strain curve varies with the critical orientation of the specimen and also depends on the width of the resulting kink. The less plastic crystals give

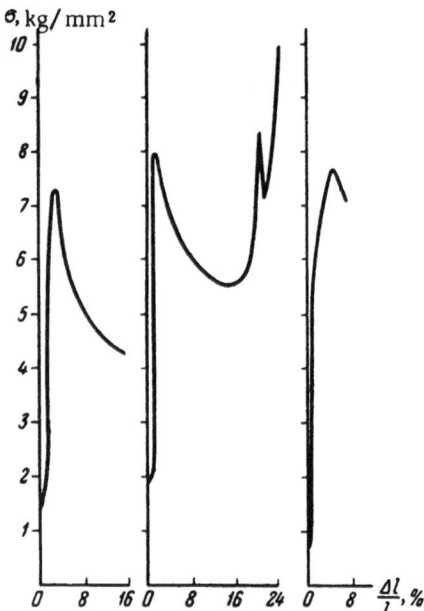

σ, kg/mm²

Fig. 184. Compression curves for zinc monocrystals of three different orientations; the (0001) slip plane makes an angle χ_0 with the compression axis. The χ_0 are respectively 2°, 4°, and 6°. The fall in stress is caused by the contraction (Gilman).

rise to narrower kink bands while the height and sharpness of the peaks increases as the kinks become narrower (for various kinks in the same specimen). If χ_0 is greater than 15°, kink formation does not lead to maxima in the stress-strain curves. Stress-strain curves for TlBr-TlI (Fig. 185a), naphthalene (Fig. 186), and CsI (Fig. 187) have been automatically recorded. Figures 185 and 186 show many very fine peaks as well as the larger ones. [*] The large peaks are not seen in the curve corresponding to the most plastic orientation of the crystal (Fig. 185b). The stress-strain curves for CsI (Fig. 187) show that kink formation consists of several stages. The first stage is elastic deformation and the second is plastic deformation by translational slip; these two stages correspond to the rising linear region OA in Fig. 187. The third stage (AB) is the formation of the kink nucleus leading to a sudden contraction of the crystal and a drop in stress; this stage is so rapid that the apparatus cannot follow it closely. The fourth stage (BC) is the growth of the kink band.

Different applied stresses are needed to produce kinks in crystals of different orientations. The required stress must be less than the yield strength of the crystal for deformation by slip (§ 15.1). The law of critical shear stress (§ 2.7) is obeyed only very approximately since kinking is preceded by plastic deformation by slip that causes hardening of the crystal (Regel' et al., 1958a).

5. Kink Bands and Irrational Twins in the Plastic Deformation of Monocrystals

Mügge's classical scheme for translational slip (§ 1.1) postulates that during plastic deformation two parts of the crystal are displaced along an atomic plane through a distance that is a multiple of the lattice parameter; such a displacement cannot alter the lattice structure. However, Ioffe, Kirpicheva, and Levitskaya (1924) found diffuse radially spread Laue spots in the x-ray patterns of deformed crystals (Fig. 188); this effect was called asterism. Similar radial spread was found with metals (Barrett, 1947) and other crystals. For many years plastic deformation was thought to be the cause of asterism. However, plastic deformation without asterism (Fig. 189a) was observed by Kochendörfer (naphthalene, 1941), Röm and Kochendörfer (aluminum, 1950), Honeycombe (cadmium, 1951), Klassen-Neklyudova and Urusovskaya (1956a). A fine banded structure is found in Laue spots from Cd (deformed by pulling) and TlBr-TlI (deformed by localized loads and longitudinal bending) (Fig. 189b).

It has been suggested by Honeycombe and by us that asterism is connected with kinking. Barrett and Levenson (1939), and Calnan (1952) have suggested that asterism is connected with the appearance of deformation bands.

The fine structure in the Laue spots shows that the coherence of the lattice is disturbed even in the absence of asterism; the nature of these microdistortions is not yet clear.

Garstone, Honeycombe, and Greetham (1956) compared metallographic studies of the surface of aluminum crystals with Laue patterns and x-ray photomicrographs. They found that optical photomicrographs taken in reflected light show only slip traces (Fig. 190a), that the Laue spots are radially spread (Fig. 190b), and that x-ray photomicrographs show macroscopic kinks in which very small lattice rotations occur (Fig. 190c) (these rotations

[*] The very fine spikes in the stress-strain curve are connected with the type of deformation. They are caused by the appearance either of thin kinked layers or of groups of dislocations.

138

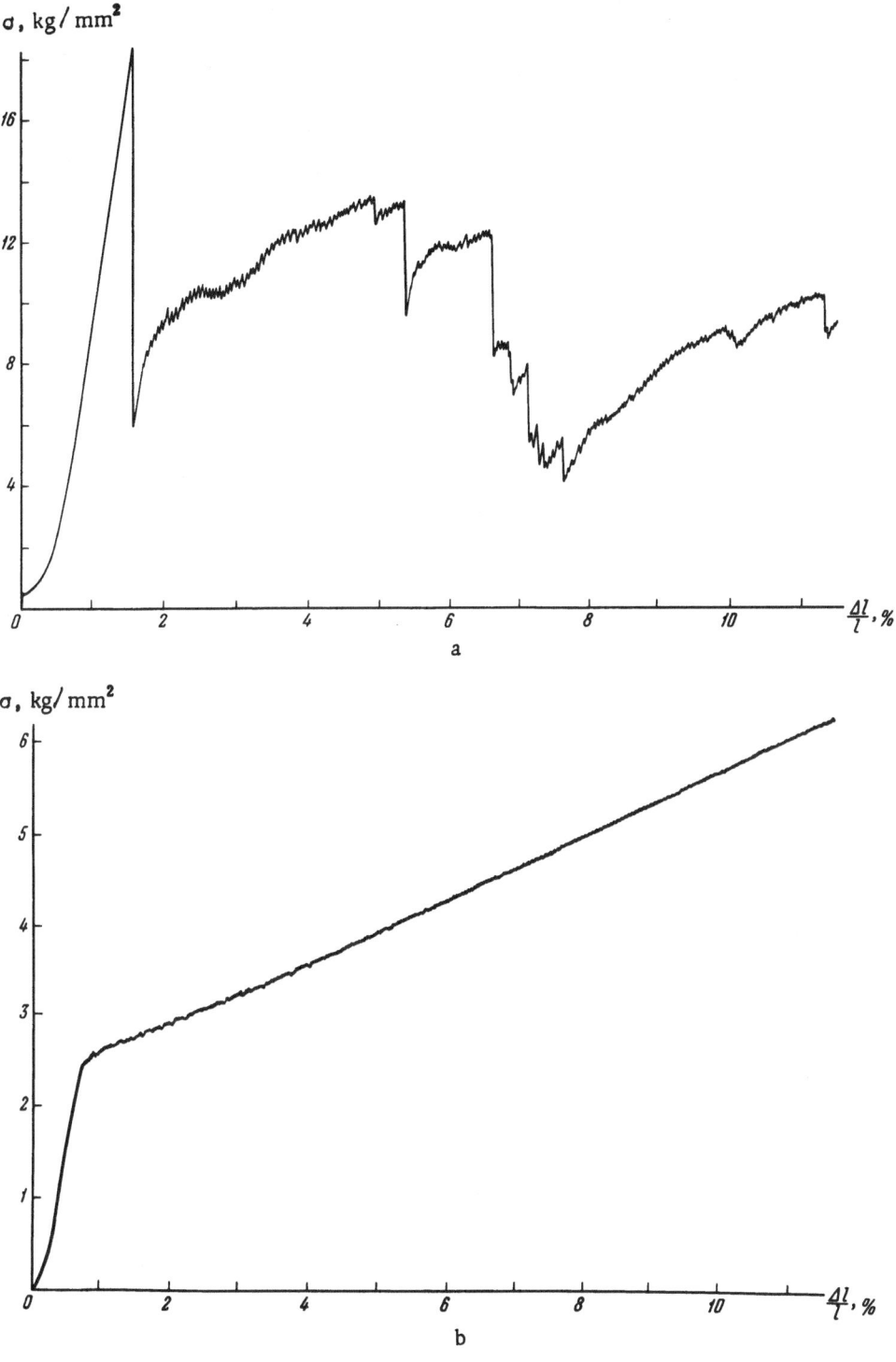

Fig. 185. Room-temperature compression curves for TlBr-TlI monocrystals recorded with an optical dynamometer and photocell: a) compression axis [100], sharp fall in stress caused by kinking (first at 18 kg/mm²); b) compression axis [110], no large jumps, no kinks evident to the eye (Regel' and Dubov).

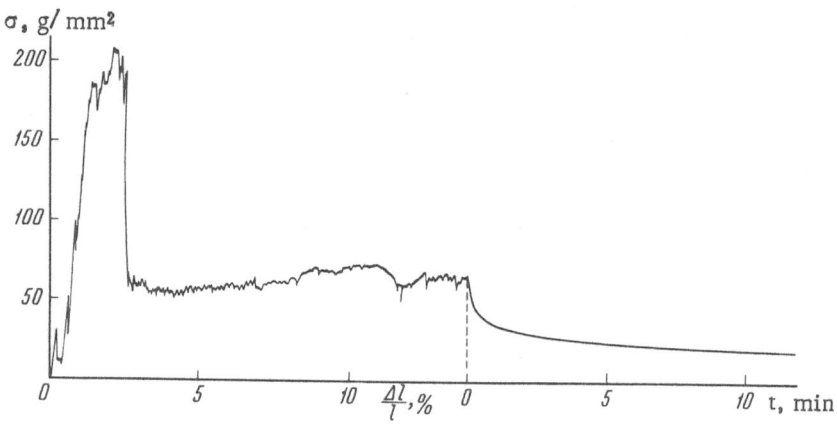

Fig. 186. Compression curves for naphthalene monocrystals; sharp fall in stress caused by kinking, orientation with $X_0 = \lambda_0 = 0$ (Perekalina, Regel', and Dubov).

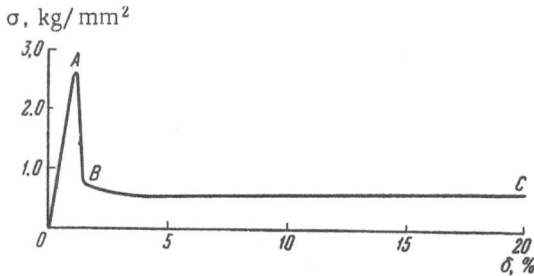

Fig. 187. Compression curve for CsI monocrystal at room temperature; compression axis [100], rate 0.15 mm/min. The fall at 2.7 kg/mm^2 is associated with the first kink band, the stress thereafter remaining almost constant at 0.9 kg/mm^2 (Regel' and Berezhkova).

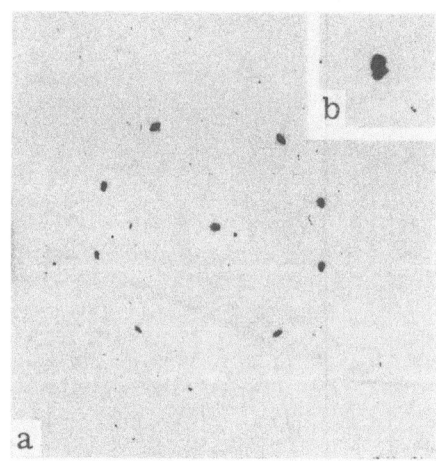

Fig. 189. Laue pattern along [110] from TlBr-TlI monocrystal deformed by localized load; beam passing through center of deformed region. No radial spread. a) Complete pattern; b) one of the spots, which has a detailed structure.

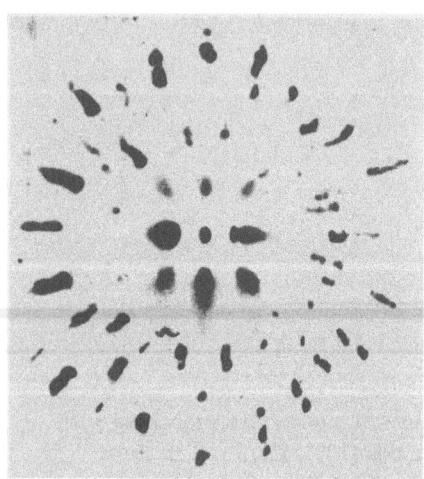

Fig. 188. Radial spread of Laue pattern from compressed rocksalt recorded along cube axis.

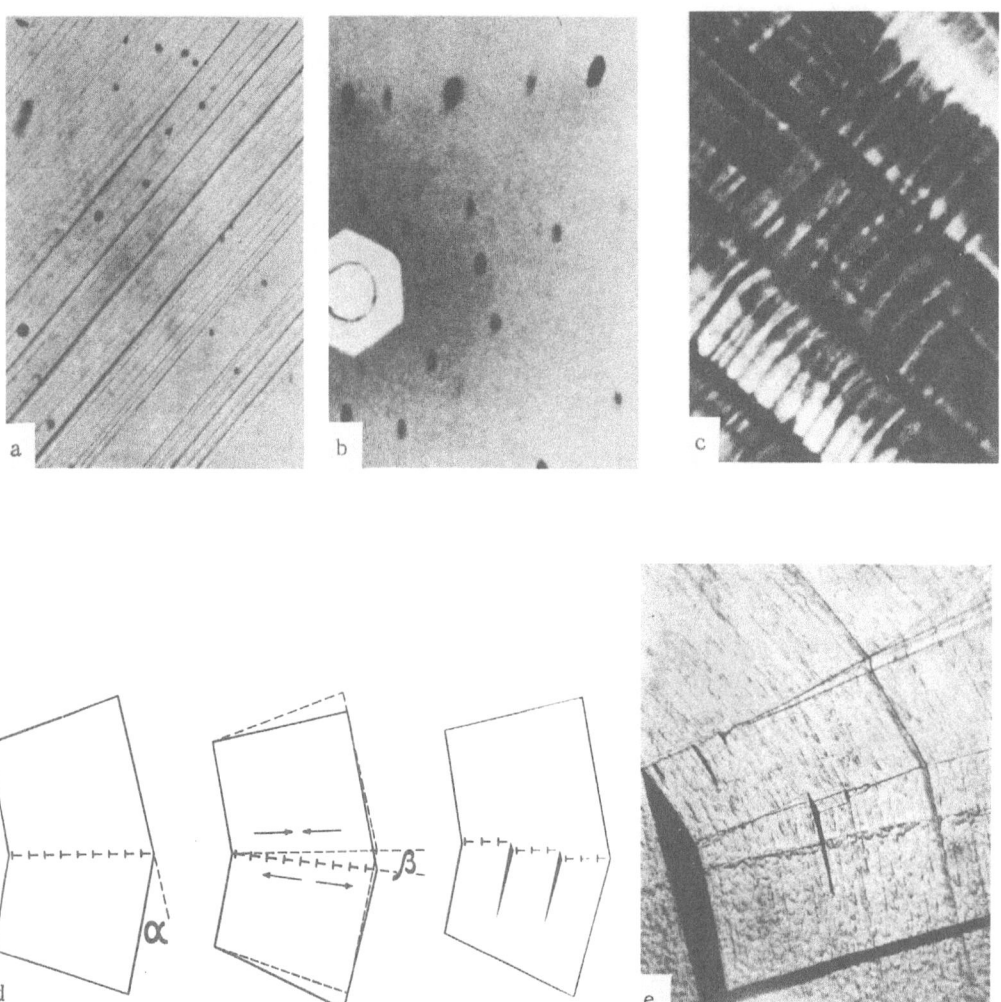

Fig. 190. Aluminum monocrystals stretched 4.4% at room temperature: a) photomicrograph, with only slip lines visible; b) x-ray pattern, spots somewhat elongated; c) x-ray photomicrograph showing series of kink bands, in each of which the lattice is turned through a very small angle (Garstone, Honeycombe, and Greetham); d) formation of cracks in kink; $\alpha\beta/2 = \varepsilon_c$ for fracture (Indenbom); e) cracks caused by kinking in a naphthalene monocrystal.

are so small that the slip traces appear straight even with a magnification of 250). They also thought that asterism is connected with the rotation of macroscopic regions.

Thus, many workers have concluded that asterism is caused by the appearance of reoriented regions (kinks, irrational twins, etc).

At present, a possible connection between hardening during plastic deformation and asterism has not been established. Honeycombe et al. observed plastic flow unaccompanied by hardening in monocrystals of copper and aluminum between -196°C and 200°C (extensions of 4-5% at 200°C and up to 10% at -196°C); however, asterism was observed in this case. Kochendörfer has observed hardening with slip but without asterism. There is no information as yet about the possibility of slip without both hardening and asterism.

6. Kinking and Fracture

High-speed cinemicrography (§ 15.1) has been used in this laboratory to examine cracking along kink boundaries in compressed naphthalene monocrystals. Such cracks are usually explained as caused by rupture of

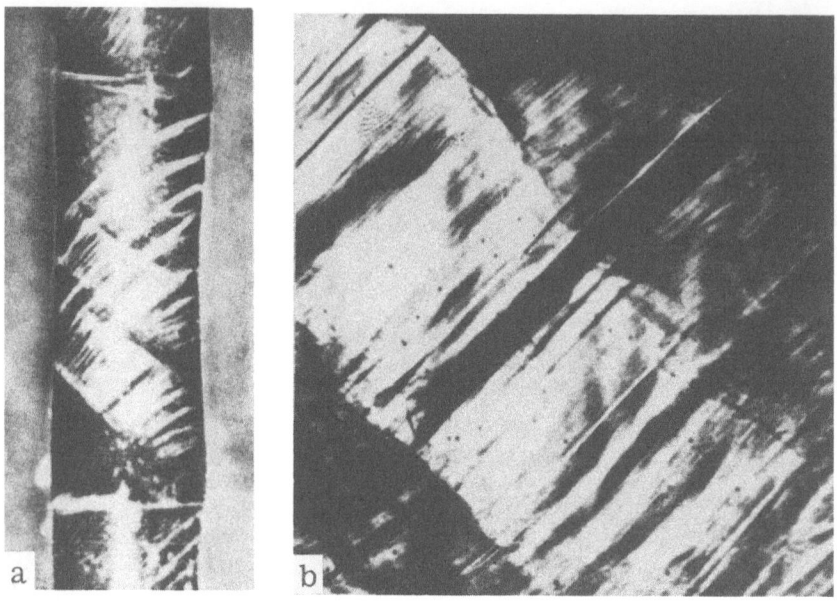

Fig. 191. a) Kink bands in coarse-grained aluminum after creep test at 600°C; b) part of one grain at high magnification (Gervais, Norton, and Grant); c) kinks in lead grains stretched 10% at 22% per min; × 200, pulling axis horizontal (Kondrat'ev).

moving dislocation walls or intersection between such walls and slip lines, but the above study showed that fracture is preceded by deviation of the kink boundaries from regular (symmetrical) positions (Indenbom, 1962), which is related to general plastic deformations and gives rise to microscopic stresses proportional to the product of α (change in lattice orientation) and β (angular deviation from the symmetrical position), as in Fig. 190d. Indenbom and Urusovskaya (1950) have calculated these stresses for irrational twins (Ch. 6, § 14.4). The boundaries fracture at a critical stress characteristic of the crystal, and the parts that break away twist back to the initial symmetrical position via the formation of connecting cracks (Fig. 190e). New dislocation walls appear at the ends as the cracks continue to develop (the converse does not occur, although those who have examined only the final states believe this to be so). Stepanov (1932 and 1937) was the first to express the view that plastic deformation (together with hardening) is the primary cause of premature fracture; the experiments described here reveal the mechanism, which is one of plastic deformation associated with kinking, and this is now generally accepted.

§ 16. Formation of Kinks and Disoriented Microregions in Polycrystalline Metals

Polycrystalline specimens contain grains of different orientations; therefore, even homogeneous pulling of these specimens can lead to inhomogeneous stress distributions. If adjacent grains are extended by different amounts, bending moments can arise at the grain boundaries and lead to reorientation inside the grains (kinks, irrational twins, and deformation bands).

This effect has been observed during pulling, fatigue testing (Forsyth, 1951) and creep testing. Forsyth found that kinks occur more often in impure metals. Kinking in coarse-grained aluminum after a creep test at 600°C is shown in Fig. 191a, b (Gervais, Norton, and Grant, 1953).

The formation of rotated regions in grains of magnesium is shown in Fig. 192 (Chaudhury et al., 1953). Figure 193 shows a series of kinks in a stretched grain of aluminum (Honeycombe, 1951).*

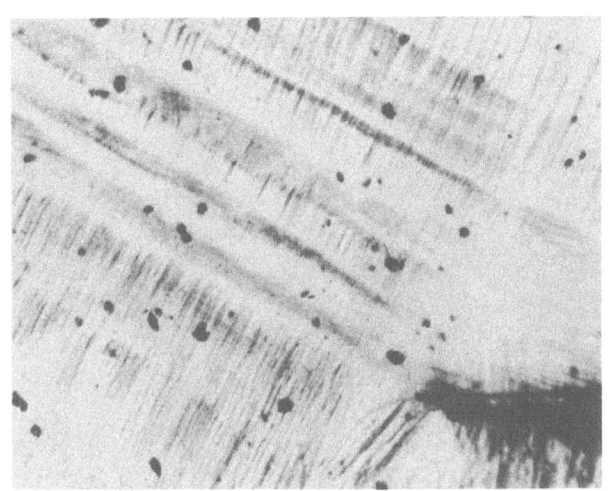

Fig. 192. Regions with twisted lattices in coarse-grained magnesium after creep test (Chaudhury, Norton and Grant).

Fig. 193. Series of kinks in a stretched grain of aluminum (Honeycombe).

Kinks are formed in monocrystals when plastic deformation by slip is difficult due to the lattice orientation (§ 14.6). It is possible that similar lattice reorientation occurs in separate grains along flow lines (Lyuders-Chernov lines) in polycrystals.

A predominant orientation (texture) in strongly deformed polycrystals (Barrett, 1947) can be caused by rotation of slip planes and directions in grains during plastic deformation, by mechanical twinning of grains, or by kinking in grains unfavorably oriented for homogeneous slip or twinning.

High-resolution microscopy has been used to detect slightly disoriented microscopic regions (substructures, mosaics, polygonization) in metal grains and monocrystals (Cahn, 1950a, b; Delise, 1953; Gifkins, 1955, etc). Some workers consider that substructure is a direct result of deformation (Klassen-Neklyudova, 1953; Liu Yi-huan and Tao Tsu-tsoong, 1956). Liu Yi-huan and Tao Tsu-tsoong confirmed that substructure in aluminum crystals that have been worked at high temperatures is due to slip and lattice rotation. Deformation by slip is now considered to be the cause of the appearance of slightly disoriented regions (polygonization) during the very early stages in the annealing of deformed monocrystals and polycrystals. The horizontal edge dislocations in the slip traces regroup when heated and form vertical dislocation walls; in this way polygonization occurs. Cracks that arise at kink boundaries are one of the sources of premature fracture of crystalline materials (see Ch. 6, § 15.6).

---

*Kondrat'ev has observed kinking prominently expressed in polycrystalline lead stretched 10%; the kinks arose on account of variation in the degree of deformation at grain boundaries (Fig. 191c).

PART III

THEORY OF TWINNING

CHAPTER 7

MACROSCOPIC THEORY OF TWINNING

Garber's experiments showed that the length of a mechanical twin as a function of load shows very little hysteresis and very little relation to the loading rate or to the temperature, which gave one reason to hope that twinning can be treated as a static problem corresponding to minimum potential energy. However, the problem cannot be solved generally within the framework of the macroscopic theory of elasticity; further simplifications are needed.

§ 17. A Thin Twin as a Fracture Surface

Vladimirskii (1947) observed that the energy of a calcite crystal containing a thin twin layer should be the sum of the energy of the twin boundaries and the usual elastic energy, the latter being proportional to the square of the deformation and evenly distributed. The lattice undergoes a finite deformation of simple-shear type near the twin. The relative displacement of the edge atomic layers in the twin is proportional to the thickness δ, the factor being s = 0.694 (Fig. 194a). Adjacent atoms remain so, and the main crystal thus experiences displacements of the order of δ on the two sides of the twin layer. The relative displacement at the end of a wedge twin is zero of course. Around the twin we have elastic strains of order δ / l, in which l is the length of the twin. These strains are localized within a volume whose linear dimensions are of the order of l, and the volume density of the energy is of the order of $G\delta^2 l$ (G is the shear modulus).

Vladimirskii estimated the specific surface energy of the boundary as the product of the shear modulus and the lattice parameter a; the total surface energy is of the order of Gal^2. The work needed to produce the twin is of the order of the load P multiplied by the height of the step resulting from the twinning.

The condition of minimum free surface energy is

$$G\ \delta^2 l + Gal + P\delta = \min. \tag{1}$$

If all terms in (1) are of the same order, we have

$$l \sim (P/G)^{2/3} a^{-1/3}; \quad \delta \sim (P/G)^{1/3} a^{1/3}; \quad \delta/l \sim (P/G)^{-1/3} a^{2/3}. \tag{2}$$

Garber's measurements (P of the order of 3 kg) then give $\delta/l \approx 10^{-3}$ from the last, which is in agreement with experiment.

Vladimirskii replaced a flat-faced wedge twin (Fig. 194b) by a surface of discontinuity (Fig. 194c). The boundary conditions for the two surfaces follow directly from the mechanism (displacement of lattice nodes along the twin direction). The nodes remain at the same distance from the twin plane. The normal displacements at the surfaces representing the twin then obey the conditions of continuity; the elastic strain within the body of the twin is assumed to be small relative to the final displacement. The tangential displacements normal to the twinning direction are continuous, but those along that direction are discontinuous. The discontinuity is proportional to the thickness.

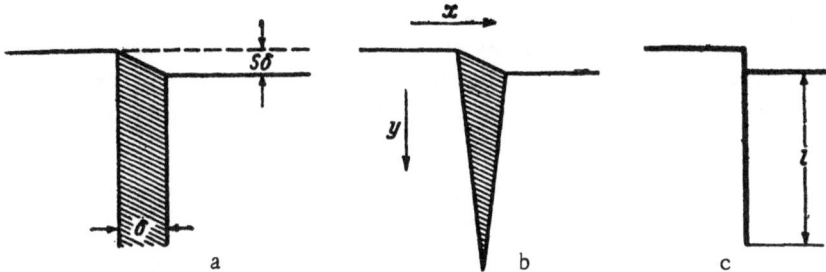

Fig. 194. Model of a twin in the macroscopic theory: a) plane twin layer (thickness δ, specific shear s, height of step sδ); b) wedge-shaped layer; c) thin layer equivalent to local shear on one plane (Vladimirskii's representation).

Variational principles applied to the displacements and energy automatically give the boundary conditions for the stresses. Calculations on the basis of the dislocation theory of grain-boundary energies enabled Vladimirskii to show that one can neglect the relation of surface energy to angle with respect to the twin plane.[*] This means that the boundary conditions for the stresses may be deduced on the basis of the energy outside the twin only. From this he derived:

1) the condition of continuity for the normal stresses;

2) the condition of continuity for the tangential stresses normal to the twinning direction;

3) the condition that the tangential stresses in the twin direction must be zero on both sides of the twin.

These boundary conditions reduce the problem to one in the linear theory of elasticity. Vladimirskii thereby solved approximately for the length of the twin produced by external forces uniformly distributed over some part of the boundary of the crystal.

If this length is small relative to the size of the crystal, the latter may be treated as a half-plane.

The twin is treated as a section of depth l in this half-plane; the problem is solved by conformal representation, the section being a radial one in a circle of unit radius. The twins are unstable for l less than $l_1 \approx c$, c being the halfwidth of the loaded region. Ones larger than c (but small relative to the crystal) give $l \sim (Pd)^{2/3}(\gamma G)^{-1/3}$, in which d is the distance from the point of emergence on the surface to the center of the loaded part and γ is the specific surface energy of the twin boundary. General arguments show that the twin again becomes unstable when its length exceeds some value l_2 comparable with the dimensions of the crystal; it then expands spontaneously. In this way he explained the basic laws of mechanical twinning discovered by Garber.

Kosevich and Bashmakov (1960) have used this model in the analysis of twinning processes in crystals of antimony, bismuth and zinc. They used δ/l to estimate the tangential stresses in the slip plane and the normal stresses in the cleavage plane. The estimates show that slip, rather than cleavage, was the probable effect.

Vladimirskii also pointed out that the boundary conditions could be stated for each boundary separately, but he did not deduce these conditions, because he considered that the stepped structure of an actual boundary (§ 22) made such treatment undesirable (the thickness of a thin twin layer is comparable with the heights of the steps).

---

[*] See § 22 for a discussion of this part of Vladimirskii's work.

§ 18. Twin Boundary on a Fracture Surface

Lifshits (1948) proposed a macroscopic theory of twinning based on the deduction of boundary conditions by means of the nonlinear relation between the stress tensor σ_{ik} and the strain tensor ε_{ik}. In two dimensions (Fig. 194b) the twin plane (x = constant) is normal to the surface of the crystal, the twin direction lying along the Y axis.

The twin angle is $\alpha = \tan^{-1}(s/2)$ (Fig. 195a) and is taken as small; deformation up to angles comparable with α is described by the usual relations from the theory of elasticity. Hooke's law applies to all components of the deformation tensor apart from ε_{xy}. The equilibrium twin position for $\varepsilon^0_{xy} = \alpha$ is used to elucidate the relation of σ_{xy} to ε_{xy}, this being symmetrical relative to the Y axis.[*] This gave the curve of Fig. 195b. The state $\eta_1 < \varepsilon_{xy} < \eta_2$ is unstable $(\frac{\partial \sigma_{xy}}{\partial \varepsilon_{xy}} < 0)$ and cannot exist, so $\varepsilon_{xy} > \eta_2$ in the twin layer, while $\varepsilon_{xy} < \eta_1$ outside it. The twin layer is thus separated from the main crystal by discontinuities in the strain tensor.

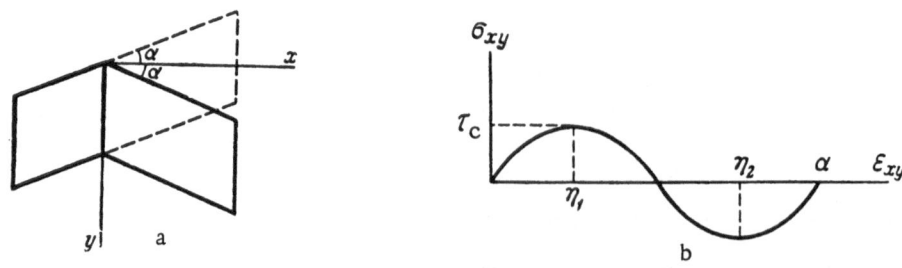

Fig. 195. a) Model of twinning; b) nonlinear relation of σ_{xy} to ε_{xy} in twinning (Lifshits's theory). The region $\varepsilon_{xy} < \eta_1$ corresponds to the main crystal, while that with $\varepsilon_{xy} > \eta_2$ corresponds to the twin layer; the states with $\eta_1 < \varepsilon_{xy} < \eta_2$ are unstable.

For convenience, the curve of Fig. 195b was replaced by straight lines over the ranges $(0, \eta_1)$ and (η_2, α), which corresponds to the assignment of an effective shear modulus.

In the main crystal

$$\sigma_{ik} = c_{iklm}\varepsilon_{lm}, \tag{3}$$

while in the twin layer

$$\sigma_{ik} = c_{iklm}\varepsilon_{lm} - \sigma^0_{ik}, \tag{3'}$$

in which c_{iklm} is the tensor for the elastic moduli and

$$\sigma^0_{ik} = c_{iklm}\varepsilon^0_{lm} = (c_{ik12} + c_{ik21})\alpha = 2c_{ik12}\alpha. \tag{4}$$

Substitution of (3) and (3') into the equation of equilibrium gives us

$$c_{iklm}\frac{\partial^2 u_l}{\partial x_k \partial x_m} = P_i + f_i, \tag{5}$$

in which P_i is the external force and f_i is the fictitious twinning force at the boundary; the latter is numerically equal to the force that the stress σ^0_{ik} would produce there.

----

[*]The author has $\varepsilon^0_{xy} = 2\alpha$, on account of a misunderstanding.

This f_i enables us to reduce the problem to the elastic theory of stresses produced by localized forces subject to additional conditions on the permissible ε_{xy} within and outside the twin layer.

The $\Delta\varepsilon$ at the boundary can be found from (5) with allowance for the continuity of the tangential derivative of the displacement vector:

$$\sin^2\psi\,\Delta\varepsilon_{xx} + \cos^2\psi\,\Delta\varepsilon_{yy} = 2\sin\psi\cos\psi\,\Delta\varepsilon_{xy} \tag{6}$$

(ψ is the angle between the twin boundary and the twin plane).

If $\psi = 0$ or $\psi = 90°$, $\Delta\varepsilon_{xy} = \alpha$ and the twin boundary produces no macroscopic stress; $\Delta\varepsilon_{xy}$ falls rapidly as the boundary deviates from the X or Y axis. But Fig. 195b shows that $\Delta\varepsilon_{xy} > \eta_2 - \eta_1$, so the boundary must form a small angle either with the X axis or with the Y axis.

The maximum possible angle ψ_{max} is roughly

$$\psi^2_{max} \approx \frac{\tau_c}{\alpha E}, \tag{7}$$

in which τ_c is the critical stress for twin shear (Fig. 195b) and E is Young's modulus. Garber gives $\tau_c \approx 10^3$ g/mm^2, which means $\psi_{max} \approx 0.003$, which agrees with experiment.

The cusp at the boundary does not cause divergence in the strain only if this is a point of reversal. Therefore the angle at the end of a wedge twin must be strictly zero.

Lifshits's boundary conditions for the plane case are readily extended to an arbitrary orientation; (5) remains valid and gives

$$n_i c_{iklm} \Delta\varepsilon_{lm} = n_i \sigma^0_{ik}, \tag{5'}$$

while (6) becomes

$$\mathbf{n} \times \Delta\varepsilon \times \mathbf{n} = 0, \tag{6'}$$

in which $\mathbf{n}$ is the normal to the boundary and $\Delta\varepsilon$ is the discontinuity in the total strain tensor.

Bullough and Bilby (1956) considered the macroscopic conditions for the contact of lattices at a plane twin boundary without assuming that the shear angle was small. The matrices for the strain and lattice rotation must be multiplied instead of added. Let the lattice undergo a homogeneous twin shear G, a homogeneous elastic strain Q (with respect to the main crystal), and a rotation R; then the total lattice transform matrix is F = RQG. Now F must not alter the intercepts in the plane of the twin boundary, so we must impose a boundary condition, which I put in the form

$$(F - I) \times \mathbf{n} = 0 \tag{8}$$

(here I deviate from the dislocation representation); I is the unit matrix and $\mathbf{n}$ is the normal to the boundary. The linear theory gives us (6) or (6') from (8).

Bullough (1957) made a special study of the case Q = I (boundary not causing macroscopic stress); then transfer from the lattice of the main crystal to that of the twin is given by a transform RG, which for twins of the first kind must be, apart from a translation U, equivalent to a reflection D in the twin plane:

$$DRG = U. \tag{9}$$

But R followed by D is equivalent to a reflection P in a plane normal to G, so DR = P and

$$PG = U. \tag{10}$$

In other words, shear G and reflection P are equivalent to lattice translation.

If we assume that G occurs in the slip plane T along the slip direction t, then (10) enables us to derive the twinning elements from the slip elements (Table IV). The relation is particularly simple to establish if the plane perpendicular to the slip direction is a symmetry plane of the lattice. Then (10) is replaced by

$$G = U \qquad (10')$$

(G restores the lattice, apart from a translation), which implies that s is determined by the ratio of the translation vector to the distance between adjacent slip planes, while the twinning angle is given by

$$2 \tan \alpha = s. \qquad (11)$$

Finally, s and 2α can be represented as a shear parallel to the twin plane K and to the twin direction η.

Twinning of the second kind implies that the lattice must be restored after twin shear and reflection in a plane normal to η, which is equivalent to s together with reflection in the T plane.

Bullough's theory indicates that the slip elements $<110>$ and $\{111\}$ in a face-centered cubic lattice correspond to mechanical twins on $\{113\}$, or in a diamond lattice to twins on $\{123\}$, $\{345\}$, and so on.

Deformation twins on $\{111\}$ planes are accounted for by the $<112>$ slip direction. Correct deductions of the slip elements for body-centered cubic lattices are obtained if the elements are taken as the $\{112\}$ planes and $<111>$ directions.

§ 19. Energy of Twinning With and Without Change of Form

We study first the case of a homogeneous stress distribution and compare the potential energy density of the crystal in the initial and twinned orientations; we do not directly study the formation and motion of the twin boundaries. For a stress distribution σ_{ik}, the potential energy of the crystal W is equal to the sum of its elastic energy[*] $\frac{1}{2}\sigma_{ik}\varepsilon_{ik}^e$, (in which ε_{ik}^e is the elastic strain corresponding to the given orientation of the crystal) and the potential energy of the crystal in the field of the external forces, $\sigma_{ik}\varepsilon_{ik}$ (in which ε_{ik} is the total strain of the specimen):

$$W = \sigma_{ik}\left(\frac{1}{2}\varepsilon_{ik}^e - \varepsilon_{ik}\right). \qquad (12)$$

The total and elastic strains are equal in the initial orientation; therefore the potential energy density

$$W' = -\frac{1}{2}\sigma_{ik}\varepsilon_{ik}^e = -\frac{1}{2}s'_{iklm}\sigma_{ik}\sigma_{lm}, \qquad (13)$$

in which s'$_{iklm}$ is the elastic constant of the crystal.

For a crystal in the twinned orientation, the total strain consists of the residual strain ε_{ik}^0, corresponding to the change of form of the crystal during twinning, and the elastic strain ε_{ik}^e, depending on the elastic constant of the twinned layer s"$_{iklm}$. Hence the potential energy density in the twinned layer is

$$W'' = -\sigma_{ik}\varepsilon_{ik}^0 - \frac{1}{2}\sigma_{ik}\varepsilon_{ik}^e = -\sigma_{ik}\varepsilon_{ik}^0 - \frac{1}{2}s''_{iklm}\sigma_{ik}\sigma_{lm}. \qquad (13')$$

We now compare W' and W".

$$\Delta W = W' - W'' = \sigma_{ik}\varepsilon_{ik}^0 + \frac{1}{2}(s''_{iklm} - s'_{iklm})\sigma_{ik}\sigma_{lm}. \qquad (14)$$

If $\Delta W > 0$, the twinned orientation is the position of stable equilibrium, and vice versa.

---

[*]The energy is referred to unit volume; summation with respect to the subscripts is to be understood.

It is necessary also to allow for the energy of the twin boundary (including the surface energy and the energy associated with the stress boundaries) in order to fully describe the twinning process.[*] If these effects can be neglected, the condition $\Delta W \geq 0$ will determine those parts of an inhomogeneously stressed crystal in which mechanical twinning can occur.

The sign of ΔW for twinning with change of form is determined by the sign of the first term in (14) at sufficiently low stresses. Therefore if the stress changes sign the effect is reversed (favorable orientations become unfavorable, the direction of motion of twin boundaries is reversed, etc). The twinning ellipsoid, corresponding to the tensor ε_{ik}^0, determines the twinning elements (the circular cross section corresponds to the twin plane, etc.).

These conditions do not apply to twinning without change of form. Here the stable orientation is that which corresponds to the maximum yielding of the crystal, and the twinning condition is governed by an equation that is quadratic in stress (i.e., the effect is unaltered if the stress changes sign). Experiments on the detwinning of quartz showed that the orientation of the final monocrystalline specimen does not depend on the sign of the applied stress (Wooster and Wooster, 1946).

For twinning with change of form, the change in yielding of the crystal can be substantial at large stresses when the elastic strain exceeds the twinning shear ε_{ik}^0. Stepanov (1947) considered only the effect of a change in the elastic constants in (14) and suggested the possibility of a link between mechanical twinning and the elastic properties of the crystal. He suggested that the effect of a point load on an elastic half-space would favor twinning only if the half-space has regions of different rigidities; this would lead to inhomogeneous loading. It is possible that reorientation of various parts of the crystal would lead to a more uniform load distribution and a reduction in the free energy of the system. The reoriented regions could be twinned.

Stepanov assumed that a crystal tends to become elastically isotropic during twinning, and also that it tries to acquire the maximum possible rigidity. This is only true, according to (14), if the crystal is very anisotropic and has a low twinning shear, and even then the crystal tends to acquire the minimum possible rigidity.

Stepanov (1950) discussed the mechanical twinning of quartz (twinning without change of form, i.e., the first term in (14) is zero) and gave a qualitative explanation (the sign of the effect was given correctly) of Shubnikov's scheme (§ 3.2) for the geometry of Dauphiné twins produced in quartz by point loads.

Thomas and Wooster (1951) analyzed the mechanical twinning of quartz in more detail and constructed (14) for the special cases of uniaxial loads and distortions. We can construct (14) for any type of loading if we remember that

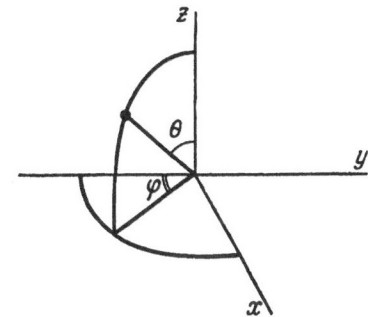

Fig. 196. Coordinate system used by Thomas and Wooster.

Dauphiné twins are rotated through 180° about the optic axis (equivalent to a change of sign of both the principal axes that are perpendicular to the optic axis). This leads to a change of sign of those components of the elastic constants tensor in which the indices corresponding to the reversed axes occur an odd number of times. These components are indicated here.

$$
\begin{array}{ccc|cc|c}
s_{11} & s_{12} & s_{13} & s_{14} & 0 & 0 \\
s_{12} & s_{11} & s_{13} & -s_{14} & 0 & 0 \\
s_{13} & s_{13} & s_{33} & 0 & 0 & 0 \\
\hline
s_{14} & -s_{14} & 0 & s_{44} & 0 & 0 \\
0 & 0 & 0 & 0 & s_{44} & 2s_{14} \\
\hline
0 & 0 & 0 & 0 & 2s_{14} & 2(s_{11}-s_{12})
\end{array}
$$

---

[*] Orowan (1954) has made estimates of this type.

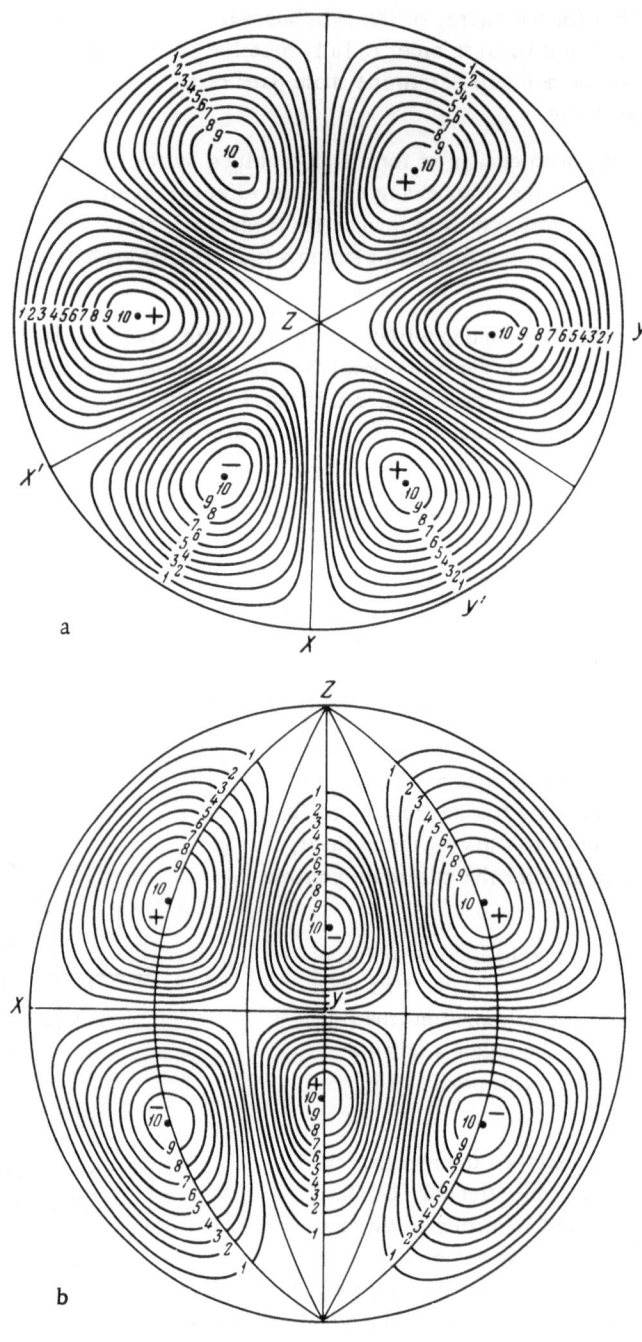

Fig. 197. Sterograms for the energy aspects of twinning
for uniaxial loading: a) projection on the (0001) plane;

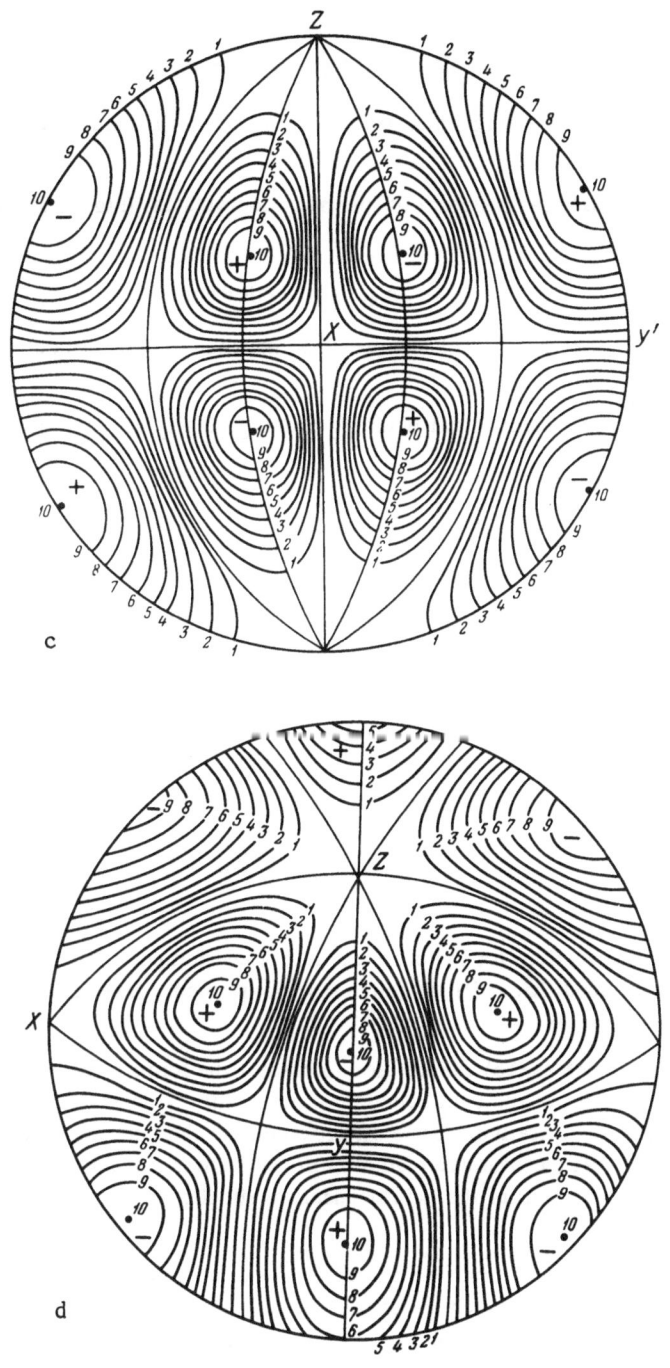

b) projection on the (01$\bar{1}$0) plane; c) projection on
the ($\bar{2}$110) plane; d) projection on the (10$\bar{1}$1) plane.

Hence, ΔW is given by

$$\Delta W = -2s_{14}[\sigma_{23}(\sigma_{11} - \sigma_{22}) + 2\sigma_{12}\sigma_{31}]. \qquad (15)$$

Thus, twinning is only affected by the tangential stresses acting along the basal plane and the plane parallel to the optic axis.

Thomas and Wooster (1951) used a system of spherical coordinates (Fig. 196) and considered a uniaxial stress of orientation (θ, φ). Then

$$\Delta W = -2s_{14}\sigma^2 \sin^3\theta \cos\theta \cos 3\varphi. \qquad (16)$$

The authors obtained (16) by using the known relation between Young's modulus for quartz and the orientation of the stress axis.

For quartz $s_{14} < 0$; therefore mechanical twinning is energetically advantageous for uniaxial stresses in directions for which

$$\Omega(\theta, \varphi) = \sin^3\theta \cos\theta \cos 3\varphi > 0.$$

Figure 197a-d shows stereograms for 10 different orientations corresponding to faces of the crystal (Thomas and Wooster); these stereograms can be used to predict the form of the twin when the stress is nearly uniaxial. The stress can be only very approximately uniaxial for a point load and yet there is definite qualitative resemblance between the stereograms of Fig. 197 and the twinning figures of Tsinzerling and Shubnikov (Fig. 114).

Thomas and Wooster assumed that for a triaxial stress (principal stresses σ_1, σ_2, σ_3; principal directions (θ_1, φ_1), (θ_2, φ_2), (θ_3, φ_3) ΔW is given by

$$\Delta W = -2s_{14} \sum_{i=1}^{3} \sigma_i^2 \Omega(\theta_i, \varphi_i). \qquad (16')$$

They calculated the stress distributions for various mechanical and temperature effects (using the theory of elastic isotropic media) and found that (16'), although it is not rigorous, gave results in good agreement with observed twin forms (§ 4.5).

The fact that formulas such as (15) can be used to predict the forms of twins produced by inhomogeneous stresses indicates that the energies of the twin boundaries are relatively unimportant in quartz; it seems that the transition zone undergoes comparatively little distortion due to the unchanged form of the twinned crystal. Thomas and Wooster assume that the structure of the transition zone is similar to that of high temperature hexagonal quartz, and that the movement of the twin boundary is similar to that of the phase boundary in the $\alpha \rightleftharpoons \beta$ transition.

Twinning without change of form must be characteristic of all rotation twins in which the axis coincides with the axis of the ellipsoid of homogeneous deformation, and of all reflection twins in which the plane of symmetry is the plane of symmetry of this ellipsoid. In these cases, the optical indicatrices of both components of the twin must also be similar.

Twin shear is absent and the elastic constants are unchanged for some reflection and rotation twins, and for all inversion twins. In these cases, (14) is identically zero and external mechanical stress cannot cause movement of the twin boundaries (triglycine sulfate, § 4.8); therefore inversion twins cannot be mechanical twins (§ 1.4).

CHAPTER 8

MICROSCOPIC THEORY OF TWINNING

The macroscopic approach treats the twin boundary as a surface of stepwise change in macroscopic quantities, whereas microscopic theories are designed to elucidate the atomic structure of the boundary and the atomic mechanism of lattice reorientation. The structure of the boundary naturally determines the mechanism, which differs very greatly as between boundaries sharp on the atomic scale and ones having broad transition regions.

§ 20. Transition Zones on Twin Boundaries

The first theoretical treatment was for ferromagnetic domains, which are twins differing in direction of spontaneous magnetization (Bloch, 1932; Landau and Lifshits, 1935). At the boundary there is a transition zone, in which the direction of magnetization varies continuously between the two opposed directions. The width of the zone is governed by the competition between the orienting forces (which are related to the magnetic anisotropy) and the exchange forces (the latter tend to make the transition more gradual).

Kontorova (1942) observed that the interaction forces between adjacent layers can give rise to a transition zone at any twin boundary; these forces tend to make the transition more gradual. On the other hand, any position other than the initial and twin ones is of higher energy (does not correspond to a state of stable equilibrium). The excess energy of the intermediate layers is the result of the orienting forces, which tend to make the transition sharper.

The total energy H of the transition zone is the sum of the interaction energy (which is inversely proportional to d, the width of the transition zone) and the energy of the orienting forces (which is proportional to d):

$$H = \frac{A}{d} + Bd. \tag{17}$$

The equilibrium value d_0 is found from the condition for minimum H(d):

$$d_0 = \sqrt{A/B}. \tag{18}$$

Then the total energy of the zone, which is equivalent to the surface energy of the boundary, is

$$\gamma = H_{\min} = 2\sqrt{AB}. \tag{19}$$

Rough estimates of the values indicate that the upper limit to the width may be 200-500 lattice parameters, the total energy per atom being 10^{-15} to 10^{-13} erg (γ of 10-100 erg/cm^2 or so).

For comparison, the 180° domains in iron have γ = 1.4 erg/cm^2 and d of 1800-2000 A; 90° domains have γ of about 1.2 erg/cm^2 and d of about 500 A (Néel, 1944).

Zhirnov (1958) has discussed the structure of the transition zone in a ferroelectric (§ 1.14); calculations analogous to the above indicate that the 180° boundaries of domains in barium titanate are 5-20 A thick and have γ = 10 erg/ cm^2. The 90° ones have d of 50-100 A and γ of 2-4 erg/ cm^2.

Rochelle salt has boundaries whose widths vary with temperature from 12 to 200 A, the surface energy varying from 0.06 to 0.012 erg/ cm^2.

Zhirnov's results show that ferroelectrics differ from ferromagnetics in that the transition zones are not so broad.

A broad zone allows one component to grow at the expense of the other by gradual change of orientation in layers parallel to the boundary. A quantitative discussion (Kontorova and Frenkel', 1938) shows that the rate of movement v cannot exceed the speed of sound c; the width varies as $\sqrt{1-v^2/c^2}$, while the energy is proportional to $\sqrt{1-v^2/c^2}$.

§ 21. Sharp Twin Boundaries: Atomic Steps on a Twin Boundary

The interaction forces may be low relative to the orienting forces, in which case the transition zone is almost absent; the boundary is sharp even on the atomic scale. Vladimirskii (1947) discussed some aspects of the structure for a boundary not coinciding with the twin plane.

A twin whose boundary is parallel to the twin plane should represent an equilibrium configuration without any stress concentration. The energy localized in the twin is simply the surface energy. Any deviation from the twin plane causes the boundary to be stepped, with parts parallel to the twin plane and sites when a step of one atomic layer occurs (Fig. 198a).

Crystallographically regular contact between the components (a stress-free configuration) is impossible in this case; each step "corresponds to a dislocation whose order is less than one lattice parameter" (Vladimirskii). Figure 198a illustrates this; it corresponds to an early stage of the dislocation theory (§ 22). The distance between steps equals the lattice parameter divided by the angle between the macroscopic interface and the crystallographic twin plane (about 10^3 atomic distances, or of the order of the thickness of the twin layer, in Garber's experiments). The total number of steps equals the thickness of the layer divided by the thickness of the atomic layer (this is about 10^4 for a layer of few microns thick).

Vladimirskii also estimated the surface energy of a stepped boundary by summing the elastic energies of the stresses caused by the steps on the assumption that these were localized in a region of outside radius R_1 (of the order of the distance between steps) and inside radius R_2 (of the order of the thickness a of an atomic layer). The result is

$$\gamma(\psi) \sim Ga\psi \ln\left(\frac{R_2}{R_1}\right), \tag{20}$$

in which $\gamma(\psi)$ is the part of the surface energy dependent on the angle ψ between the planes.

By analogy with the boundary energy of blocks (Cottrell, 1953c), we can calculate $\gamma(\psi)$ more rigorously; allowance may be made for the overlap between stress fields. The calculation is as for the distortion energy of a slip plane (Indenbom, 1960) and gives

$$\gamma(\psi) = \frac{s^2 Ga\psi}{4(1-v)}\left(-1+\ln\frac{R_2}{2\pi R_1}\right), \tag{20'}$$

in which ν is Poisson's ratio.

Garber's experiments were for $\psi < 10^{-3}$, in which case the surface energy differed little from that of a coherent boundary parallel to the twin plane.

Lifshits and Obreimov (1948) also considered the stepped structure of twin boundaries; they treated the growth as a result of movement of atomic steps (dislocations) along the interface. However, this idea was fully

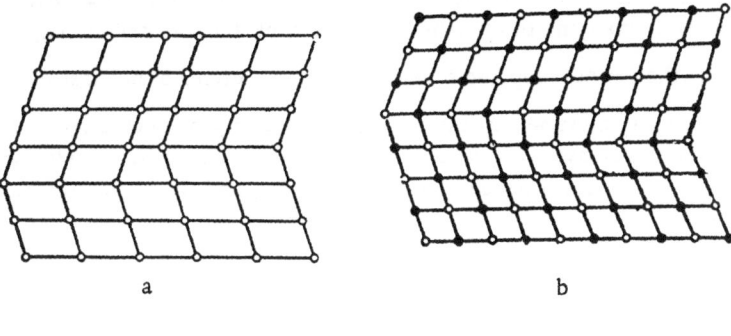

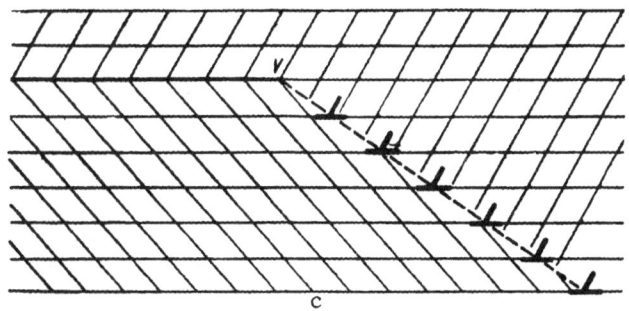

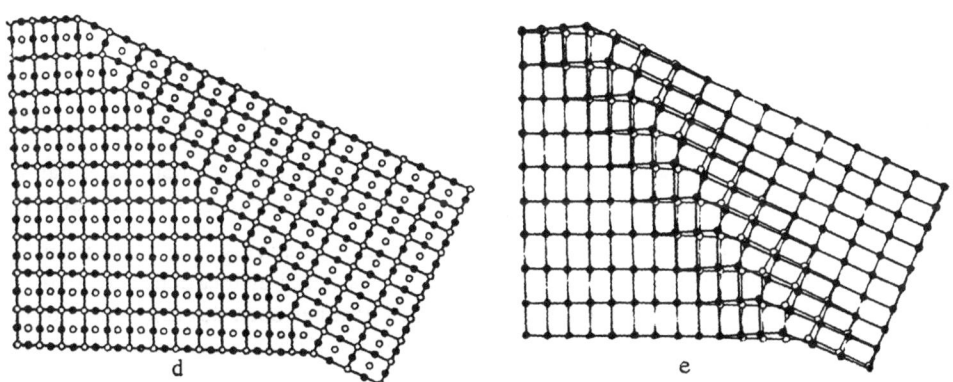

Fig. 198. Stepped structure of a twin boundary: a) Vladimirskii (1947); b) twinning dislocation in a body-centered cubic crystal (Read, 1953), the (101) plane of the figure being normal to the dislocation and the (111) twin plane horizontal; the white and black
circles correspond to atoms lying in different planes; c) Siems and Haasen's (1958) scheme
for coherent and incoherent twin boundaries, each step at the boundary corresponding to
a twinning dislocation; d) two twinning dislocations (edge ones) on a (101) twin boundary in a face-centered tetragonal lattice (Basinski and Christian, 1954b), the plane of
the figure being (010) and the white and black circles denoting atoms lying in different
planes; e) screw twinning dislocation in a simple tetragonal lattice; notation as for d)
(Basinski and Christian, 1954).

developed only in the dislocation theory of twinning, where the twinning dislocation appears naturally as the edge of an incomplete plane of atoms lying in the twin position (Frank and van der Merve, 1949).

Figure 198, b to e, shows the various proposed schemes for the structures of boundaries.

The growth on a step boundary must be entirely analogous to layer growth on a face (tangential growth of close-packed layers). Further, two cases are possible. If both ends of a step lie on side faces, the step is essentially lost (Fig. 199); a fresh nucleus is needed for the next layer, and the nucleation probability is governed by the supersaturation.

The nucleation difficulty is completely overcome if one end of the step lies on the surface (Fig. 200).

The step may end on the surface at an ordinary screw dislocation, which in effect converts the entire crystal to a single atomic plane. Figure 200 shows that growth does not cause loss of the step in this case; the step rotates, and each layer is formed regularly above the previous one.

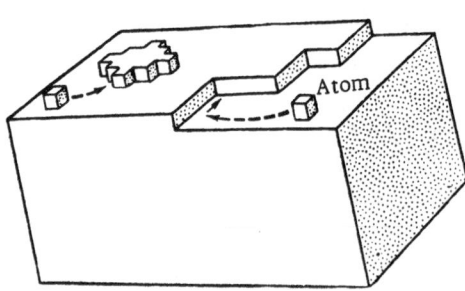

Fig. 199. Growth steps on the surface of an ideal crystal. The steps are lost as growth proceeds, and new nuclei are needed.

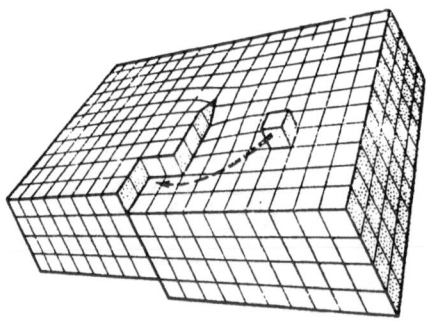

Fig. 200. Growth steps on the surface of a crystal containing a screw dislocation; growth does not lead to loss of the step, and characteristic spiral figures are formed.

Mechanical twinning may be treated as growth of a crystal in the solid state, so both cases can occur. Nucleation becomes probable if the stress at the boundary is high, and the twin can be built up layer by layer.

Steps (twinning dislocations) give most of the growth if the stress is relatively low; these terminate on the surface at points where ordinary dislocations emerge (Fig. 201). Twinning dislocations with fixed points can produce growth without requiring fresh nuclei. This pole mechanism has been discussed for body-centered cubic crystals by Cottrell and Bilby (1951) [*] and by Thompson and Millard (1952); Okawa (1957) has done the same for face-centered cubic crystals.

§ 22. Twinning Dislocations and Stresses

Ordinary (complete) dislocations are line defects analogous to current lines in magnetostatics or vortex lines in hydrodynamics. The translation vectors in an ideal crystal form a closed figure, but the figure formed around a dislocation is open. The vector needed to close the figure is also a translation vector; it is called the Burgers vector of the dislocation. This vector is normal to the dislocation line for an edge dislocation (Fig. 202a) and is parallel to it for a screw dislocation (Fig. 202b).

---

[*] Price (1961) was unable to detect the pole mechanism in his electron-microscope study of the motion of twin boundaries in whiskers (see p. 63).

The Burgers vector of a twinning dislocation may be defined similarly (Frank, 1951); here we must start from a figure closed for a coherent boundary (Fig. 203). The Burgers vector for an edge twinning dislocation is

$$b = sa, \qquad (21)$$

in which s is the twin shear and a is the distance between planes parallel to the twin plane.

The dislocation representation enables us to use the theory to calculate stresses and lattice rotations associated with twin boundaries. Figure 204 shows examples of dislocation schemes corresponding to various types of twin layers.

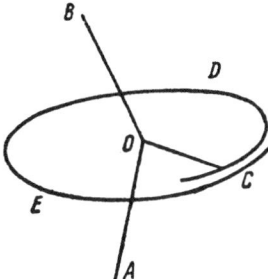

Fig. 201. Pole growth mechanism for a twin layer, analogous to spiral growth as in Fig. 200. AOB is the screw dislocation and OC is the twinning dislocation (step at the boundary), which rotates about O (the pole) and rises one layer per turn.

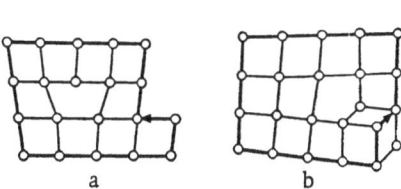

Fig. 202. Construction of the Burgers vector and contour for complete dislocations: a) edge dislocation, with Burgers vector normal to dislocation line; b) screw dislocation, Burgers vector parallel to dislocation line.

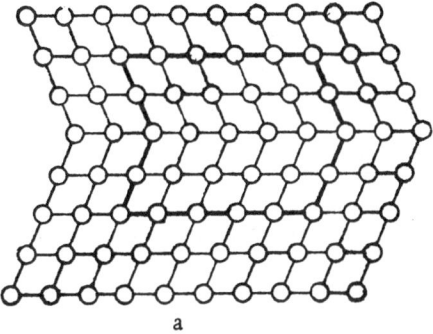
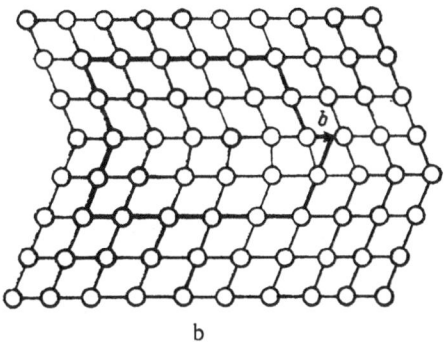

Fig. 203. Construction of the contour and Burgers vector b for a twinning dislocation with an edge orientation: a) coherent twin boundary; b) step at boundary.

A wedge-shaped layer (Fig. 204a) has its boundary conditions for macroscopic stresses simply as (5') and (6'), which were deduced without allowance for the dislocation structure. Allowance may be made for the microstructure of the stress at a boundary close to the twin plane by analogy with the treatment for slip planes (Indenbom and Tomilovskii, 1958).

Reducing polysynthetic twins (Fig. 204b) produces an effect analogous to that of a vertical series of edge dislocations; the lattice is rotated through

$$\vartheta = nb = \frac{s\delta}{\delta + \delta'},$$ (22)

in which n is the mean density of dislocations in the plane, δ is the mean thickness of one component, and δ' is the mean thickness of the second one (Basinski and Christian, 1954a and 1954b).

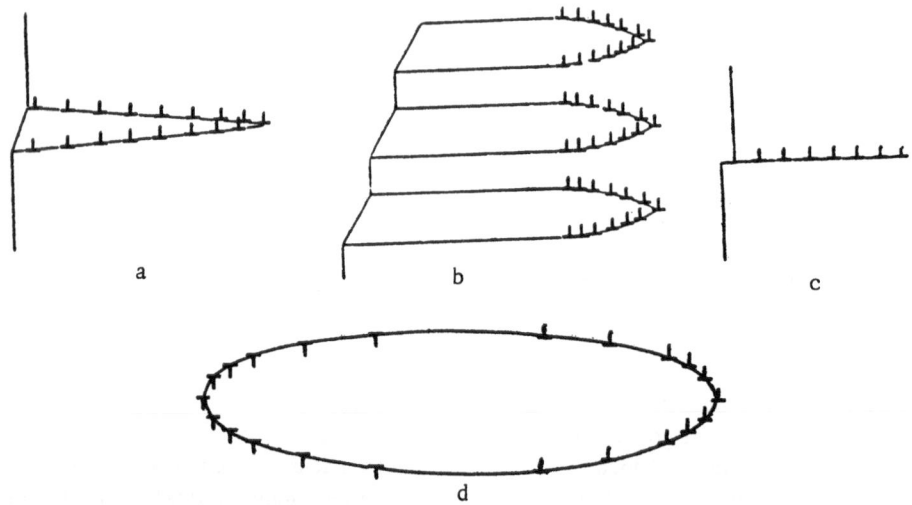

Fig. 204. Dislocation schemes for the structures of twin layers; the dislocations are shown as ⊥: a) wedge-shaped layer; b)"reducing" polysynthetic twins; c) nucleus of a layer; d) thin twin layer corresponding to Vladimirskii's approximation (Fig. 194c).

Nuclei (Fig. 204c) produce stresses acting in the twin plane in opposition to the direction of twinning; each loop of diameter L causes a stress Gb/2L (Nabarro, 1952), so a nucleus of diameter L and thickness δ consisting of δ/a loops must produce a stress

$$\tau \approx \frac{Gb}{2L} \cdot \frac{\delta}{a} = \frac{sG}{2} \cdot \frac{\delta}{L}.$$ (23)

If this stress is balanced by the external forces, we obtain the following estimate for the ratio of the dimensions (Friedel, 1956):

$$\frac{\delta}{L} \sim \frac{2\tau}{sG}.$$ (23')

For calcite, with $\tau/G \approx 10^{-3}$, the thickness must be about 1 μ for a twin layer 1 mm long.

A thin wedge layer may be represented roughly as a sequence of dislocations lying in a plane (Fig. 204d). The energy per unit length for the two-dimensional case is

$$W = \frac{s^2 G \delta^2}{\pi (1 - \nu)} \left(2\ln 2 - \frac{1}{2} \right)$$ (24)

(only the energy of the macroscopic stresses in the matrix is considered). The numerical values for calcite show that this latter energy was of the order of the external energy in the experiments of Obreimov and Startsev (1958), because the latter cannot be a measure of the surface energy of the interface.

A flat nucleus of elliptic cross section may be compared with the distribution of dislocations on a bounded area of a glide plane (Leibfried, 1951). The ratio of the axes is

$$\frac{\delta}{L} = (1 - \nu)\frac{\tau}{sG}, \tag{25}$$

in which τ is the external stress. The nucleus produces in the matrix macroscopic stresses corresponding to the following energy per unit length:

$$W = \frac{\pi s^2 G \delta^2}{8(1 - \nu)}. \tag{26}$$

Recently, Kosevich (1961 and 1962) has discussed several problems by means of the above conception of a thin twin as a sequence of dislocations lying in the slip plane.

§ 23. Stacking Faults as Monolayer Twins

A close-packed structure can be considered as a succession of layers of spheres as shown in Fig. 205. The next layer can be packed on top of layer A in two possible ways, B or C (similarly, A or B can be packed on C and A or C on B). The face-centered cubic structure has the sequence ... ABCABC ..., the hexagonal close-packed structure the sequence ... ABABAB ....

It is convenient to use the symbols ∇ and $\triangle$ to represent the sequence of atomic planes; ∇ represents the sequence A→B→C→A, etc., while $\triangle$ represents the opposite sequence. Thus...$\nabla\nabla\nabla$...represents a face-centered cubic structure and... $\nabla\triangle\nabla\triangle$... represents a hexagonal close-packed structure.

The normal sequence of atomic planes is reversed at a twin boundary. Thus ... $\nabla\nabla\nabla\downarrow\triangle\triangle\triangle$ ... represents a twin boundary along {111} in a face-centered cubic crystal, and ... $\nabla\triangle\nabla\downarrow\nabla\triangle\nabla$ ... represents a twin boundary along the basal plane in a hexagonal close-packed crystal (the arrow denotes the position of the boundary). The packing sequence is reversed again at the second boundary of the twinned layer.

In a monolayer twin such as ... $\nabla\nabla\nabla\downarrow\triangle\uparrow\nabla\nabla\nabla$ ... or ... $\nabla\triangle\nabla\downarrow\nabla\uparrow\nabla\triangle\nabla$ ..., there is only one incorrectly packed layer (stacking defect). The two-dimensional lattice can terminate inside the crystal forming a linear defect (partial dislocation). Read (1953) and Cottrell (1953c) have reviewed the various types of stacking defects and partial dislocations in closed-packed structures.

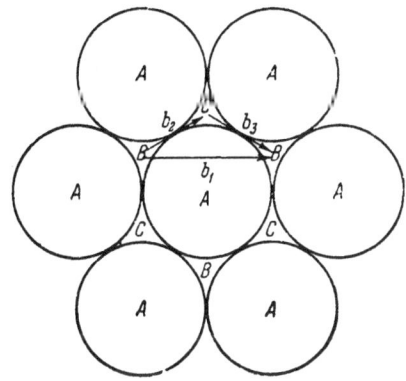

Fig. 205. Array of atoms in close packing.

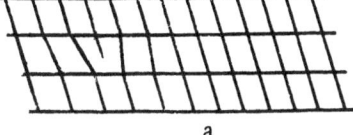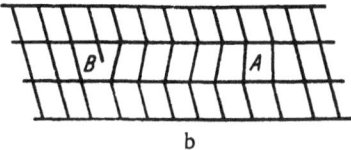

Fig. 206. Splitting of complete dislocation a) into two partial ones, b) joined by a stacking defect (monolayer twin) AB. The plane of the stacking defect is (111) for face-centered crystals; the Burgers vector of the complete dislocation is (a/2) <110>, and that of the partial one is <112>.

161

In some cases a complete dislocation in a close-packed plane can split into two partial dislocations forming an interstitial stacking defect (Fig. 206). The stacking defect (monolayer twin) can act as twinning dislocation if the Burgers vector of the partial dislocation lies in the plane of the defect.

If the Burgers vectors of both partial dislocations lie in the plane of the defect, then the split dislocation is more mobile than the complete dislocation. However, a low-mobility defect (Lomer-Cottrell barrier) can be formed by the fusion of split dislocations lying in intersecting slip planes (Fig. 207); this blocks both slip planes and hampers further plastic deformation (Lomer, 1951; Cottrell, 1952; Seeger and Schoek, 1953; Green, 1956; Stroh, 1956).

Modern theories of the hardening of face-centered cubic crystals are based on calculations of the formation and destruction of Lomer-Cottrell barriers (Friedel, 1955; Seeger, 1957). A stacking defect is the thinnest possible twin layer with two coherent boundaries; the energy of the defect is about twice the energy of a coherent boundary. Recent electron-microscope and x-ray methods have provided direct means of observing these defects and of estimating their energies by experiment (Whelan, 1959; Howie and Swann, 1961; Aerts and Delavignette, 1962; Siems, Delavignette, and Amelinckx, 1961).

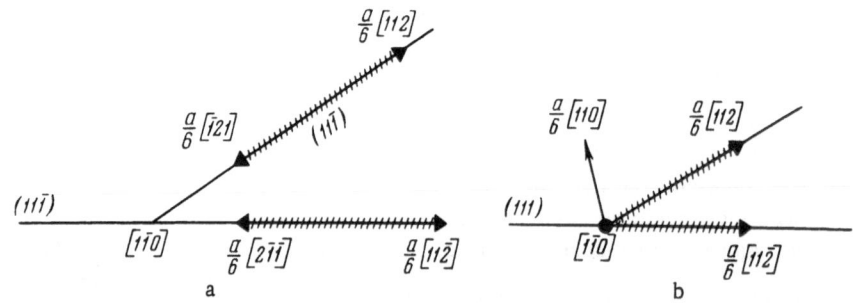

Fig. 207. Fusion of split dislocations a) lying in intersecting slip planes to give b) a Lomer-Cottrell barrier. The stacking defects (monolayer twins) are shown cross-hatched. The Burgers vector is shown for each dislocation.

§ 24. Atomic Model of Twinning

The above microscopic theory of twin boundaries does not consider the nature of the interatomic forces that determine the character and form of the boundary.

The analysis of mechanical twin formation involves nonlinear atomic interactions (Lifshits and Obreimov, 1948). The authors simplified the calculations by considering a simple lattice and assuming that all displacements are parallel to a certain axis Z.

If the dissociation energy equation contains no terms in which the atomic displacement is raised to powers higher than the second, then the equation for lattice equilibrium is

$$\sum_{k'} A_{k-k'} \xi_{k'} = F_k , \qquad (27)$$

in which ξ_k is the displacement of the k-th atom along the Z axis, and F_k is the force acting on this atom in the same direction (the position of an atom in the lattice is specified by the radius vector **k**, drawn from its original position).

Thus the force between atoms k and k' is given by

$$F_{k, k'} = A_{k-k'} (\xi_k - \xi_{k'}). \qquad (28)$$

The interatomic force between atoms k and k' varies with the relative displacement η of the atoms approximately as shown by the full line in Fig. 208; there is still one equilibrium position. Lifshits and Obreimov simplified the calculations by replacing this curve (full line) with the broken line; this is equivalent to the introduction of an additional twinning force f [equivalent to an external force in (27)] that arises when η exceeds a critical value d.

The nonlinear relation $F(\eta)$ is only calculated for adjacent atoms in the X direction as a first approximation, and it is possible to neglect the effects of more remote atoms and of relative displacements in other directions; these effects need only be considered for much larger deformations.

Thus, the equilibrium equation becomes

$$\sum_{k'} A_{k-k'}(\xi_k - \xi_{k'}) - F_k = f\sum_{r}(\delta_{k,r} - \delta_{k,r+a}). \tag{27'}$$

Here the summation on the right is with respect to all $\mathbf{r}$ for which the relative displacement exceeds a critical value d, i.e., for all twinned atoms

$$\xi_r - \xi_{r+a} > d, \tag{29}$$

in which a is the lattice parameter along the X axis, $\delta_{k,r} = 1$ if $\mathbf{k} = \mathbf{r}$, and $\delta_{k,r} = 0$ if $\mathbf{k} \neq \mathbf{r}$.

The first atom to pass over into the twinned position (under the influence of an external point force F) experiences a force $F + f$, while its neighbor to the left experiences a force $-f$. Thus, atoms in the twinned position exert an additional force on the remaining atoms. The inequality (29) must be satisfied before a certain atom will pass into the twinned position. The displacements in (29) must be calculated by allowing for the effect of the external force F and of the atoms that are already in the twinned position; the effect of these atoms is calculated by using Green's functions to describe the displacements caused by the point force.

Lifshits and Obreimov showed how similar calculations can be made for any lattice and for one-, two-, or three-dimensional twin layers.

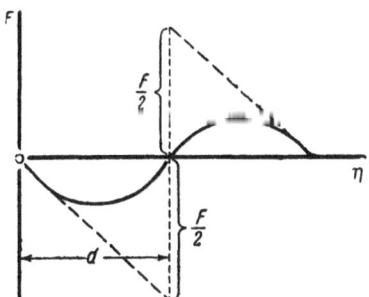

Fig. 208. Interatomic force F as a function of displacement η of atoms in the Lifshits-Obreimov model.

If we assume that n atoms have passed into the twinned position, and that the n-th and (n + 1)-th atoms undergo displacements of $\eta_n^{(n)}$ and $\eta_{n+1}^{(n)}$ respectively (due to the action of the force f on the n-th atom, the force $-f$ on its neighbors to the left, and the external force F), then the twin boundary is given by the condition

$$\eta_{n+1}^{(n)} < d < \eta_n^{(n)}. \tag{30}$$

There are three possible cases:

1) $\lim\limits_{n\to\infty} \eta_n^{(n)} < d$, i.e., the twinning force f is not large enough to maintain a twin layer of more than a certain thickness. Twinning ceases when $\eta_n^{(n)} = d$. The twin returns to its initial position when the force F is removed; therefore, it can be called an elastic twin.

2) $\lim\limits_{n\to\infty} \eta_{n+1}^{(n)} < d < \lim\limits_{n\to\infty} \eta_n^{(n)}$, i.e., the twinning force is capable of maintaining a twin layer of any thickness above a certain length but is incapable of spontaneously continuing the twinning process. Such a twin layer is stable even if the force F is removed; therefore the twin is not elastic.

3) $\lim_{n \to \infty} \eta_{n+1}^{(n)} > d$, i.e., once a layer of a certain thickness is produced, the twinning force is capable of continuing the growth of the twin layer which spontaneously spreads through the crystal.

Thus the nature of the twinning process depends on the functions $\eta_n^{(n)}$ and $\eta_{n+1}^{(n)}$. Lifshits and Obreimov are trying to develop a method of constructing these functions in the general case. For a given load F, the relative displacement of atoms in the X direction in a three-dimensional twin decreases rapidly along both the axes lying in the twin plane; therefore, an elastic twin must be narrow in the direction perpendicular to the twin plane. Lifshits (1948) has considered the case of a half-space bounded by a plane, and has given a general method of constructing Green's functions for an arbitrary unbounded lattice.

The Lifshits-Obreimov theory and Garber's experiments both indicate that lattice reorientation occurs in stages. In the first stage, the crystal lattice is deformed elastically (according to Hooke's law). In the second stage, an elastic twin forms and develops; the deformation is still reversible but varies nonlinearly with stress. In the third stage, a permanent twin is formed which can develop further either spontaneously or by the action of an external force.

§ 25. Difficulties: Problems of Nucleation

Most authors treat the structure of twin boundaries and the mechanism of lattice reorientation involving movement of these boundaries in similar ways despite some differences in approach and in the type of model used in quantitative calculations. For example, the Lifshits-Obreimov theory consists of a dislocation scheme for the structure of the twin boundary together with calculations of the atomic structure of the nuclei of dislocations.[*] The stability of the dislocations is deduced from the asymptotic behaviour of the functions $\eta_n^{(n)}$ and $\eta_{n+1}^{(n)}$.

The usual models for the atomic structure of dislocations (Peierls, 1940; Nabarro, 1947; Huntington, 1955; Indenbom, 1958; Lehmann and Leibfried, 1958) lead to conclusions about the symmetrical nature of dislocation nuclei and the equilibrium of dislocations along the close-packed direction in inorganic crystals.

If the surface energy of the coherent twin boundary is ignored (following Lifshits and Obreimov), it is shown that a monolayer twin is unstable (elastic) if the resistance of the lattice to the motion of dislocation is less than the force of attraction between the dislocations and the crystal surface, and also that a residual twin will grow spontaneously if the attraction of the dislocation to the surface is greater than the resistance of the lattice.

All theories of abrupt twin boundaries assume that the boundaries change position because of tangential motion of dislocations. The formation of new dislocations is either treated as in the dislocation theory for the spiral growth of crystals (polar mechanism) or explained by the presence of locally overstressed regions in which the deformation reaches the twinning shear.

It is possible to calculate the probability of the formation of flat twinning nuclei at twin boundaries by analogy with the theory for the spontaneous formation of nuclei of translational slip (Cottrell, 1953b). The energy barrier for the formation of a flat nucleus at the twin boundary is approximately $G^2 b^3 / \tau$ while the critical radius of a nucleus is $Gb / 2\tau$ for an external tangential stress τ acting along the twin plane in the twin direction. The barrier is approximately $10^4 S^3$ eV if Ga^3 is about 1 eV and τ is about $10^{-4} G$; therefore, the twinning shear S can be reached at thermal energies for small twinning angles.

The formation of twin layer nuclei in monocrystals depends on the surface energy γ of the coherent twin boundary.

The force $b\tau$ acting per unit length of the dislocation due to the external stress τ must be greater than 2γ. The stress must be about 20 kg/mm^2 if γ is about 1 erg/cm^2 and b is about 0.1 A. The stress required for the formation of twinning nuclei is determined by the energy of the twin boundary for small twinning angles, and by the elastic energy of the matrix of twinning dislocations for large twinning angles.

---

[*] The model of Kontorova and Frenkel' (1938) can also be considered as an atomic model of twinning dislocations.

In most cases, the required stress concentration can be produced either by macroscopic defects in the crystal or by the plastic flow or fracture that precedes twinning (Bilby and Entwisle; 1954; Bilby and Bullough 1954; Friedel, 1956; Siems and Haasen, 1958).

Thus the formation of twinning nuclei is closely linked with the formation of stress concentrations in deformed crystals; therefore, further progress in the theory of twinning can only be achieved by linking it with the general theory of plastic flow and fracture of crystals.

APPENDIX 1*

SELECTIVE ETCHING AS A MEANS OF STUDYING TWINNING

This has become widely used with deformed crystals. It is usual to employ a slow-acting reagent, which produces pits at the points where dislocation lines emerge.

Konstantinova (1960) has used the method with deuterated triglycine sulfate and has shown that regular series of pits remain in the monodomain region behind twin boundaries (domain boundaries) displaced by an electric field; these relate to edge dislocations.

Keith and Gilman (1960) and Startsev (1960) have observed dislocations lagging behind twin boundaries displaced by mechanical stresses in calcite. This effect may be related to the 'memory' of quartz and Rochelle salt observed by Tsinzerling (p. 97) and Chernysheva (p. 74).

Startsev has observed the etching of coherent and incoherent boundaries in calcite, bismuth, and antimony. The etch pits in the grooves indicate that the dislocation density at an incoherent boundary is much the higher. Plane-parallel boundaries of very thin twins have dislocations arranged in linked pairs. The density increases near regions of parting on twin planes. A study has been made of the effects of annealing on the motion and distribution of dislocations in layers.

Selective etching has also been applied to elastic twin layers; dislocations near such layers are found to vanish when the crystal is unloaded. Removal of a residual layer by force reversal leaves behind a widely spaced series of dislocations. New dislocations are formed by a twin layer penetrating a boundary between adjacent blocks.

The above papers contain many other very interesting observations, whose interpretation is as yet far from clear. Further work should provide a complete picture of the atomic mechanism.

---

*To Russian edition.

TABLE Ia

Mechanical Twinning Elements for Metals

Metal	K_1	η_1	K_2	η_2	s	Notes
			Cubic System			
			Face-centered Lattice			
Cu	(111)	[11$\bar{2}$]	(11$\bar{1}$)	[112]	0.707	Only recrystallization twins (with these elements) occur in Ag, Al, Au, γ-Fe, and Co
			Body-centered Lattice			
Cr, α-Fe, Mo, Na	(112)	[11$\bar{1}$]	(11$\bar{2}$)	[111]	0.707	
	(441)	–	–	–		
	(332)	–	–	–		
Pb	(221)?	–	–	–	0.354	
	(112)	–	–	–		
W	(112)	[11$\bar{1}$]	(11$\bar{2}$)	[111]	–	
			Lattice of Diamond Type			
Ge	(111)	[11$\bar{2}$]	(11$\bar{1}$)	[112]	0.707	
			Hexagonal System			
			Hexagonal Close Packing			
Be		[$\bar{1}$011]			0.199	c/a = 1.568
Cd		[10$\bar{1}$1]			0.171	c/a = 1.886
Mg	(10$\bar{1}$2)	[$\bar{1}$011]	(10$\bar{1}$2)	[10$\bar{1}$1]	0.129	c/a = 1.624
Ti		[$\bar{1}$011]			0.189	c/a = 1.587
Zn		[10$\bar{1}\bar{1}$]			0.139	c/a = 1.856
Be	(10$\bar{1}$1) (10$\bar{1}$3)					Additional Forms
Mg	(10$\bar{1}$1)	[$\bar{1}$012]	(0001)	[10$\bar{1}$0]	1.066	Ditto
	(11$\bar{2}$1)	[$\bar{1}\bar{1}$26]	(0001)	[11$\bar{2}$0]	0.638	"
	(11$\bar{2}$2)	[11$\bar{2}\bar{3}$]	(11$\bar{2}$2)	[11$\bar{2}$3]	0.957	"
	(11$\bar{2}$3)	[$\bar{1}\bar{1}$22]	(0001)	[11$\bar{2}$0]	1.194	"
Ti	(11$\bar{2}$4)	[2$\bar{2}$43]	(11$\bar{2}$4)	[2$\bar{2}$43]	0.468	"
	{30$\bar{3}$4}	<20$\bar{2}$3>	{$\bar{1}$012}	<$\bar{1}$011>		"
			{30$\bar{3}$2}*	<$\bar{1}$013>*		*150 and 286°C
	{10$\bar{1}$3}*	<30$\bar{3}$2>*	or	or		
			{$\bar{1}$011}**	<$\bar{1}$012>**		**At any temperature apart from -190°C

TABLE Ia (Continued)

Metal	K_1	η_1	K_2	η_2	s	Notes
			Rhombohedral Lattice			
As	$(\bar{1}012)$?	$[10\bar{1}\bar{1}]$	$(10\bar{1}\bar{1})$	$[\bar{1}012]$	0.256	
Bi	$(\bar{1}012)$	$[10\bar{1}\bar{1}]$	$(10\bar{1}\bar{1})$	$[1012]$	0.118	
Hg	$(\bar{1}012)$	$[10\bar{1}\bar{1}]$	$(10\bar{1}\bar{1})$	$[\bar{1}012]$	0.447	
Sb	$(\bar{1}012)$	$[10\bar{1}\bar{1}]$	$(10\bar{1}\bar{1})$	$[\bar{1}012]$	0.146	
α-Zr	$(10\bar{1}2)$					
	$(11\bar{2}1)$					
	$(11\bar{2}2)$					
	$(11\bar{2}3)$					
			Tetragonal System			
			Tetragonal Lattice			
β-Sn	(301)	$[\bar{1}03]$	$(\bar{1}01)$	[101]	0.119	c/a = 0.541
In	(101)	$[10\bar{1}]$	$(\bar{1}01)$	[101]	0.150	c/a = 1.078
			Orthorhombic System			
			Orthorhombic Lattice			
α-U	$(1\bar{7}2)$	[312]	(112)	X	0.228	
	(112)	$X[3\bar{7}\bar{2}]$	$(1\bar{7}2)$	[312]	0.228	X = irrational twin
	(121)	X[100]	X(111.4)	[311]	0.329	
	(111)	$[12\bar{3}]$	$(1\bar{7}6)$	$[5\bar{1}\bar{2}]$	0.214	
	(130)	$[3\bar{1}0]$	$(1\bar{1}0)$	[110]	0.299	

TABLE Ib

Twinning Element for Alloys

Alloys	K_1	η_1	K_2	η_2	
F.c. Alloys Cu–Al, Al–Cu Cu–Zn, Al–Zn Au–Ag	(111)	$[11\bar{2}]$	$(\bar{1}\bar{1}1)$	[112]	
B.c. Alloys β-Cu–Zn α-Fe–Si	(112)	$[\bar{1}\bar{1}1]$	$(11\bar{2})$	[111]	Recryst. only

TABLE II

Mechanical Twinning Elements for Nonmetallic Crystals

Crystal	Formula	K_1	η_1	K_2	η_2	s	Notes
Cubic System							
Gallium* and indium** antimonides		(111) (123)	$[11\bar{2}]$ $[41\bar{2}]$			0.408 0.655	* 100–500°C ** 80–320°C
Silver bromide	AgBr				[111]		
Silicon	Si	(111) (123)	$[11\bar{2}]$	$(11\bar{1})$		0.408	
Magnetite	Fe_3O_4	(111)		$(11\bar{1})$			
Galena	PbS	$(113)^*$ $(441)^{**}$ $(332)^{***}$ $(221)?$ (112)	$[11\bar{1}]$ $[011]$ $[11\bar{2}]$ $[2\bar{2}5]$ $[33\bar{2}]$	$(11\bar{1})$ (011) $(11\bar{2})$ $(2\bar{2}5)$ $(33\bar{2})$	$[113]$ $[441]$ $[332]$ $[221]$ $[112]$	0.354	* ⎫ 1-3% ** ⎬ Bi *** ⎭
Rocksalt	NaCl	(111)					Spinel law 400–500°C
Sphalerite	ZnS	(111)	$[11\bar{2}]$			0.408	300–400°C
Hexagonal System							
Graphite	C	$(11\bar{2}1)$					
Trigonal System							
Hematite	Fe_2O_3	(0001) $(10\bar{1}0)$	$[02\bar{2}1]$ $[\bar{1}012]$	$(02\bar{2}1)$ $(\bar{1}012)$	$[0001]$ $[10\bar{1}0]$	0.634 0.205	
Dolomite	$CaMg(CO_3)_2$	$(02\bar{2}1)?$	$[01\bar{1}1]$	$(01\bar{1}1)$	$[02\bar{2}1]$	0.588	
α-Quartz	SiO_2						Dauphiné twins, axial law
Calcite	$CaCO_3$	$(\bar{1}012)$	$[10\bar{1}1]$	$(10\bar{1}1)$	$[\bar{1}012]$	0.693	
Corundum	Al_2O_3	(0001)	$[02\bar{2}1]$	$(02\bar{2}1)$	$[0001]$	0.635	
Magnesite	$MgCO_3$	$(10\bar{1}1)$	$[\bar{1}011]?$	$(10\bar{1}1)$	$[\bar{1}012]$	0.799	
Millerite	NiS	$(\bar{1}012)$	$[10\bar{1}0]$	$(10\bar{1}0)$	$[\bar{1}012]$	0.380	
Pyrargyrite	$Ag_3Sb_4S_3$	$(10\bar{1}4)?$	$[0001]$	(0001)	$[10\bar{1}4]$	0.456	
Sodium nitrate	$NaNO_3$	$(\bar{1}012)$	$[10\bar{1}1]$	$(10\bar{1}1)$	$[\bar{1}012]$	0.753	
Siderite	$FeCO_3$	$(\bar{1}012)$	$[10\bar{1}1]$	$(10\bar{1}1)$	$[\bar{1}012]$	0.781	
Tetragonal System							
Hausmannite	Mn_3O_4	(101)	$[\bar{1}01]$	$(\bar{1}01)$	$[101]$	0.323	
β-Leucite	$K(AlSi_2O_6)$	(110)	$[1\bar{1}0]$	$(1\bar{1}0)$	$[110]$		Below 620°C
Tinstone	SnO_2	(101)	$[\bar{3}01]$	$(\bar{3}01)$	$[101]$	0.265	
Rutile	TiO_2	(101) (101)	$[\bar{1}01]$ $[\bar{3}01]$	$(\bar{1}01)$ $(\bar{3}01)$	$[101]$ $[101]$	0.908 0.190	

TABLE II (Continued)

Crystal	Formula	K_1	η_1	K_2	η_2	s	Notes
			Orthorhombic System				
Ammonium sulfate	$(NH_4)_2SO_4$	$(1\bar{3}0)$		(110)		0.409	
		(110)		$(1\bar{3}0)$			
Anhydrite	$CaSO_4$	(101)	$[\bar{1}01]$	$(\bar{1}01)$	[101]	0.228	
Aragonite	$CaCO_3$	(110)	$[1\bar{3}0]$	$(1\bar{3}0)$	[110]	0.130	
Baryta	$BaSO_4$	(110)?	$[1\bar{1}0]$	$(1\bar{1}0)$	[110]	0.411	
Bournonite	$PbCuSbS_3$	(110)	$[1\bar{1}0]$	$(1\bar{1}0)$	[110]	0.128	
Maleic anhydride	$\begin{array}{c}CHCO\\ \| \quad\diagdown O\\ CHCO\diagup\end{array}$	(101)		$(10\bar{1})$		0.583	
Potassium nitrate	KNO_3	(110)	$[1\bar{1}0]$	$(1\bar{3}0)$	[310]	0.041	
		(110)		$(1\bar{3}0)$			
Potassium sulfate	K_2SO_4	$(1\bar{3}0)$?		(110)		0.041	
Potassium chromate	K_2CrO_4	(110)		$(1\bar{3}0)$		0.024	
Carnallite	$KClMgCl_2 \cdot 6H_2O$	(110)	$[1\bar{1}0]$	$(1\bar{3}0)$	[130]	0.048	
Copper blende	Cu_2S	(201)	[010]	(010)	[201]	0.600	
		(131)?			[110]		
Mercury bromide	$HgBr_2$	(110)		$(1\bar{3}0)$		0.048	
Magnesium tartrate	$Mg(C_4H_4O_6)_2 \cdot 4H_2O$	(110)		$(1\bar{1}0)$		0.106	
Nickel ammonium chloride	$NH_4 \cdot ClNiCl_2 \cdot 6H_2O$	(111)				0.049	
			Monoclinic System				
Barium bromide	$BaBr_2 \cdot 2H_2O$	(001)		(100)		0.8696	
		(100)		(001)			
Barium iodide	$BaI_2 \cdot 2H_2O$	(100)		(001)		0.845	
Barium chloride	$BaCl_2 \cdot 2H_2O$	(001)		(100)		0.038	
		(100)		(001)			
Bischoffite	$MgCl_2 \cdot 6H_2O$		[111]	(111)	[112]	0.404	
Diopside	$CaMgSi_2O_6$	(001)	[100]	(100)	[001]	0.567	
Triammonium disulfate	$(NH_4)_3 \cdot H(SO_4)_2$	(310)	$[1\bar{3}0]$	$(\bar{1}10)$	[110]	0.0245	
Tripotassium disulfate	$K_3H(SO_4)_2$	(310)	$[1\bar{3}0]$			0.0115	
Jordanite	$Pb_4As_2O_7$	(101)	$[\bar{1}01]$	$(\bar{3}01)$	[301]	0.051	
		$(\bar{1}01)$	[101]	(301)	$[\bar{3}01]$	0.137	
		(100)	[001]	(001)	[100]	0.016	
Artificial mineral	$(ClO_3)K$	(100)	$[0\bar{0}1]$	(001)	[100]	0.716	
		(110)	$[\bar{1}10]$	$(\bar{1}10)$	[110]	0.669	
Calcium chloraluminate	$(CaCl_2) \cdot Ca_3Al_2O_6$	(310)	$[1\bar{3}0]$			0.080	
Aminoethanoic acid	$C_5H_{10}N_2O_2$	(100)		(001)		0.231	*Rare
		(001)*		(100)			
Cryolite	$Na_3Al_5F_6$		[110]	(110)		0.069	
Cobalt chloride	$CoCl_2 \cdot 6H_2O$		[011]	$(11\bar{1})$		0.232	
Leadhillite	$2PbCO_3 \cdot PbSO_4 \cdot$ $\cdot Pb(OH)_2$	(310)	$[1\bar{3}0]$	(110)	$[\bar{1}10]$	0.0141	

TABLE II (Continued)

Crystal	Formula	K_1	η_1	K_2	η_2	s	Notes
Lithium sulfate	$LiSO_4 \cdot H_2O$	$(1\bar{2}1)$	$[0\bar{1}1]$	(012)		0.473	
Nickel chloride	$NiCl_2 \cdot 6H_2O$		$[0\bar{1}1]$	$(11\bar{1})$		0.217	
Nickel sodium uranylacetate	$NiNa(UO_2)_3 \cdot$	(110)	$[\bar{1}10]$	$(\bar{3}10)$	$[130]$	0.1285	
	$\cdot(CH_3COO)_9 \cdot 9H_2O$	$(1\bar{3}0)$	$[310]$	(110)	$[1\bar{1}0]$		
Sphene (titanite)	$CaTi(SiO_4)O$		$[110]$	(131)		0.598	

Triclinic System

Crystal	Formula	K_1	η_1	K_2	η_2	s	Notes
Albite	$Na(AlSi_3O_8)$	(010)	$[001]$	(001)	$[010]$	0.142	
Bromonitrobenzene	$C_6H_4Br(NO_3)$		$[100]$	(100)			
Iodonitrobenzene	$C_6H_4I(NO_3)$		$[100]$	(100)		0.645	
Plagioclase	$Na(AlSi_3O_8)$	$(010)?$	$[001]$	(001)	$[010]$	0.151	
Sodium platinocyanide	$Na_2Pt(CN)_4 \cdot 3H_2O$	(001)	$[010]$			0.018	
				(010)	$[001]$	0.018	
Barium cadmium double chloride	$BaCl_2CdCl_2 \cdot 4H_2O$	(010)	$[001]$	(001)	$[010]$	0.0956	
Potassium manganese double chloride	$KMnCl_3 \cdot 2H_2O$	(111)	$[\bar{1}01]?$	$(\bar{1}\bar{1}1)$	$[101]$		
Potassium cadmium double sulfate	$K_2Cd(SO_4)_2 \cdot 2H_2O$	(010)	$[001]$	(001)	$[010]$	0.069	
Potassium manganese double sulfate	$K_2Mn(SO_4) \cdot 2H_2O$	(010)	$[001]$	(001)	$[010]$	0.1887	
Ammonium cadmium selenate	$(NH_4)_2Cd \cdot (SeO_4)_2 \cdot$ $\cdot 2H_2O$	$(010)?$	$[001]$	(001)	$[010]$	0.3548	

TABLE III

Crystallographic Data on Martensite Transitions

Structure change on cooling	Comp.	Rel. of phases		Habit planes	Notes	Lit. ref.
		old	new			
F.c.c. → hex.	Pure Co Fe-Mn(15— —20% Mn)	(111) ∥ (0001) $[1\bar{1}0]$ ∥ $[11\bar{2}0]$		(111) ?	Four types of mutual orientation, each having its own habit plane; relationships known only approximately	[1—4]
F.c.c. → b.c.c.	Fe—Ni (27— —34% Ni)	(111) ∥ (101) $[1\bar{2}1]$ ∥ $[10\bar{1}]$ (Nishiyama's relation)		∼(259)	Exact relations unknown; Nishiyama's relations are only approximate representations of irrational ones. The Kurdyumov-Sachse relations occur above room temperature. Much variation in data on habit planes.	[5—8]

TABLE III (Continued)

Structure change on cooling	Comp.	Rel. of phases		Habit planes	Notes	Lit. ref.
		old	new			
F.c.c. → b.c.t.	Fe−C (0.04% C)	?		Sets ‖ to $\langle 1\bar{1}0 \rangle$ on {111} planes	Bowles states that the sets of needles are traces of plates on the plane of section	
	Fe−C (0.5− 1.4% C)	(111) ‖ (101) [1$\bar{1}$0] ‖ [11$\bar{1}$] (Kurdyumov- Sachse relation)		∼(225)	24 relations occur, which consist of 12 pairs of twins with the same habit planes for each pair	[5, 6 10−12]
	Fe−C (1.5− −1.8% C)	?		∼{259}		
	Fe−Ni−C (0.8% C 22% Ni)	(111) ‖ (101) [1$\bar{2}$1] ‖ [10$\bar{1}$]		<1° between (3, 10, 15) and (9, 22, 23)	Orientations established precisely; habit plane and orientation determined for the individual plates	[5] [13]
	Fe−C Fe−Ni−C, Fe−Cr−C, Fe−Mn−C (with much C)			∼{225} and ∼{259} occur together	The relative number of the habit planes is not dependent on the proportion of C, but it varies with temperature from 90% {225} at 30°C to about 90% {259} at 40°C	[14]
F.c.c. → f.c.t.	In−Tl (18−20% Tl)	(111) ‖ (111) [01$\bar{1}$] ‖ [011] Orientation with twins on (110) planes		(101)	36 possible orientations for the tetragonal lattice, of which 12 are forms of the relations given, the other 24 being twins of these 12	[15−18]
					One system of parallel martensite plates can contain up to 8 orientations with 4 oppositely inclined plates in each. One theory indicates that this degeneracy is only apparent and that there are 24 types of habit plane containing 2 twins in each. 4 habit planes lie at 0.5° to {110} and cannot be distinguished by experiment	

TABLE III (Continued)

Structure change on cooling	Comp.	Rel. of phases		Habit planes	Notes	Lit. ref.
		old	new			
B.c.c. → hex	Pure Li	Slight deviations from the relations of Burgers vectors: $(10\bar{1}) \parallel (0001)$ $[111] \parallel [\bar{1}\bar{1}20]$		{441} very roughly	The Burgers relations take 12 forms; slight deviations provide 24 forms with common indices for the habit planes. The data for Ti are from [22]. Two special habit planes	[19, 20] [2, 21] [2, 23]
	Pure Zr	Two directions differ by ∠3° in Li and by ∠2° in Zr		$(8, \bar{12}, 9)$		
B.c.c. → hex	Pure Ti	Planes in Ti inclined at < 0.5°		$(9, \bar{12}, 8)$ ∠4° to $(3\bar{4}4)$	These correspond to two orientations with the basal plane inclined at 1°. The Ti alloys show two systems of habit planes: 1) close to {334} or (8, 9, 12) at 87° to the basal plane of the hexagonal lattice and 2) close to {344} at 40° to the basal plane	[24]
	Ti—Mn (4.3 −5.2% Mn)	?		{334} and {344}	Both habit planes may exist together. The martensite plates are often twinned; the twin plane is (110) for Ti—Mo	[25]
B.c.c. → f.c.t.	Cu—Zn (40% Zn)	?		~{2, 11, 12}	Martensite structure unknown. The 1% Sn—Pb alloy has a f.c.t. structure, but this explains the x-ray patterns of Cu—Zn only in part. The martensite structure of Cu—Zn is unknown	[26, 27]
	Cu—Sn (25.6% Sn)	?		~{133}		[27]
B.c.c. → only hex.	Cu—Al (11 −13.1% Al)	$(10\bar{1}) \angle 4°$ to (0001) $[111] \parallel [10\bar{1}0]$		∠2° to {133}	The martensite has a distorted hexagonal structure. [0001] lying at 2° to the normal to (0001). Two orientations occur in the system of oppositely inclined plates	[28—30]

TABLE III (Continued)

Structure change on cooling	Comp.	Rel. of phases old	Rel. of phases new	Habit planes	Notes	Lit.ref.
	Cu–Al (12.9 –14.7% Al)	$(10\bar{1}) \parallel (0001)$ $[111] \parallel [10\bar{1}0]$ $(1\bar{1}01)$ twins in which $(0\bar{1}1)$ lies at $\angle 4°$ to $(000\bar{1})$		$\angle 3°$ to $\{122\}$	The martensite has a hexagonal close-packed structure, which goes over to a distorted ortho-rhombic one as the Al content increases. Two orientations of parallel plates occur in each system	[28–31]
B.c.c.→ orthorhombic	Au–Cd (47.5% Cd)	$(100) \parallel (100)$ $[11\bar{1}] \parallel [110]$		$(\bar{3}31)$	There are 12 types of orientation, with their own habit planes	[32–33]
Tetr.→ ortho-rhombic	U–Cr (0.4 at. % and 1.4 at. % Cr)	a)$(410) \parallel (0\bar{2}1)$ $[001] \parallel [124]$ b)$(410) \parallel (0\bar{2}1)$ $[001] \parallel [\bar{1}24]$ c)$(\bar{1}40) \parallel (001)$ $[001] \parallel [320]$		Between $(\bar{4}\bar{4}1)$ and (321) Close to (321)	There are 8 forms of a) and b); the orientations of b) are not equivalent but are mirror images of those of a) in the (410) plane. In a), $(001) \parallel (817)$, while in b) $(001) \parallel (212)$; orientations a) and b) are obtained for indi-vidual plates. The polar figures indicate that orientations close to c) also exist, though this is only very rough and is obtained by twinning of martensite on $(1\bar{1}2)$. The results are for 1.4 at. % Cr; an orientation analo-gous to a) occurs in an alloy with 0.4 at. %Cr.	[34–36]

LITERATURE CITED

1. T. R. Anantharaman and J. W. Christian "Philos. Mag.", 1952, 43, p. 1338.
2. W. G. Bürgers "Physica", 1934, 1, p. 561.
3. Z. Nishiyama "Sci. Rep. Tohoku Imp. Univ.", 1936 (1), 26, p. 77.
4. J. G. Parr "Acta crystallogr.", 1952, 5, p. 842.
5. A. B. Greninger and A. R. Troiano "Trans. AIMME", 1940, 140, p. 307.
6. G. Wassermann and K. W. Mitt "Inst. Eisenforsch.", 1935, 17, p. 149.
7. Z. Nishiyama "Sci Rep. Tohoku Imp. Univ.", 1934–1935, (1), 23, p. 638.
8. R. F. Mehl and G. Derge "Trans. AIMME", 1937, 125, p. 482.
9. J. S. Bowles "Acta crystallogr.", 1951, 4, p. 162.
10. R. F. Mehl, C. S. Barrett, and D. W. Smith "Trans. AIMME", 1933, 105, p. 215.
11. G. V. Smith and R. F. Mehl "Trans. AIMME", 1942, 150, p. 211.
12. G. V. Kurdyumov and G. Sachs "Z. Phys.", 1930, 64, p. 325.
13. A. B. Greninger and A. R. Troiano "Trans. AIMME", 1949, 185, p. 590.

14. R. F. Mehl and S. W. van Winkle "Rev. metallurgie", 1953, 50, p. 465.

15. Z. S. Basinski and J. W. Christian "Acta metallurgica", 1954, 2, p. 148.

16. J. S. Bowles, C. S. Barrett, and L. Guttman "Trans. AIMME", 1950, 188, p. 1478.

17. Z. S. Basinski and J. W. Christian "Acta metallurgica", 1953, 1, p. 759.

18. Z. S. Basinski and J. W. Christian "J. Inst. Metals", 1951–1952, 80, p. 659.

19. C. S. Barrett and O. R. Trautz "Trans. AIMME", 1948, 175, p. 579.

20. J. S. Bowles "Trans. AIMME", 1951, 191, p. 44.

21. A. J. J. van Ginneckin and W. G. Bürgers "Acta crystallogr.", 1952, 5, p. 548.

22. A. K. Williams, R. W. Cahn, and C. S. Barrett "Acta metallurgica", 1954, 2, p. 117.

23. C. J. McHargue "Acta crystallogr.", 1953, 6, p. 529.

24. S. Weining and E. S. Machlin "Trans. AIMME", 1954, 200, p. 1280.

25. Y. C. Liu and H. Margolin "Trans. AIMME", 1953, 197, p. 667.

26. D. Hull and R. D. Garwood Inst. Metals: Symposium on the mechanism of phase transformations in metals, 1955, p. 219.

27. A. B. Greninger and V. G. Mooradian "Trans. AIMME", 1938, 128, p. 337.

28. A. B. Greninger "Trans. AIMME", 1939, 133, 204.

29. I. Isaichev, E. Kaminsky, and G. V. Kurdyumov "Trans. AIMME", 1939, 133, p. 361.

30. V. Gavranek, E. Kaminsky, and G. V. Kurdyumov "Metallwirtschaft", 1936, 15, p. 370.

31. G. V. Kurdyumov "Trans. AIMME", 1939, 139, p. 22 (discussion).

32. L. C. Chang and T. A. Read "Trans. AIMME", 1951, 191, p. 47.

33. L. C. Chang "Acta crystallogr.", 1951, 4, p. 320.

34. A. N. Holden "Acta metallurgica", 1953, 1, p. 617.

35. R. F. Mehl and S. W. van Winkle "Rev. metallurgie", 1953, 50, p. 465.

36. B. R. Butcher and A. H. Rowe Inst. Metals: Symposium on the mechanism of phase transformations in metals, 1955, p. 229.

TABLE IV *

Glide Elements for Metal and Other Crystals

Crystal	Lattice	Slip plane	Slip direction	Notes
Cu, Au, Ag, Ni, CuAu, α-CuZn, AlCu, AlZn	Cubic f.c.	{111}	<110>	
Al	Cubic f.c.	{111} {100}	<110> <110>	Above 450°
α-Fe	Cubic b.c.	{110} {112} {123}	<111> <111> <111>	
α-Fe + 4% Si	Cubic b.c.	{110}	<111>	
Mo, Nb	Cubic b.c.	{110}	<111>	Simultaneous slip on two {110} planes is sometimes erroneously considered as slip on {112} or {123}

* Taken from Seeger's article in the Handbuch der Physik (1958), with correction of misprints and later additions.

TABLE IV (Continued)

Crystal	Lattice	Slip plane	Slip direction	Notes
Cd, Zn, ZnCd	Hexagonal close-packed c/a > 1.85	(0001) $(10\bar{1}0)$ $(11\bar{2}2)$	$<2\bar{1}\bar{1}0>$ $[11\bar{2}0]$ $[\bar{1}\bar{1}23]$	In Zn mainly for T > 250°C In Zn at 20°C
Mg	Hexagonal close-packed c/a = 1.623	(0001) $\{10\bar{1}1\}$	$<2\bar{1}\bar{1}0>$ $<2\bar{1}\bar{1}0>$	Occurs above 225°C and also under other conditions
		$\{10\bar{1}0\}$	$<2\bar{1}10>$	Begins to become dominant at room temperature and below in specially oriented monocrystals and in poly-crystals. A trace of lithium facili-tates the process and greatly in-creases the low-temperature plasticity
Be	Hexagonal close-packed c/a = 1.568	(0001) $\{10\bar{1}0\}$	$<2\bar{1}\bar{1}0>$ $<2\bar{1}\bar{1}0>$	
Ti	Hexagonal close-packed c/a = 1.587	$\{10\bar{1}0\}$ $\{10\bar{1}1\}$ (0001)	$<2\bar{1}\bar{1}0>$ $<2\bar{1}\bar{1}0>$ $<2\bar{1}\bar{1}0>$	Pyramid planes and basal planes less active than prism planes
C, Ge, Si, ZnS (cubic)	Diamond structure	$\{111\}$		<101> is the slip direction in Ge
NaCl, KCl, KBr, KI, AgCl, LiF	Rocksalt structure	$\{110\}$	<110>	
PbTe, PbS	The same	$\{110\}$	<100>	Slip on $\{110\}$ and <110> occurs also in PbS
MgTl, LiTl, AuZn, AuCd, NH_4Cl, NH_4Br, CsI, CsBr, TlCl − TlBr (KPC-6)	Cesium iodide structure	$\{110\}$	<100>	
TlBr − TlI(KPC-5)		$\{110\}$ $\{100\}$	<100> <100>	For > 250°C
β'-CuZn	The same	$\{110\}$	<111>	
AgMg	The same	$\{3\bar{2}\bar{1}\}$	<111>	
$CaCO_3$	Hexagonal system, trigonal system	$\{111\}$ $\{100\}$	<011> <011>	At 20°C From 20 to 300°C

APPENDIX 2*

SELECTIVE ETCHING APPLIED TO THE DISLOCATION MECHANISM OF TWINNING WITH CHANGE OF FORM**

Selective etching has given valuable results here, as in the case of slip. Most of them have been obtained from calcite, though some interesting results have also been obtained on metal crystals.

Rais (1957), Watts (1959), and Stanley (1959) have observed etch figures on calcite crystals, which are related to dislocations within the crystal.

Kolontsova et al. (1959) used 10% hydrochloric acid to produce etch figures on cleaved surfaces; the same pattern was produced on both sides of a plate 2 mm thick, which shows that faults run for considerable distances. It was also found that dynamically generated twins contain far more dislocations than the parent crystal.

Startsev has examined the dislocation mechanism of twinning for calcite and some metals; Startsev, Bengus, et al. (1960) used 2% aqueous tartaric acid as the etching agent, which was allowed to act for 10-30 sec. This gave pits with sharp edges; the shape did not alter even on very deep etching. They also discovered other good agents for calcite, but tartaric acid was the best for revealing dislocations at twin boundaries.

Figure 1 shows the etch pattern on calcite cleaved and etched as above. The symmetrical patterns on the mating surfaces indicate that the pits represent points of emergence of dislocations; the same is to be concluded from the localization at points of large deformation and the displacement under load. Figure 2 shows an axial section of a calcite plate after rapid bending at 700°C: The crystal was cooled while remaining under load and then was etched. The pattern clearly corresponds to the stress distribution produced by bending.

The structure of twin boundaries is of particular interest; etching reveals dislocations localized there (Fig. 3). No boundary is free from dislocations throughout its length, but very slow loading with the force precisely in the shear plane and along the twin direction can give large coherent regions. Figure 4 shows coherent and incoherent boundaries. The dislocation density was observed to range from 30 per cm (for a nearly coherent boundary) to 5000 per cm; the latter corresponds to a deviation equivalent to a rotation of 1' in the (100) plane. This very small rotation would be difficult to detect by eye, so the boundaries are always virtually parallel to [100]. Interferometry (for wedge-shaped twins) gives roughly the same result.

Bengus et al. (1961) have shown that pits of two types occur at twin boundaries (Fig. 5). Those of the first type etch less readily and correspond to dislocations that cause less lattice disturbance. Those of the second type etch very vigorously. The first type was shown to represent twinning dislocations, which are extremely mobile in response to stress or annealing, although the latter sometimes leaves some dislocations unmoved. Moreover, the first type is always displaced along the twin plane (Fig. 6).

---

*Added for the English edition.
** Compiled by V. I. Startsev.

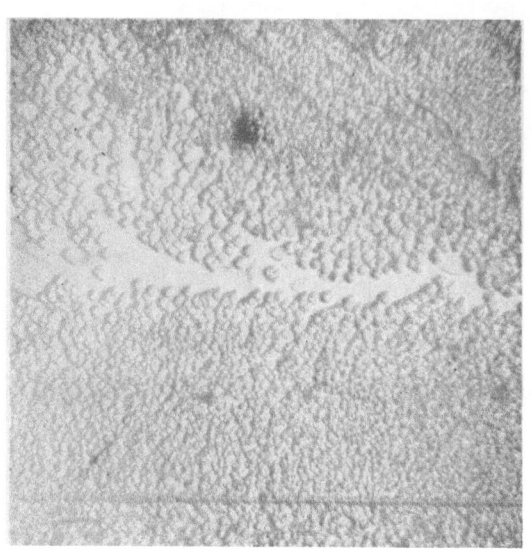

Fig. A-1. Etch pattern on cleavage-plane surface of calcite. × 600.

Fig. A-2. Formation of dislocations by the bending of a calcite plate at a high temperature; the neutral line is seen. Etching time 15 sec. × 115.

Fig. A-3. Etch pattern on cleavage-plane surface of calcite showing twin boundaries. M = main crystal, T = twin, B = boundary. × 200.

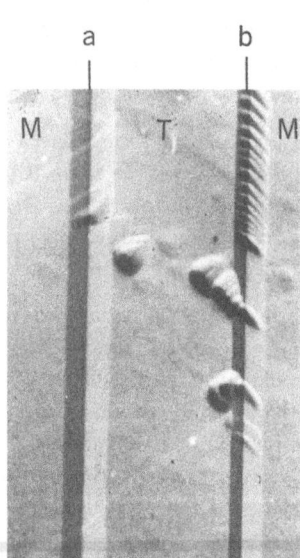

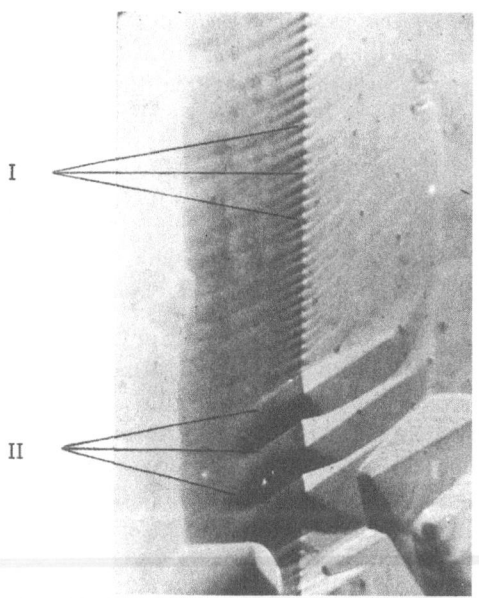

Fig. A-4. Etch pattern on calcite having twin boundaries of different types. M = main crystal, T = twin, a = almost coherent boundary, b = incoherent boundary. × 200.

Fig. A-5. Part of the boundary of a twin layer showing pits of two types: I) first type, II) second type. × 650.

The most convincing proof that these are twinning dislocations was given by Bengus et al. (1961a) from the mechanical displacement along the twin plane. It is difficult to observe the motion directly, for displacement of one dislocation along the entire twin plane displaces the twin boundary by only one lattice constant. The load and loading rate must be such that the twin layer does broaden and the dislocations are displaced only a short distance. The previously etched calcite crystal was placed on the stage of a microhardness tester and was loaded with a diamond pyramid normal to the cleavage plane. Figure 7 shows the result for a part of the twin boundary near the pyramid (to the right under the boundary). The flat-bottomed pits correspond to the initial positions; the dislocations were displaced leftward by the pressure from the pyramid, and the new positions are shown by the sharp-pointed pits. The load on the pyramid provides a means of measuring the stress needed to start the motion (starting stress).

It is difficult to establish the stress acting on a dislocation here, because the dislocations are not isolated; each is acted on by the stresses set up by adjacent ones. The starting stress may be calculated on the assumption that the latter effect is negligible (which is certainly untrue), but this gives a large range of values. For instance, for the dislocations of Fig. 7 the range is 60 ± 15 g/mm^2. The precise value varies from experiment to experiment; it is affected by the number and nature of the adjacent dislocations.

Sometimes the starting stress can be deduced much more simply for certain arrays of dislocations.

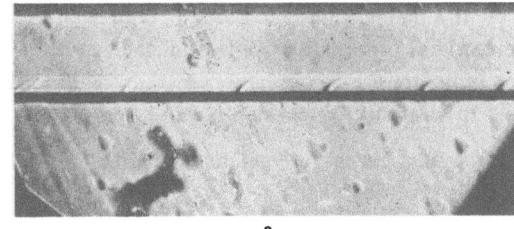

a

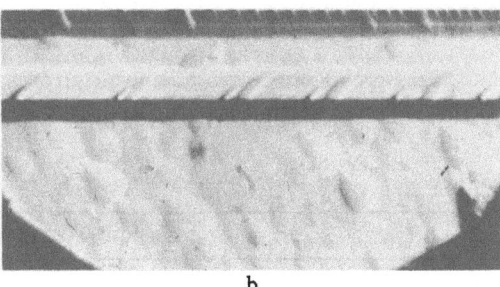

b

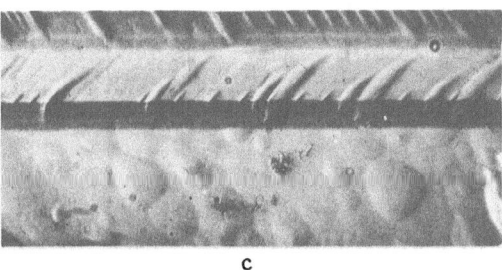

c

Fig. A-6. Redistribution of dislocations at a twin boundary by annealing: a) before; b) after first annealing; c) after second annealing. Etching time 3 sec in each case. × 306.

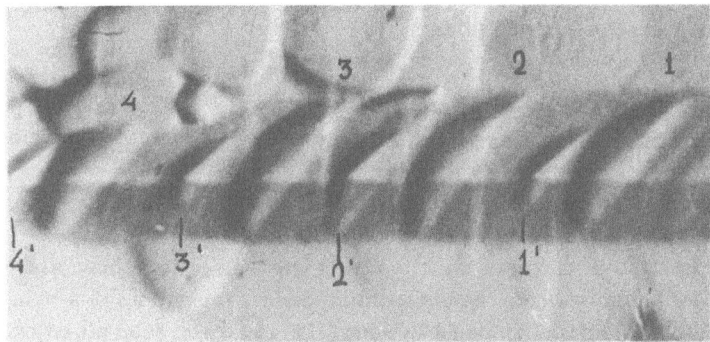

Fig. A-7. Displacement of twinning dislocations by mechanical stress (diamond pyramid, to the right outside the field of view). The flat-bottomed pits 1-4 denote the initial positions of dislocations; the sharp-pointed ones 1'-4' denote the positions after displacement. × 1040.

If a dislocation is retarded by an obstacle, it retards those following after it in adjacent twin planes. If the external force is then removed, the repulsion between the dislocations causes them to take a definite pattern that indicates the forces between them. Each dislocation moves until the repulsion becomes equal to the resistance to motion in the twin plane.

Results given by Frank and Nabarro (1951) have been used to calculate this resistance (starting stress), which is given by

$$\sigma_0 = (nGb/L\pi)/(1-\nu),$$

in which n is the number of dislocations in a group, G is the shear modulus (3.2×10^{11} dyne/cm^2), b is the Burgers vector, L is the total length of the group, ν is Poisson's ratio (0.3) and σ_0 is the starting stress.

These groups occur fairly frequently; ones are found along [00$\bar{1}$] and along [001]. Figure 8 shows a typical group. Results taken from typical photographs are listed in Table A-I.

TABLE A-I

n	L,cm	σ_0,g/mm^2	n	L,cm	σ_0,g/mm^2
130	0.1126	45	26	0.011	91
96	0.0635	59	357	0.1500	94
26	0.0170	60	27	0.0112	95
23	0.0134	67	33	0.0127	102
15	0.0068	86	45	0.0128	138

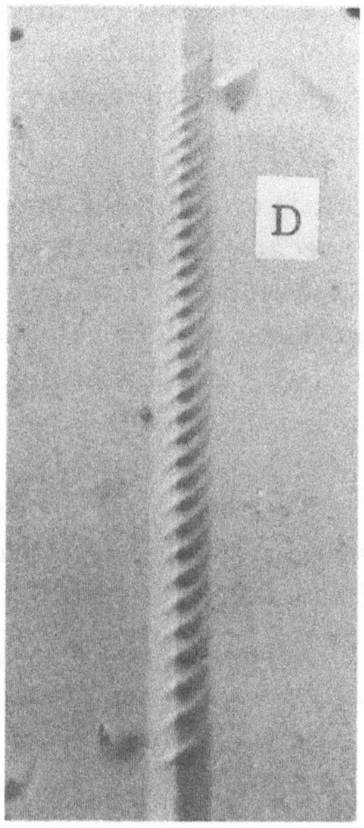

The spread in the σ_0 may be caused by curvature in the dislocations (which may not be of purely edge type) or by the distance of the obstacle from the surface.

The mean value is 87 g/mm^2, the minimum being 45 g/mm^2 and the maximum 138 g/mm^2. The force needed to start the expansion of a layer (per dislocation) should not exceed the starting force if the crystal contains no obstacles; the dislocations would then move more freely in the twin planes. In fact, obstacles of various types (complete dislocations, sessile dislocations, inclusions, and so on) are present; the force needed to overcome these is seen as a contribution to the resistance.

The second type of pit (II in Fig. 5) corresponds to dislocations of low mobility, whose tendency to etch is comparable with that of dislocations away from the boundary. These dislocations are not displaced along the twin plane by external forces, but they migrate together with the boundary and in the same direction (twinning dislocations move along the boundary). This transverse motion is seen in Fig. 9, which shows the initial and final positions of a twin boundary with pits of the second type. These dislocations often occur in pairs (Fig. 10); each deep pit on one boundary corresponds to a similar one on the other.

Etching following detwinning shows that these paired pits remain flat-bottomed in most cases (Fig. 11) for thin layers. It is difficult to determine the pairs in thick twins, although Fig. 12 shows that such pairs are present. Such twins often give isolated pits clearly corresponding to deep pits at the boundaries (Fig. 13). During detwinning, these pits at the boundaries approach those within the layer, and on completion they become

Fig. A-8. Set of twinning dislocations along a twin boundary at an obstacle: D = complete dislocation retarding the twinning ones. × 600.

180

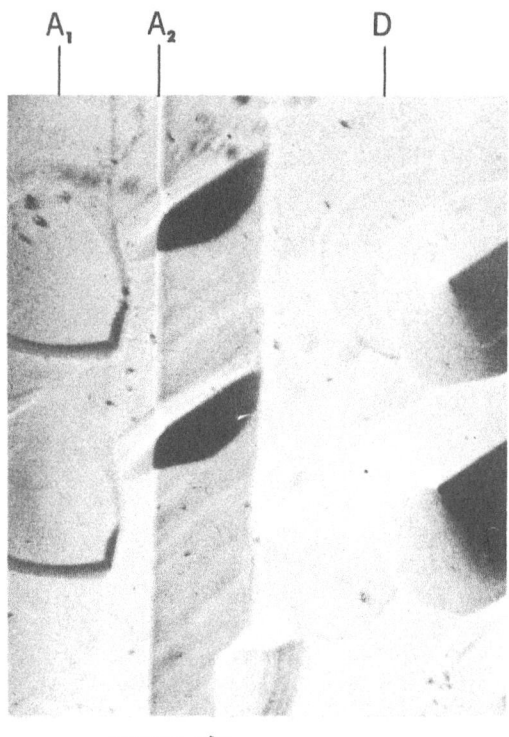

Fig. A-9. Motion of a boundary in reversal of twinning (shown by arrow): A_1 is the initial position (flat-bottomed pits) and A_2 is the later position (sharp-pointed pits); D = twin. The sharp-pointed pits in the body of the twin represent the emergence of the other end of the half-loops. × 540.

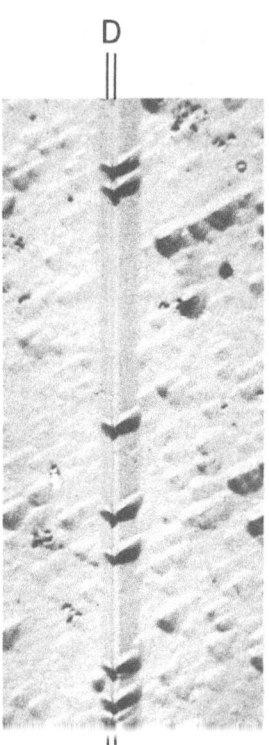

Fig. A-10. Paired dislocations at two boundaries of a thin twin layer; D = twin. × 340.

flat-bottomed pairs. This by itself is not surprising if we consider that a pair corresponds to the emergence of a dislocation loop that intersects the surface. Such a loop is called a transverse one to distinguish it from loops lying in the interface; such loops do not expand along the boundary. These points of emergence are not always detectable if one end of the loop lies at the boundary, but annealing (which alters the distribution) usually makes the relation clear.

The properties and interactions of dislocations indicate that the low mobility dislocations always present in calcite are sessile dislocations whose Burgers vectors lie along [110].

Bengus et al. (1961b) found that many other lattice defects arise during twinning, quite apart from the production of twinning dislocations. These defects block the twinning dislocations and hinder the

Fig. A-11. Retention of a half-loop after detwinning; D = twin. b) site as of a) after detwinning. × 600.

expansion of twin layers; they cause hardening. Many of these defects persist within the twin layer, as Kolontsova et al. have (1959) shown.

The dislocation arrays in a twin are not randomly arranged; they may almost always be correlated as pairs of pits along definite crystallographic directions. Very often the straight line between two pits will etch as a groove of triangular shape, which resembles the groove at a twin boundary but which etches much less readily (Fig. 14a). Here the paired pits differ in shape, which indicates that the dislocation lines differ in direction. Figure 14 shows also the effects of varying the etching time; comparison shows that the distance between the vertices decreases as the time is increased, and finally the pits fuse to form a single flat-bottomed one. This result shows that the dislocation lines are branches of a loop intersecting the surface.

a

b

c

d

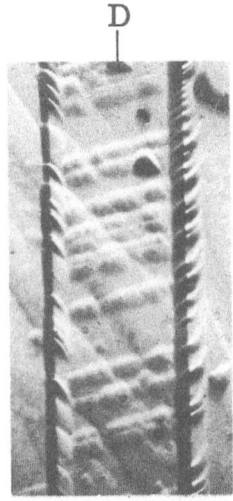

Fig. A-12. Paired dislocations at the boundary of a thick twin. × 306.

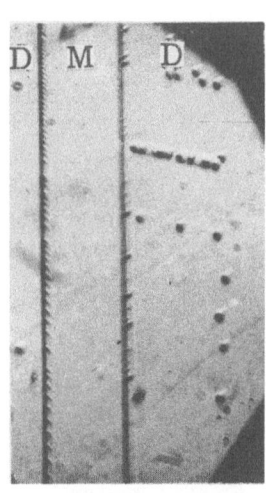

Fig. A-13. Paired dislocations at the boundary and in the twin. M = main crystal, D = twin. × 340.

Fig. A-14. Patterns of loops from etching of calcite; etching times (sec): a) 3; b) 6; c) 9; d) 30. × 600.

Brief annealing at low temperatures reduces the size of the loops. It is supposed that the grooves represent packing defects and that the dislocation lines around such defects are the lines of partial dislocations.

It is usual to find more dislocations near the surface of a twinned crystal; this may be a result of the method of loading, which produces maximum stress near the surface. It may also mean that the stress needed to produce a stable loop is less at the surface. It is also found that low-temperature annealing reveals numerous dislocations at points of stress concentration.

The production of dislocations in a twin is often related to a halt in the motion of the boundary (Fig. 15); here each row of dislocations corresponds to a halt during the growth of the twin. Keith and Gilman (1960) have observed a similar effect.

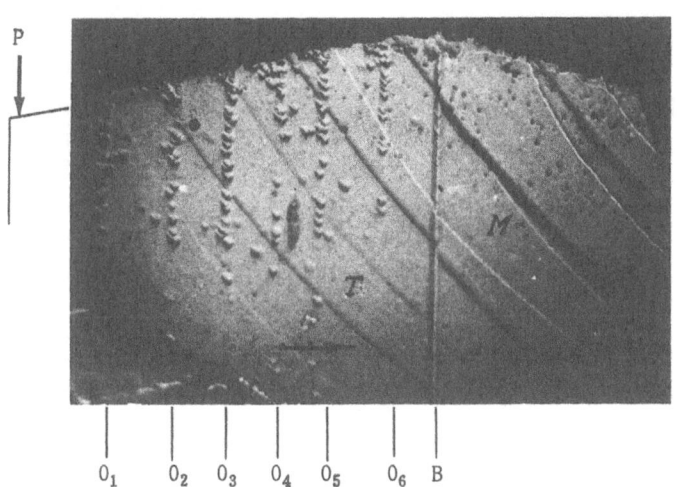

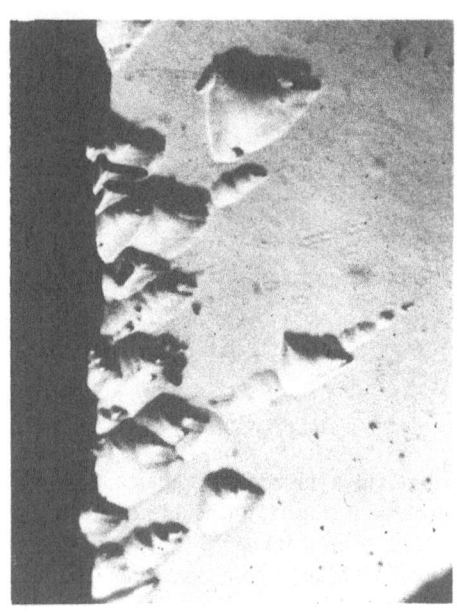

Fig. A-15. Dislocations in a twin layer in calcite produced in stepwise motion of the boundary. M = main crystal, T = twin, B = boundary. The horizontal arrow indicates the direction of motion; the vertical one, the direction of the force P. 0_1, 0_2, and so on are the points where the boundary halted. × 200.

Fig. A-16. Motion of dislocations starting at a twin boundary in antimony crystals. × 450.

Soifer and Startsev (1961) have applied selective etching to antimony monocrystals; Lavrent'ev and Startsev (1961) have done the same for zinc. Observations have also been made on dislocations at twin boundaries, which show some of the effects previously reported for calcite. The site of the twin layer bears many dislocations after a metal crystal has been detwinned. It has been found that the twin boundaries in antimony and bismuth are stressed and can act as sources of dislocations. This is well seen in Fig. 16, which shows the effects of continuous etching under load, although the same effect is often seen when there are no external forces; this may be the result of removal of obstructions at the surface.

APPENDIX 3*

SELECTIVE ETCHING APPLIED TO TWINNING
WITHOUT CHANGE OF FORM

This type occurs in quartz, triglycine sulfate, and barium titanate. Twinning without macroscopic displacement (Fig. 119) should not produce twinning dislocations, although local lattice distortion may occur (Ch. 2, § 3.3), though this does not give dislocations. Prolonged etching of quartz with hydrofluoric acid causes the whole surface to become covered with etch figures, which are associated with differential attack on the faces. The reoriented regions are revealed by the altered orientation of the figures; steps occur at the boundaries.

Tsinzerling (1963) has used saturated aqueous boric acid with hydrofluoric (10:1) at 20°C to reveal dislocations in quartz; those appearing on cleavage surfaces on the rhombohedron and basal plane were seen as separate figures against a smooth background in monocrystals and in twinned crystals. Dislocations were not seen at the junctions. Dislocations may represent the earliest stage of slip in the case of quartz.

Konstantinova (1962) has found several agents that reveal the domains (positive and negative) in triglycine sulfate, as well as the dislocations. Twins were revealed on (010) cleavages and on (100) polished surfaces.

Fig. B-1. Twin (domain) boundaries and dislocations; no dislocations at boundaries. (010) section in triglycine sulfate (Konstantinova).

The signs of the domains were established from the black-and-white patterns seen in reflection. The twin boundaries are revealed as a result of differences in the rates of attack on the positive and negative components; a step is seen. The polarity is absent on (100), but the regions still stand out clearly (Fig. 131a). Some modes of etching reveal the boundaries as narrow grooves.

Individual dislocations and ones along slip lines are revealed in both types of component; the dislocations differ in shape and orientation as between components. No dislocations are seen along boundaries on (010), as one would expect (Fig. B-1), but (unexpectedly) line defects are revealed along the boundaries of wedge-shaped domains. The density and disposition of these defects are not the same on the two boundaries, but the orientation of the etch pits is the same.

Another interesting effect has been observed by Konstantinova. Cleaved surfaces show shallow dislocation loops in response to etching; these are pairs of pits that soon become flat-bottomed and then vanish (Fig. B-3). New domains (twins) often appear in the region of these residual pits.

* Added for the English edition.

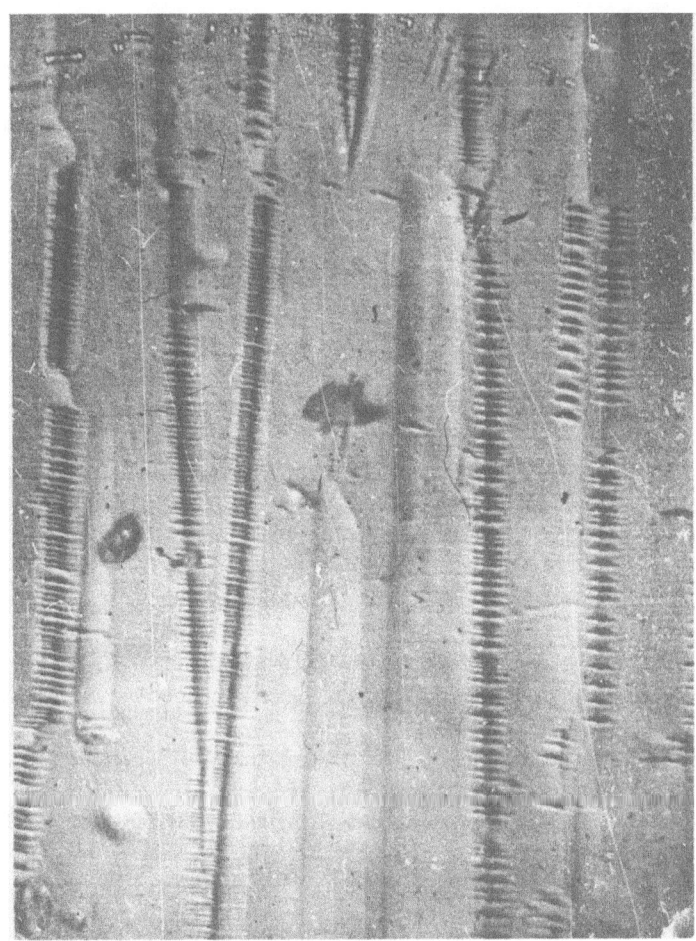

Fig. B-2. Series of line defects at twin boundaries in a (100) section of triglycine sulfate (Konstantinova).

Fig. B-3. Appearance of fresh domains in flat-bottomed etch pits; (010) section of triglycine sulfate (Konstantinova).

LITERATURE CITED

Aerts, E., Delavignette, P., Siems, R, and Amelinckx, S. "Stacking-fault energy in silicon," J. Appl. Phys. 33(10): 3078-3080 (1962).

Akulov, N.S. Ferromagnetism, Moscow-Leningrad, GIZ, 1939, p. 94.

Akulov, N.S. and Dekhtyar, M. V. Phys. Rev. 5: 750 (1932).

Akulov, N.S. and Raevskii, S. Ann. phys. 5(2): (1934).

Allen, N.P., Hopkins, B.E., and McLennan, J.E. "Tensile properties of single crystals of high-purity iron," Proc. Roy. Soc. (London) A 234(1197): 221-246 (1956).

Aminoff, G. and Broomé, B. "Strukturtheoretische Studien über Zwillinge, I," Z. Krist. 80: 355-376 (1931).

Ancker, B. "Präzisionsuntersuchungen am Zinkgitter 1-2," Ann. Phys. 12(1-3): 121-144, 145-154 (1953).

Anderson, A., Jillson, J., and Dunbar, B. "Deformation mechanisms in alpha-Ti," J. Metals 5, Sec. 2(9): 1191-1197 (1953).

Anderson, W.A. and Mehl, R.F. "Recrystallization of aluminum in terms of rate of nucleation and rate of growth," Trans. AIMME 161: 140 (1945).

Andrade, E.N. da C., and Hutchings, P.J. "The mechanical behavior of single crystals of mercury," Proc. Roy. Soc. (London) A 148(863): 120-146 (1935).

Andrews, K.W. and Johnson, W. "Formulas for the transformation of indices in twinned crystals," Brit. J. Appl. Phys. 6(3): 92-96 (1955).

Ansel, G. Metals Handbook, Cleveland, Am. Soc. Metals, 1948, p. 975.

van Arkel, A.E. and van Bruggen, M.G. "Rekristallisations Erscheinungen bei Aluminium, II," Z. Physik 51: 520(1928).

Bain, E.C. and Jeffries, Z. "Mixed orientation by plastic deformation," Chem. & Met. Eng. 25: 775 (1921).

Bakarian, P.W. and Mathewson, C.H. Trans. AIMME 194: 865 (1952).

Barrett, C.S. "Structure of iron after compression," Trans. AIMME 135: 246-322 (1939).

Barrett, C.S. Trans. AIMME 161: 15 (1945).

Barrett, C.S. Structure of Metals, McGraw-Hill Book Co., Inc., New York, 1943.

Barrett, C.S. Trans. AIMME 200: 1003 (1954).

Barrett, C.S. Imperfections in Nearly Perfect Crystals, Symposium Held at Pocono Manor, 1950, John Wiley & Sons, Inc., New York, Chapman and Hall, London, 1952.

Barrett, C.S. "Metallurgy at low temperatures." Trans. Am. Soc. Metals, 49: 53-117 (1957).

Barrett, C.S. and Heller, C.T. Trans. AIMME 171: 246 (1947).

Barrett, C.S., and Levenson, L. "Structure of iron after drawing, swaging, and elongation in tension," Trans. AIMME 135: 327-343 (1939).

Basinski, Z.S., and Christian, J.W. "Crystallography of deformation by twin-boundary movements in indium-thallium alloys," Acta Met. 2(1): 101-116 (1951a).

Basinski, Z.S., and Christian, J.W. J. Inst. Metals 80: 659 (1951b).

Basinski, Z.S., and Christian, J.W. "Experiments on martensitic transformation in single crystals of indium-thallium alloys," Acta Met. 2: 148 (1954a).

Basinski, Z.S., and Christian, J.W. "Crystallography of deformation by twin-boundary movement in indium-thallium alloys," Acta Met. 2: 101-116, 148 (1954b).

Basinski, Z.S., and Sleeswyk, A. "Ductility of iron at 4.2°K," Acta Met. 5(3): 176-179; (12): 762-764 (1957).

Baumhauer, H. "Über künstliche Kalkspatzwillinge nach -1/2 R," Z. Krist. 3: 588 (1879).

Baumhauer, H.Z. Krist. 49: 113 (1911).

Bechtold, I.H. Trans. AIMME 197: 1469 (1953).

Bechtold, I.H., and Shewmon, P.G. Trans. Am. Soc. Metals 46: 397 (1954).

Beck, P.A. "Annealing of cold-worked metals," Advances in Physics 3(11): 245-325 (1954).

Beck, P.A. and Hu, H. "Annealing textures in rolled face-centered cubic metals," Trans. AIMME 194: 83 (1952).

Beck, P.A. and Sperry, P.R. "Strain-induced grain boundary migration in high-purity aluminum," J. Appl. Phys. 21: 150 (1950).

Becker, R. "Recrystallization texture in aluminum after compression," Z. tech. Physik 7: 547 (1926).

Bell, R.L. and Cahn, R.W. "The nucleation problem in deformation twinning," Acta Met. 1(6): 752-753 (1953).

Bell, R.L. and Cahn, R.W. "The dynamics of twinning and the interrelation of slip and twinning in zinc crystals," Proc. Roy. Soc. (London) A 239: 494-521 (1957).

Bell, R.L. and Cahn, R.W. "The initiation of cleavage fracture at the intersection of deformation twins in zinc single crystals," J. Inst. Metals 86: 433-438 (1958).

Belov, N.V. and Klassen-Neklyudova, M.V. "Character of damage to crystals," Zhur. Tekh. Fiz. 18(3): 265-278 (1948).

Bengus, V.Z., Komnik, S.N., and Startsev, V.I. "Formation of dislocation in calcite crystals," Kristallografiya 6(4): 599-604 (1961).

Bengus, V.Z., Komnik, S.N., and Startsev, V.I. "Some effects observed at the boundaries of a twin band in calcite," Kristallografiya 6(4): 614-620 (1961).

Bengus, V.Z. and Startsev, V.I. "A study of the dislocation mechanism of twinning in ionic crystals," In: Physics of Alkali-Halide Crystals, Latvian State University, Riga, 1962, pp. 475-477.

Berg, W. "Mechanical twinning in bismuth crystals," Nature 134: 143 (1934).

Berndt, G.V. Verhandl. deut. physik. Ges. 19: 314-327 (1917); 21: 110-117 (1919).

Betteridge, W. "The crystal structure of cadmium-indium alloys rich in indium," Proc. Phys. Soc. (London) 50: 519-523 (1938).

Biggs, W.D. and Pratt, P.L. "The deformation and fracture of alpha-iron at low temperatures," Acta Met. 6(11): 694 (1958).

Bilby, B.A. and Bullough, R. "The formation of twins by a moving crack," Phil. Mag. 45(365): 631-646 (1954).

Bilby, B.A. and Christian, J.W. Martensitic Transformations. The Mechanism of Phase Transformations in Metals, London, 1956, pp. 121-172.

Bilby, B.A. and Entwisle, A.R. "The formation of mechanical twins," Acta Met. 2(1): 15-19 (1954).

Billig, E. and Ridout, M.S. "Transmission of electrons and holes across a twin boundary in germanium," Nature 173, 490 (1954).

Bitter, F. Phys. Rev. 38: 1930; 41: 4 (1932).

Blank, F. "Über die Kohäsionsgrenzen des Steinsalzkristallen," Z. Physik 61(11-12): 727-749 (1930).

Blewitt, T.H., Coltman, R.R., Klabunde, C.E., and Noggle, T.S. "Low-temperature reactor-irradiation effects in metals," J. Appl. Phys. 28(6): 639-644 (1957).

Blewitt, T.H., Coltman, R.R., and Redman, J.K. "Work-hardening in copper crystals," Defects in Crystalline Solids: Report of the Conference on Defects in Crystalline Solids, held at the University of Bristol, July, 1954, Phys. Soc. (London) (1955), pp. 369-382.

Bloch. Z. Physik, 74: 295 (1932).

Boas, W. and Schmidt, B. Plasticity of Crystals, Especially Metallic Ones [Russian translation], GONTI, 1938; [English translation] 1950.

Bond, W.L. and Andrus, I. "Structural imperfections in quartz crystal," Am. Mineralogist, 37: 622-632 (1952).

Bowden, F.P. and Cooper, R.E. "Velocity of twin propagation in crystals," Nature 195(4846): 1091-1092 (1962).

Bragg, W. and Gibbs, R.E. Proc. Roy. Soc. (London) A 109: 401 (1925).

Brauns, R. "Über die Bedeutung der Morphotropie für Mineralogie," Neues Jahrb. Mineral. 1: 113 (1889).

Brick, R.M. Trans. AIMME, 1953.

Brick, R.M. and Steijn, R.P. Trans, Am. Soc. Metals 46: 1406 (1954).

Bridgman, P.W. Proc. Am. Acad. Arts Sci. 60: 306 (1925).

Bridgman, P.W. Physics of High Pressure. Macmillan Co., New York, 1931.

Bridgman, P.W. Studies in Large Plastic Flow and Fracture. Effects of High Hydrostatic Pressures on the Mechanical Properties of Materials, McGraw-Hill Book Co., Inc., 1952.

Brilliantov, N.A. and Obreimov, I.V. "Plastic deformation in rocksalt," Zhur. Eksptl. i Tekh. Fiz. 5(3-4): 330-339 (1935).

Brilliantov, N.A. and Obreimov, I.V. "Plastic deformation, IV," Zhur. Eksptl. i Tekh. Fiz. 7(8): 978-986 (1937).

Bullough, R. "Deformation twinning in the diamond structure," Proc. Roy. Soc. (London) A 241(1227): 568-577 (1957).

Bullough, R. and Bilby, B.A. "Continuous distribution of dislocations. Surface dislocation and the crystallography of martensitic transformations," Proc. Phys. Soc. (London) B 69(12): 1276-1286 (1956).

Bunshah, R.F. and Mehl, R.F. Trans. AIMME 197: 1251 (1953).

Bürgers, W.G. "Neuere Untersuchungen und Auffassungen auf dem Gebiete der Rekristallisationsforschung," Berg- und hüttenmänn. Monatsh. Montan. Hochschule Leoben 101(7): 151-154 (1956).

Bürgers, W.G. and Louwerce, P.C. "Über den Zusammenhang zwischen Deformations-Vorgang und Rekristallisationstexture bei Aluminium," Z. Physik 67: 605 (1931).

Bürgers, W.G., Meijs, J.C., and Tiedema, T.J. "Frequency of annealing twins in copper crystals grown by recrystallization," Acta Met. 1: 75-78 (1953).

Burkart, M.W. and Read, T.A. "Diffusionless phase change in the indium-thallium system," J. Metals 197: 1516-1524, (1953).

Burke, J.E. Trans. AIMME 188: 1324 (1950).

Burke, J.E. and Turnbull, D. "Recrystallization and grain growth," Progr. in Metal Phys. 3: 220 (1952).

Burke, J.E. and Turnbull, D. "Recrystallization and grain growth," Progr. in Metal Phys. 4: (1953).

Cahn, R.W. "A new theory of recrystallization nuclei," Proc. Phys. Soc. (London) A 63: 323 (1950a).

Cahn, R.W. "Internal strains and recrystallization," Progr. in Metal Phys. 2: 151 (1950b).

Cahn, R.W. "Plastic deformation of α-uranium: twinning and slip," Acta Met. 1: 49-70 (1953a).

Cahn, R.W. "Soviet work on mechanical twinning," Nuovo cimento, Ser. 9-10 (4): 350-384 (1953b).

Cahn, R.W. "Twinned crystals," Advances in Phys. 3(12): 363-446 (1954).

Cahn, R.W. "Mechanical twinning in molybdenum," J. Inst. Metals 83(11): 493-496 (1954/1955).

Calnan, E.A. "Laue asterism and deformation bands," Acta Cryst. 5: 557-564 (1952).

Calnan, E.A. and Clews, C.I. "Deformation textures in face-centered cubic metals," Phil. Mag. 41(322): 1085-1100 (1950).

Calnan, E.A. and Clews, C.I. Phil. Mag. 42: 616, 919 (1951).

Calnan, E.A. and Clews, C.I. "The prediction of uranium deformation textures," Phil. Mag. 43: 93-104 (1952).

Carlile, S.J., Christian, J.W., and Hume-Rothery, S.I. J. Inst. Metals 77: 169 (1949).

Carrington, W.E. "Lamellae in chromium," J. Inst. Metals 82: 170 (1953/1954).

Chalmers, B. "The twinning of single crystals of tin," Proc. Phys. Soc. (London) 47(Pt.4): 733 (1935).

Chang, L.C. and Read, T.A. Trans. AIMME 47(189): 191 (1951).

Chaudhury, A., Grant, N., and Norton, J. "Metallographic observation on the deformation of high-purity magnesium in creep at 500°F," J. Metals 5(5): 71-76 (1953).

Chen, N.K. and Maddin, R. Trans. AIMME 191: 937 (1951).

Chentsova, L.G. "Nature of the color centers in smoky quartz," Kristallografiya 1(4): 484-485 (1956).

Chernysheva, M.A. "Mechanical twinning in crystals of Rochelle salt," Doklady Akad. Nauk SSSR 74: 247-249 (1950).

Chernysheva, M.A. "Effects of an electric field on the twinned structure of crystals of Rochelle salt," Doklady Akad. Nauk SSSR 81(6): 1065-1068 (1951).

Chernysheva, M.A. Twinning Phenomena in Crystals of Rochelle Salt, Dissertation, 1955.

Chernysheva, M.A. "Some details of the domain structure of crystals of Rochelle salt (from optical studies)," Izvest. Akad. Nauk SSSR, Ser. Fiz. 21(2): 289-292 (1957).

Churchman, A.T. "The formation and removal of twins in titanium during deformation," J. Inst. Metals 83: 39-40 (1954/1955).

Churchman, A.T. "The yield phenomena, kink band and geometric softening in titanium crystals," Acta Met. 3(1): 22-29 (1955).

Churchman, A.T., Geach, G.A., and Winton, J. "Deformation twinning in materials of the A4 (diamond) crystal structure," Proc. Roy. Soc. (London) A 238: 194-203 (1956).

Clogh, W.R. and Pavlovic, A.S. "The flow, fracture and twinning of commercially pure vanadium," Trans. Am. Soc. Metals 52: 948-970 (1960).

Collection. Elasticity and Inelasticity of Metals, edited by S.V. Vonsovskii, Moscow, Izd-vo inostr. lit., 1952.

Collection. Aspects of the Physics of Plasticity of Crystals, edited by M.V. Klassen-Neklyudova, Moscow, Izd. Akad. Nauk SSSR, 1960.

Conrad, H. and Schoek, G. "Cottrell locking and the flow stress in iron," Acta Met. 8(11): 791 (1960).

Cottrell, A.H. "The formation of immobile dislocations during slip," Phil. Mag. 43: 645-647 (1952).

Cottrell, A.H. Dislocations and Plastic Flow in Crystals, Oxford University Press, Fairlawn, N.J., 1953a.

Cottrell, A.H. "Theory of dislocations," Progr. in Metal Phys. 4: 205-264 (1953b).

Cottrell, A.H. and Bilby, B.A. "A mechanism for the growth of deformation twins in crystals," Phil. Mag. 42: 573-581 (1951).

Cross, L.E. and Nicholson, B.J. Phil. Mag. 46: 453 (1955).

Czochralski, J. Moderne Metallkunde, Berlin, 1924.

Davidenkov, N.N. The Problem of Impact in Metal Science, Moscow-Leningrad, Izd-vo AN SSSR, 1938.

Davidenkov, N.N. and Chuchman, T.N. "A review of current theories of cold-shortness," Studies on Heat-Resisting Alloys, Vol. 2, Moscow, Izd-vo AN SSSR (1957), pp. 9-34.

Davidenkov, N.N. and Chuchman, T.N. "Twinning and cold-shortness," Zhur. Tekh. Fiz. 28(11): 2502-2513 (1958).

Deicha, G. "Sensibilité des macles mecaniques aux defauts primaires de structure ($BaCl_2 \cdot 2H_2O$)," Bull. min. et cristallogr. 72(10-12): 543-548 (1949).

Delise, L. "Electron microscopic study of the effect of cold working on the subgrain structure of copper," J. Metals 5(5): 733-735 (1953).

Dereyttere, A. and Greenough, G.B. "The markings in the cleavage surfaces of zinc single crystals," Phil. Mag. 45(365): 624-630 (1954).

Dichl, J. "Kristallen. I. Verfestigungskurven und Oberflächenerscheinungen," Z. Metallk. 47(5): 331-343 (1956).

Dijkstra, L.J. and Martins, U.M. "Domain pattern in silicon-iron under stress," Rev. Modern Phys. 25(1): 146-150 (1953).

Dijkstra, L.J., Martins, U.M., Chalmers, B., and Cavanagh. "Internal microstrains and the deformation and failure of metals," J. Soc. Nondestructive Testing 12(1): 13-18 (1954).

Dubov, G.A. and Regel', V.R. "Methods of testing small specimens in compression and in stress relaxation," Kristallografiya 2(6): 746-755 (1957).

Dunn, C.G., Daniels, F.W., and Bolton, M.I. Trans. AIMME 188: 368 (1950).

Elam, C.F. "Discussion of paper by Adcock," J. Inst. Metals 27: 94 (1922).

Elam. C.F. "An investigation of some banded structures in metal crystals," Proc. Roy. Soc. (London) A 121(787): 237-247 (1928).

Elam, C.F. "Distortion of beta-brass and iron crystals," Proc. Roy. Soc. (London) A 153, 273 (1936)

Ellis, W.C. and Treuting, R.G. Trans. AIMME 191: 53 (1951).

Eshelby, J., Frank, F., and Nabarro, F. "The equilibrium of linear arrays of dislocations," Phil. Mag. 42: 351 (1951).

Fairbairn, H.W. "Correlation of quartz deformation with its crystal structure," Am. Mineralogist 24(6): 351 (1939).

Fine, M.E. "Evidence for domain structure in antiferromagnetic CoO from elasticity measurements," Phys. Rev. 87(6): 1143 (1952).

Fine, M.E. and Kenneey, N.T. "Moduli and internal friction of magnetite as affected by the low-temperature transformation," Phys. Rev. No. 6: 1573-1576 (1954).

Fisher, E. Naturwissenschaften 41: 117 (1954).

Fisher, J.C., Johnston, W.G., Thomson, R., and Vreeland, T. (Editors). Dislocations and Mechanical Properties of Crystals, John Wiley & Sons, Inc., New York, 1956.

Fisher, R.M., Darken, C.S., and Curroll, K.G. Acta Met. 2(3): 368-373 (1954).

Flunkert, H. "Über das Zustandekommen der optischen Abbildung der ferroelektrischen Domänen von Seignetterkristallen," Z. Physik, 3: 253-271 (1960).

Fonrie, I.T., Weinberg, F., and Boswell, F.W.C. "The growth of twins in tin single crystals as observed by transmission electron microscopy." Acta Met. 8: 851 (1960).

Forster, F. and Scheil, E. "Untersuchung des zeitlichen Ablaufes von Umklappvorgängen in Metallen," Z. Metallk. 32: 165 (1940).

Forsyth, P. "Some metallographic observations on the fatigue of metals," J. Metals 80: 181 (1951).

Frank, F.C. "Crystal dislocation: elementary concepts and definitions," Phil. Mag. 42(362): 809-819 (1951).

Frank, F.C. and van der Merve, J.H. "One-dimensional dislocations, I, Static theory," Proc. Roy. Soc. (London) A 198(1053): 205-216 (1949).

Frank, F.C. and Thompson, N. "On deformation by twinning," Acta Met. 3(1): 30-33(1955).

Frank, N. "Stable dislocations in the common crystal lattices," Phil. Mag. 44: 1213 (1953).

Franks, I. Metal whiskers, Nature 177(4517): 984 (1956).

Friedel, G. Leçons de Cristallographie, Paris, Berger-Levzault, 1926.

Friedel, J. "On the linear work-hardening rate of f.c.c. single crystals," Phil. Mag. 46(382): 1169-1186 (1955).

Friedel, J. Les Dislocations, Paris, 1956, § 6.5.3.

Friedel, J. "The mechanism of work-hardening and slip-band formation," Proc. Roy. Soc. (London) 242(1229): 147-158 (1957).

Frocht, M.M. Photoelasticity, John Wiley & Sons, Inc., New York, 1941.

Frondel, A.O. "Symposium on quartz oscillator plates," Am. Mineralogist, 1945.

Fullman, R.L. "Formation of annealing twins during grain growth," J. Appl. Phys. 22: 1350 (1951); 21: 1069 (1950).

Garber, R.I. "Residual stresses in plastically deformed crystals of rocksalt," Zhur. Eksptl. i Theoret. Fiz. 8(6): 746-752 (1938).

Garber, R.I. "Mechanical twinning of calcite, III," Zhur. Eksptl. i Teoret. Fiz. 10(3): 354-357 (1940).

Garber, R.I. "Increase in the yield point during annealing of twinned calcite," Zhur. Eksptl. i Teoret. Fiz. 16(10): 923-927 (1946).

Garber, R.I. "Mechanism of twinning of calcite and sodium nitrate during plastic deformation," Zhur. Eksptl. i Teoret. Fiz. 17(1): 48-62 (1947a).

Garber, R.I. "Annealing of twinned crystals of calcite and sodium nitrate," Zhur. Eksptl. i Teoret. Fiz. 17(1): 63-68 (1947b).

Garber, R.I., Gindin, I.A., Kogan, V.S., and Lazarev, B.G. "Study of plastic behavior of monocrystals of beryllium," Fiz. Metal. i Metalloved. 1(3): 528-537 (1955).

Garber, R.I., Gindin, I.A., and Konstantinovskii, M.G. "Effects of grain size on the initiation and development of twin bands in iron," Zhur. Tekh. Fiz. 23(12): 2127-2135 (1953).

Garber, R.I., Gindin, I.A., Konstantinovskii, M.G., and Startsev, V.I. "Annealing of twinned crystals of iron," Doklady Akad. Nauk SSSR 74(2): 343-344 (1950).

Garber, R.I., Obreimov, I.V., and Polyakov, L.M. "Formation of ultramicroscopic inhomogeneities in the plastic deformation of rocksalt," Doklady Akad. Nauk SSSR 108(3): 425-427 (1956).

Garber, R.I. and Stelina, E.I. "Etch figures of tapering elastic twins," Kristallografiya 5(5): 811 (1960).

Garber, R.I. and Stelina, E.I. "Defects at the boundaries of twin bands." In: Physics of Alkali-Halide Crystals, Latvian State University, Riga (1962), pp. 479-481.

Garber R.I., Zalivadnyi, S.Ya., and Startsev, V.I. "Effects of mosaic structure on resistance to mechanical twinning in sodium nitrate," Doklady Akad. Nauk SSSR 58(4): 571-572 (1947).

Garstone, J., Honeycombe, R., and Greetham, G. "Easy glide of cubic metal crystals," Acta Met. 4(5): 485-494 (1956).

Geil, G.W. and Garwille, N.L. Welding J. (N.Y.) 32: 273 (1953).

Gervais, A., Norton, J., and Grant, N. "Kink-band formation in high-purity aluminum during creep at high temperatures," J. Metals 5, Sec. 2(11): 1487-1492 (1953).

Gifkins, R. "Grains within grains," Australian Engr., February, pp. 41-50 (1955).

Gilman, J.J. "Mechanism of ortho kink-band formation in compressed zinc monocrystals," J. Metals 6, Sec. 2(5): 621-629 (1954).

Gilman, J.J. and Read T.A. "Bend-plane phenomena in the deformation of zinc monocrystals," J. Metals 5(1): 49-55 (1953).

Gilman, J.J. and Read, T.A. Trans. AIMME, 194: 875 (1952).

Gindin, I.A., Lazarev, B.G., and Khotkevich, V.I. "Low-temperature polymorphism of metals." Zhur. Eksptl. i Teoret. Fiz. 23(3): 892-904 (1953).

Gindin, I.A. and Startsev, V.I. "Twinning of bismuth," Zhur. Eksptl. i Teoret. Fiz. 20(8): 738-741 (1950).

Gogoberidze, D.B. "Mechanical twinning of crystals," Uspekhi Fiz. Nauk 16: 1104-1109 (1936).

Gogoberidze, D.B. Mechanical Twinning, Gos. nauchno-tekh. izd-vo Ukrainy, 1938.

Gogoberidze, D.B. Some Bulk Defects of Crystals, Leningrad, Izd-vo Leningr. Gos. un-ta, 1952.

Gogoberidze, D.B. and Ananiashvili, E.G. "Über die mechanische Zwillingsbildung beim Kalkspat, 1-2," Phys. Z. Sov. Un. 7(5-6): 547-552 (1935); 10(6): 820-825 (1936).

Gogoberidze, D.B. and Ananiashvili, E.G. "Twinning of calcite," Zhur. Eksptl. i Teoret. Fiz. 5(7): 654-657 (1952).

Golovchiner, Ya.M. and Kurdyumov, G.V. "Microstructure studies for the austenite-martensite transition in alloy steels at low temperatures," Problemy Metalloved. i Fiz. Metal. 2: 98-118 (1951).

Goucher, F.S. "Studies on the deformation of tungsten single crystals under tensile stress," Phil. Mag. 48: 800-819 (1924a).

Goucher, F. S. "On the strength of tungsten single crystals and its variation with temperature," Phil. Mag. 48: 229-249 (1924b).

Gough, H. J. and Cox, H. L. "Behavior of a single crystal of zinc," Proc. Roy. Soc. (London) A 123(791): 143 (1929).

Grasc, C. and Baudeau, M. "Sur les differents processus de recristallisation du beryllium," J. Nucl. Mater. 6(1): 120-122 (1962).

Green, R. B. "Intersection faulting mechanism theory of flow and fracture of f.c.c. metals," Phys. Rev. 102(2): 376-380 (1956).

Greninger, A. B. and Troiano A. R. Trans. AIMME 185: 590 (1949).

Griggs, D. T. Am. Mineralogist 23: 28 (1938).

Griggs, D. T. and Miller, F. B. Bull. Geol. Soc. Am. 62: 853 (1951).

Griggs, D. T., Turner, T. J., Borg, T., and Sosoka. Bull. Geol. Soc. Am. 62: 1385 (1951); 64: 1327 (1953).

Grühn, A. and Johnsen, A. "Künstliche Schiebung in Rutil," Zentr. Mineral. Geol. 1917a, p. 370.

Grühn, A. and Johnsen, A. Zentr. Mineral. Geol. 1917a, p. 366; 1917b, p. 370.

Haasen, P. "Twinning in indium antimonide," J. Metals 9: 30 (1957).

Haasen, P. "Plastic deformation of nickel single crystals at low temperatures," Phil. Mag. No. 3: 384-418 (1958).

Hall, E. O. Twinning and Diffusionless Transformations in Metals, London, Butterworths, 1954.

Hall, E. O. "Twinning in cobalt," Acta Met. 5: 110 (1957).

Hall, M. J. and Thompson, M. W. "Epitaxy and twinning in foils of some noble metals condensed upon lithium fluoride and mica," Brit. J. Appl. Phys. 12(9): 495-498 (1961).

Harker, D. and Parker, E. R. "Grain shape and grain growth," Trans. Am. Soc. Metals 46: 156 (1954).

Harnecker, K. and Rassow, E. Z. Metallk. 16: 312 (1924).

Hasiguti, R. R. Acta Met. 3(2): 200-201 (1955).

Hatwell, H. and Votava, E. "Germination preferentielle du carbure de chrome sur les dislocations de macles dans les alliages inoxydables du type 18/8," Acta Met. 9(10): 945-948 (1961).

Hausner, H. H. and Pinto, N. P. Trans. Am. Soc. Metals 43: 1152 (1951).

Heide, F. "Über Deformationen an Kristallen bei erhöhtem Druck und erhöhter Temperatur," Z. Krist. 78: 257-278 (1931).

Herenguel, J. and Lelong, P. Compt. Rend. 233: 53-55 (1951).

Herring, W. C. and Galt, J. K. "Elastic and plastic properties of very small metal specimens." Phys. Rev. No. 6: 1060-1061.

Hess, J. and Barrett, C. S. "Structure and nature of kink bands in zinc," Trans. AIMME 185: 599 (1949).

Hiroshi, Furnichi. "Some observations on the persistent slip markings near grain and twin boundaries," J. Phys. Soc. Japan 17(9): 710 (1962).

Hirsch, P. B. "A study of cold-worked aluminum by an x-ray microbeam technique," Acta Cryst. 5(2): 168 (1952).

Holden, A. and Kunz, F. "Dimension and orientation effects in the yielding of carburized iron sheet crystals," Acta Met. 1(5): 495-502 (1953).

Holt, D. B., "Defects in the sphalerite structure," J. Phys. Chem. Solids 23(October): 1353-1362 (1962).

Honeycombe, R. "Inhomogeneities in the plastic deformation of metal crystals," J. Inst. Metals 80: 45-56 (1951/1952).

Hornbogen, E. "Dynamic effects during twinning in alpha-iron," Trans. Met. Soc. AIME 221(4): 711-715 (1961).

Hornstra, J. Dislocations, stacking faults and twinning the spinel structure, J. Phys. Chem. Solids 15(3/4): 311-323 (1960).

Howie, A. and Swann, P. R. "Direct measurements of stacking-fault energy from observation nodes," Phil. Mag. 6(70): 1215-1226 (1961).

Hull, D. "Twinning and fracture of single crystals of silicon iron," Acta Met. 8(1): 11-18 (1960).

Hull, D. "The initiation of slip at the tip of a deformation twin in iron," Acta Met. 9(9): 909-912 (1961).

Huntington, H. B. "Modification of the Peierls-Nabarro model for edge dislocation core," Proc. Phys. Soc. (London) B 68, Pt. 12(432): 1043-1048 (1955).

Indenbom, V. L. "Domain mobility in the Frenkel'-Kontorova model," Kristallografiya 3(2): 197-205 (1958).

Indenbom, V. L. "Dislocation description of simple effects in plastic deformation," in collection: Problems of the Physics of Plasticity, Moscow, Izd-vo AN SSSR, 1960.

Indenbom, V.L. "Dislocations in ionic crystals," In: Physics of Alkali-Halide Crystals, Latvian State University, Riga (1962), p. 448.

Indenbom, V.L. and Chernysheva, M.A. "Significance of optical studies of domains in Rochelle salt for the theory of ferroelectricity," Kristallografiya 2: (4): 526-535 (1957a).

Indenbom, V.L. and Chernysheva, M.A. "Derivation of the thermodynamic potential of Rochelle salt from optical studies of domains," Zhur. Eksptl. i Teoret. Fiz. 32(4): 697-701 (1957b).

Indenbom, V.L. and Tomilovskii, G.E. "Microstructure of the stresses in slip lines and dislocations," Doklady Akad. Nauk SSSR 123: (4): 673-676 (1958).

Indenbom, V.L. and Urusovskaya, A.A. "What are irrational twins?" Kristallografiya 4(1): 90-98 (1959).

Ioffe, A.F., Kirpicheva, M.V., and Levitskaya, M.A. "Deformation and strength of crystals," Zhur. Russ. Fiz.-Khim. Ob-va, Chast' Fiz. 56: 489 (1924).

Isely, F.C. "The relation between the mechanical and piezoelectrical properties of a Rochelle salt crystal," Phys. Rev. 24: 569-574 (1924).

Jackson, A. and Chalmers, B. "Influence of striations on the plastic deformation of single crystals of tin," Can. J. Phys. 31 (6): 1017-1018 (1953).

Jaffe, H.R. "Polymorphism of Rochelle salt," Phys. Rev. 51: 43-47 (1937).

Jaswon, M.A. "Twinning properties of lattice planes," Acta Cryst. 9(8): 621-626 (1956).

Jaswon, M.A. and Dove, D.B. "The prediction of twinning modes in metal crystals," Acta Cryst. 10(1): 14-19 (1957).

Jeffries, Z. and Archer, R. The Science of Metals (Fig. 60), McGraw-Hill Book Co., Inc., New York, 1924.

Jillson, D.C. "An experimental survey of deformation and annealing processes in zinc," J. Metals 188: 1009-1017 (1950).

Johnsen, A. "Untersuchungen über Kristallzwillinge und deren Zusammenhang mit anderen Erscheinungen," Neues Jahrb. Mineral. Geol. 23: 237-344 (1907a).

Johnsen, A. Ibid., 23: 237 (1907b).

Johnsen, A. Ibid., 39: 500 (1914a).

Johnsen, A. "Geometrical treatment of mechanical twinning," Jahrb. Radioakt. u. Elektronik 11: 226 (1914b).

Johnsen, A. Atombewegungen im Bi, Zentr. Mineral. Geol. p. 385 (1916a).

Johnsen, A. "Condition of twinning," Ibid., p. 121 (1916b); p. 385 (1916c); p. 433 (1917).

Johnson, F. "Influence of cold rolling upon mechanical properties of oxygen-free copper," J. Inst. Metals 21: 335 (1919).

Johnson, F. "Discussion of paper by Adcock," J. Inst. Metals 27: 94 (1922).

Jones, E.R.W. and Munro, W. J. Mech. and Phys. Solids 1: 182 (1953).

Kaczer, Jan. "A new method for investigating the domain structure of ferromagnetics," Czechoslov. J. Phys. 5(2): 239-244 (1955).

Kaiser, W. and Kohn, J.A. "Birefringence and twinning in silicon," Acta Met. 4(2): 220-221 (1956).

Känzig, W. "Ferroelectrics and antiferroelectrics," Solid State Physics, Vol. 4 (Editors: Seitz and Quinbull), Academic Press, New York, 1957, pp. 97-124.

Kauffmann, A.K., Gordon, P., and Lillie, D.W. Trans. Am. Soc. Metals 42: 785 (1950).

Kay, H.P. "Preparation and properties of crystals of barium titanate $BaTiO_3$," Acta Cryst. p. 229 (1948).

Kê, T.S. Science Record 3(1): 61 (1950).

Kear, B.H. "Clustering of slip bands in Cu_3Au crystals," Trans. Metal. Soc. AIME 224(4): 669-673 (1962).

Keith, R.E. and Gilman, J.J. "Dislocation etch pits and plastic deformation in calcite," Acta Met. 8(1): 1-10 (1960).

Kelly, A. "Neumann bands in pure iron," Proc. Phys. Soc. (London) A 66(4): 403-405 (1953).

King, R. Nature 169: 543 (1952).

Kitchingman, W.I. Crystallographic aspects of transformation in the β, β' and ξ phases of the silver-zinc alloys, Acta Met. 10(9): 799-802 (1962).

Klassen-Neklyudova, M.V. "Nature of plastic deformation," Zhur. Russ. Fiz.-Khim. Ob-va, Chast' Fiz. 59(5-6): 509-515 (1927).

Klassen-Neklyudova, M.V. "Laws of stepwise deformation," Zhur. Russ. Fiz.-Khim. Ob-va, Chast' Fiz. 60(5): 373-381 (1928).

Klassen-Neklyudova, M. V. "Über die sprungartige Deformation," Z. Physik 55(7-8): 555-568 (1929).

Klassen-Neklyudova, M. V. "The yield point," Vestnik Metalloprom. Nos. 9-10: 145-150 (1930).

Klassen-Neklyudova, M. V. Plastic Behavior and Strength of Crystals, GTTI, 1933, p. 130.

Klassen-Neklyudova, M. V. "Spinel-law mechanical twinning of rocksalt crystals," Zhur. Eksptl. i Teoret. Fiz. 12(9): 349-357 (1942a).

Klassen-Neklyudova, M. V. "Mechanical properties of corundum crystals, parts I and II," Zhur. Tekh. Fiz., 12(9): 519-534, 535-551 (1942b); Collection Dedicated to A. F. Ioffe, Moscow, 1950, p. 551-560.

Klassen-Neklyudova, M. V. "Results from studies of the mechanical properties of corundum in conjunction with optical tests," Trudy Inst. Krist. Akad. Nauk SSSR No. 8: 151-164 (1953).

Klassen-Neklyudova, M. V., Chernysheva, M. A., and Shternberg, A. A. "The real structure of crystals of Rochelle salt," Doklady Akad. Nauk SSSR 63: 527-530 (1948).

Klassen-Neklyudova, M. V., Tomilovskii, G. E., and Chernysheva, M. A. "Process of formation of kinks," Kristallografiya 8(4): (1960).

Klassen-Neklyudova, M. V. and Urusovskaya, A. A. "Transverse impact and pressure figures in cubic crystals," Trudy Inst. Krist. Akad. Nauk SSSR No. 11: 146-151 (1955).

Klassen-Neklyudova, M. V. and Urusovskaya, A. A. "Effects of a state of inhomogeneous stress on the mechanism of plastic deformation for the halides of thallium and cesium," Kristallografiya 1(4): 410-418 (1956a).

Klassen-Neklyudova, M. V. and Urusovskaya, A. A. "A study of the structure of kink bands in crystals of thallium halides," Kristallografiya 1(5): 564-571 (1956b).

Klassen-Neklyudova, M. V. and Urusovskaya, A. A. "Plastic deformation of crystals caused by lattice rotation without the formation of slip lines," Kristallografiya 2(1): 134-139 (1957).

Klassen-Neklyudova, M. V. and Urusovskaya, A. A. "Deformation of rocksalt at high temperatures," Kristallografiya 8(5): (1960).

Knapp, H. and Dehlinger, U. "Mechanik und Kinetik der diffusionslosen Martensitbildung," Acta Met. 4(3): 300-307 (1956).

Kochendörfer, A. Plastische Eigenschaften von Kristallen und metallischen Werkstoffen, Berlin (1941), p. 12.

Kohra, Kazutave. "X-Ray observations of lattice defects in particular stacking faults in the neighborhood of a twin boundary in silicon single crystals," J. Phys. Soc. Japan 17(6): 1041-1052 (1962).

Kolesnikov, G. "Geometrie mechanischer Zwillingsbildung von Zinn und Zink," Phys. Z. Sov. Un. 3: 651-667 (1933).

Kolontsova, E. V., and Plavnik, G. M. "Structures of the kink bands of some ionic crystals," Kristallografiya 1(4): 419-424 (1956).

Kolontsova, E. V., Sorokina, Yu. G., and Telegina, I. V. "A study of twinning in calcite crystals by means of x-ray microbeams and etch figures," Kristallografiya 4(5): 742-748 (1959).

Kolontsova, E. V. and Telegina, I. V. "Mechanism of formation of kink bands," Doklady Akad. Nauk SSSR 116(4): 605-608 (1957).

Konstantinova, V.P. "Domain boundaries and etch-pit networks in crystals of deuterated triglycine sulfate," Izvest. Akad. Nauk SSSR, Ser. Fiz. No. 11 (1960).

Konstantinova, V.P. "Use of selective etching in studying twin and dislocation structures in triglycine sulfate," Kristallografiya 7(4): 348-354 (1962).

Konstantinova, V.P., Sil'vestrova, I.M., and Aleksandrov, K.S. "Preparation and physical properties of crystals of triglycine sulfate," Kristallografiya 4(1): 69-73 (1959).

Konstantinova, V.P., Sil'vestrova, I.M., and Yurin, V.A. "Twinning and dielectric behavior of crystals of triglycine sulfate," Kristallografiya 4(1): 125-129 (1959).

Kontorova, T.A. "Existence of transition zones in twinned crystals," Zhur. Eksptl. i Teoret. Fiz. 12(1-2): 68-78 (1942).

Kontorova, T.A. and Frenkel', Ya. I. "Theory of plastic deformation and twinning," Zhur. Eksptl. i Teoret. Fiz. 8(1): 89-95; 8(12): 1340-1358 (1938).

Kosevich, A.M. "Dislocation theory of hysteresis effects in twinning and shearing in an inorganic medium," Fiz. Tverd. Tela 3(11): 3263-3271 (1961).

Kosevich, A.M. "Some aspects of the dislocation theory of twins," Fiz. Tverd. Tela 4: 1103 (1962).

Kosevich, A.M. and Bashmakov, V.I. "A study of the elastic stages of twinning in metal monocrystals," Kristallografiya (in press).

Kosevich, A.M. and Pastur, L.A. "Dislocation model of a twin," Fiz. Tverd. Tela 3: 1290 (1961).

Kosevich, A.M. and Pastur, L.A. "Shape of a thin twin lying at an angle to a surface," Fiz. Tverd. Tela 3(6): 1870-1875 (1961).

Kosevich, A.M. and Pastur, L.A. "Dislocation model of a thin twin at the surface of a crystal," In: Physics of Alkali-Halide Crystals, Latvian State University, Riga, 1962, pp. 482-485.

Kosevich, A.M., Startsev, V.I., and Pastur, L.A. "Distribution of dislocations at the boundaries of twinned bands in calcite," Kristallografiya (in press).

Kozu, S. "Japanese twins of quartz," Am. J. Sci. Bowen volume, pp. 281-292 (1952).

Krafft, I.M., Sullivan, A.M., and Tipper, C.W. "The effect of static and dynamic loading and temperature on the yield stress of iron and mild steel in compression," Proc. Roy. Soc. (London) A 221: 114-127 (1953).

Kronberg, M.L. "Plastic deformation of single crystals of sapphire," Acta Met. 5(9): 507-524 (1957).

Kulin, S.A. and Cohen, M. Trans. AIMME 188: 1139 (1950).

Kurdyumov, G.V. "General laws of phase transitions in eutectic alloys," Izvest. Akad. Nauk SSSR, Otdel. Mat. i Estestven. Nauk, Ser. Khim. No. 2: 271 (1936).

Kurdyumov, G.V. "Diffusionless (martensite) transformations in alloys," Zhur. Tekh. Fiz. 18: 999-1025 (1948).

Kurdyumov, G.V. and Khandros, L.G. "Thermoelastic equilibrium in martensite transformations," Doklady Akad. Nauk SSSR 66: 211 (1949).

Kuznetsov, V.D. and Zolotov, V.A. Compt. rend. acad. sci. URSS 2(1): 13 (1934).

Kuznetsov, V.D. and Zolotov, V.A. "Role of mechanical twinning in the recrystallization of deformed zinc monocrystals," Zhur. Eksptl. i Teoret. Fiz. 5(1): 75-86 (1935).

Landau, L.D. and Lifshits, E.M. Sov. Phys. 8: 1533 (1935).

Landau, L.D. and Lifshits, E.M. Statistical Physics, Gostekhizdat, 1948.

Landau, L.D. and Lifshits, E.M. Electrodynamics of Continuous Media, 1958, § 39.

Laves, F. "Über den Einfluss von Ordnung und Unordnung auf mechanische Zwillingsbildung," Naturwiss. 39(23): 546 (1952a).

Laves, F. "Mechanische Zwillingsbildung in Feldspäten in Abhängigkeit von Ordnung-Unordnung der Si/Al-Verteilung innerhalb des (Si, Al)$_4$O$_8$-Gerüstes," Naturwiss. 39(23): 546 (1952b).

Laves, F. and Baskin, L.V. "On the formation of the rhombohedral graphite modification," Z. Krist. 107: 337-365 (1956).

Laves, F. and Goldsmith, J.R. "Long-range—short-range order in calcic plagioclases as a continuous and reversible function of temperature," Acta Cryst. 7: 465 (1954).

Lavrent'ev, F.F. and Startsev, V.I. "Structure of accommodation regions in monocrystals of zinc and bismuth," Fiz. Metal. i Metalloved. 13(3): 441-450 (1962).

Lean, B., Plateau, J., Bachet, C., and Crussard, C. "Sur la formation d'ondes sonores, au cours d'essais de traction dans des éprouvettes metalliques," Compt. rend. acad. sci. Paris 246: 2845 (1958).

Lehman, O. "Mikrokristallographische Untersuchungen," Z. Krist. 10: 321 (1885).

Lehmann, Chr. and Leibfried, G. "Gittertheoretische Behandlung einer Schraubenversetzung in Alkalimetallen," J. Phys. Chem. Solids 6(2/3): 195-204 (1958).

Leibfried, G. Verteilung von Versetzungen in statischem Gleichgewicht, Z. Physik 130: 214-226 (1951).

Lifshits, I.M. "Macroscopic description of twinning phenomena in crystals," Zhur. Eksptl. i Teoret. Fiz. 18(12): 1134-1143 (1948).

Lifshits, I.M. and Obreimov, I.V. "Some considerations on the twinning of calcite," Izvest. Akad. Nauk SSSR, Ser. Fiz. 12(2): 65-80 (1948).

Liu, Yi-huan and Tao, Tsu-tsoong. "Formation and development of substructure in aluminum crystals strained at high temperatures," Acta Phys. Sinica 12(6): 550-558 (1956).

Lomer, W.M. "A dislocation reaction in the f.c.c. lattice," Phil. Mag. 42: 645-647 (1951).

Lyubov, B.Ya. and Osip'yan, Yu.A. "Kinetics of martensite transformations at temperatures close to absolute zero," Doklady Akad. Nauk SSSR 101: 853 (1955).

Madsen, P.E. "The behavior of interfaces of lightly worked uranium during recrystallization," J. Inst. Metals, Pt. II, 71-76 (1956).

Marcinkowski, M.I. and Lippsitt, H.A. "The plastic deformation of chromium at low temperatures," Acta Met. 10(2): 95 (1962).

Mason, W.P., McSkimin, H.J., and Shockley, W. "Ultrasonic observation of twinning in tin," Phys. Rev. 73: 1213 (1948).

Mathewson, C.H. "Twinning in metals," Trans. AIMME 76: 555-601 (1928).

Mathewson, C.H. Trans. Am. Soc. Metals 32: p. 38 (1944).

Mathewson, C.H., Kent, R., and von Horn. "Directed stress in copper crystals," Inst. of Metals Division, AIMME, p. 59 (1930).

Mathewson, C.H. and Phillips, A.I. "Recrystallization of cold-worked alpha-brass on annealing," Trans. AIMME 54: 608 (1916).

Mathewson, C.H. and Phillips, A.I. "Be, Mg, Zn, Cd: Sekundäre Translation in Zwillingslamellen," Trans. AIMME, No. 53 (1928).

Matthews, J.W. "The observation of dislocations to accommodate the misfit between crystals with different lattice parameters," Phil. Mag. 6(71): 1347-1349 (1961).

McHargue, C.J. "Deformation stacking faults in face-centered cubic thorium and cerium," Acta Met. 9(9): 851 (1961).

McHargue, C.J. and Hammond, J.T. Acta Met. 1: 70 (1953).

Megaw, H.D. "Origin of ferroelectricity in barium titanate and other perovskite-type crystals," Acta Cryst. 5(6): 739-749 (1952).

Megaw, H.D. Ferroelectricity in Crystals, London, 1957.

Mehl, R.F. and van Winkle, S.W. Rev. Met. 50: 465 (1953).

Merz, W.J. "Domain formation and domain-wall motion in ferroelectric $BaTiO_3$ single crystals," Phys. Rev. 95: 690-698 (1954).

Mitsui, T. and Furuichi, G. "Domain structure of Rochelle salt and KH_2PO_4," Phys. Rev. 90: 193-212 (1953).

Moore, A.I.W. "Accommodation kinking associated with the twinning of zinc," Proc. Phys. Soc. (London) B, No. 12: 956-958 (1952).

Moore, A.I.W. "Twinning and accommodation kinking in zinc," Acta Met. 3(2): 163-169 (1955).

Mügge, O. "Über Umlagerungen in Zwillingsstellungen am Chlorbaryum, $BaCl_2 \cdot 2H_2O$," Neues Jahrb. Mineral. Geol. 1: 131-146 (1881).

Mügge, O. "Beiträge zur Kenntnis der Strukturflächen des Kalkspates und über die Beziehungen derselben untereinander und zur Zwillingsbildung am Kalkspat und einigen anderen Mineralen," Neues Jahrb. Mineral. Geol. No. 1: 32 (1883).

Mügge, O. "Über künstliche Zwillingsbildung durch Druck am Antimon, Wismuth und Diopsid," Neues Jahrb. Mineral. Geol. No. 1: 183-187 (1886).

Mügge, O. "Über Umlagerungen in Zwillingstellung am Chlorbaryum $BaCl_2 \cdot 2H_2O$," Neues Jahrb. Mineral. Geol. No. 1: 131-146 (1888).

Mügge, O. "Über die Kristallform des Brombaryums $BaBr_2 \cdot 2H_2O$ und verwandter Salze und über Deformationen derselben," Neues Jahrb. Mineral. Geol. No. 1: 145 (1889a).

Mügge, O. "Formulas for transformation of indices in simple glide," Neues Jahrb. Mineral. Geol. No. 2: 98 (1889b).

Mügge, O. "Über Translation und verwandte Erscheinungen in Kristallen," Neues Jahbr. Mineral. Geol. No. 1: 71-158 (1898).

Mügge, O. "β-Tin." Zentr. Mineral. Geol., p. 233 (1917); Z. Krist. 65: 603 (1927).

Mügge, O. "Einfache Schiebung an einigen künstlichen Kristallen," Z. Krist. 75(1/2): 32-41 (1930).

Mügge, O. "Die Gleitfläche als Ursache gewisser Verzerrungen an Kalkspat," Z. Krist. 82: 59-71 (1932).

Nabarro, F.R.N. "Dislocation in simple cubic lattices," Proc. Phys. Soc. (London) Pt. II, 59(332): 256-272 (1947).

Nabarro, F.R.N. "The mathematical theory of stationary dislocations," Advances in Physics 1: 271-394 (1952).

Nakamura, T. "On the correlation between screw dislocations and twin structure in WO_3 crystals," J. Phys. Soc. Japan 11(4): 467-468 (1956).

Nakamura, T. and Nakamura, H. "Domain wall caught in dislocation in ferroelectric glycine sulfate crystals," Japanese Jour. of Appl. Phys. 1(5): 253 (1962).

Néel, L. J. Phys. Radium, pp. 241-265 (1944).

Niggli, P. Lehrbuch der Mineralogie, 2nd edition, Berlin, Borntraeger, 1924, p. 131.

Nye, J.F. "Plastic deformation of silver chloride, I. Internal stresses and the glide mechanism," Proc. Roy. Soc. (London) A. 198(1053) (1949).

Obreimov, I.V. "Treatment strength of mica," Proc. Roy. Soc. (London) A. 127: 290-297 (1930).

Obreimov, I.V. and Shubnikov, A.V. "Eine Methode zur Herstellung einkristalligen Metalle," Z. Physik 25: 31-36 (1924).

Obreimov, I.V. and Shubnikov, A.V. "A study of the plastic deformation of rocksalt by an optical method," Zhur. Russ. Fiz.-Khim. Ob-va 58: 817-828 (1926); Z. Physik 41: 907 (1927).

Obreimov, I.V. and Startsev, V.I. "Work of formation for an elastic twin in calcite," Zhur. Eksptl. i Teoret. Fiz. 35(5): 1065-1074 (1958).

Okada, K. and Koda, Sh. "The influence of quenching of twinning in zinc single crystals," J. Inst. Metals 89(12): 479-480 (1961).

Okawa, A. "On the mechanism of deformation twin f.c.c. crystals," J. Phys. Soc. Japan 12(7): 825 (1957).

Orlov, L.G. "An electron-microscope study of the dislocation structure and brittle fracture of iron," Author's abstract of dissertation, Nauch.-Issled. Institut Chernoi Metallurgii, 1962.

Orlov, L.G., Usikov, M.P., and Utevskii, L.M. UFN 76(1): 109 (1962).

Orowan, E. "A type of plastic deformation new in metals," Nature 149(3788): 643-644 (1942).

Orowan, E. "Dislocations in metals," Trans. Met. Soc. AIME, p. 116 (1954).

Owen, W.S., Averbach, B.L., and Cohen, M. "Brittle fracture of mild steel in tension at -196°C," ASTM 50: 634 (1958).

Pabst, A. "Transformation of indices in twin gliding," Bull. Geol. Soc. Am. 66: 89-912 (1955).

Palache, C. "Morphology of graphite, arsenopyrite, pyrite and arsenic," Am. Mineralogist 26: 709-717 (1941).

Paxton, H.W. "Experimental verification of the twin system in alpha-iron," Acta Met. 1(2): 141-143 (1953).

Pearson, G.L. and Feldmann, W.L. "Powder-pattern techniques for delineating ferroelectric domain structures," J. Phys. Chem. Solids, 9(7): 28-30 (1959).

Pearson, G.L., Read, W.T., and Feldmann, W.L. "Deformation and fracture of small silicon crystals," Acta Met. 5(4): 181-191 (1957).

Peierls, R. "The size of a dislocation," Proc. Phys. Soc. (London) 52: 34-87 (1940).

Perekalina, Z.B., Regel', V.R., and Dubov, G.A. "Results from compression tests on naphthalene," Kristallografiya 3(1): 71-79 (1958).

Pérez, I.P. "Variation de la constante diélectrique du quartz," Ann. Physik 7, Sér. 12: 238-283 (1952).

Pfeil, L.B. "Effect of cold work on structure and changes, produced by subsequent annealing," Carnegie School, Mem. Iron. and Steel Inst. 16: 153-210 (1927).

Pfeil, L.B. "Twinning of α-Fe," Ibid. 15: 319 (1926).

Plateau, J., Henry, G., and Crussard, C. "Complements à l'interpretation des images fournies par la microfractographie," Metaux, Corrosion, Industries 33(392): 141 (1958).

Platt, J.R. "Atomic arrangements and bonding across a twinning plane in graphite," Z. Krist. 109(5): 226-230 (1957).

Pratt, P.L. "Cleavage deformation in zinc and sodium chloride," Acta Met. 1(6): 692-699 (1953).

Pratt, P.L. and Pugh, S.F. "Twin accommodation in zinc," J. Inst. Metals, 80: 653-658 (1952).

Pratt, P.L. and Pugh, S.W. "The movement of twins, kinks, and mosaic walls in zinc," Acta Met. 1(2): 218-222 (1953).

Price, P.B. "Nucleation and growth of twins in dislocation-free zinc crystals," Proc. Roy. Soc. (London) 260 (1301): 251-262 (1961).

Price, P.B. "Twinning in cadmium dendrites," Phil. Mag. 4(47): 1229-1242 (1959).

Rais, G.B. "Dislocations in mechanically twinned crystals of calcite," Doklady Akad. Nauk SSSR 117: 419-421 (1957).

Rais, G.B. "Lattice distortion at the interface in a mechanically twinned crystal of calcite," Kristallografiya 3(3) 325-328 (1958).

Ramsey, J.A. J. Inst. Metals 80: 167 (1951).

Read, W.T. Dislocations in Crystals, McGraw-Hill Book Co., Inc., 1953.

Reed-Hill, R.E. "A Study of the [10$\bar{1}$1] and [10$\bar{1}$3] twinning modes in magnesium," Trans. Met. Soc. AIME 218: 554-557 (1960).

Reed-Hill, R.E. and Buchanan. "Zigzag twins in zirconium," Acta Met. 11(1): 73-75 (1963).

Reed-Hill, R.E. and Robertson, W.D. "Additional modes of deformation twinning in magnesium," Acta Met. 5: 717-727 (1957).

Reed-Hill, R.E. and Robertson, W.D. "The crystallographic characteristics of fracture in magnesium single crystals," Acta Met. <u>5</u>: 728-737 (1957).

Regel', V.R. and Berezhkova, G.V. "Relation of limit of kink formation to crystallographic orientation for monocrystals," Kristallografiya <u>4</u>(5): 761-767 (1959).

Regel', V.R. and Govorkov, V.G. "Relation of critical cleavage stress for a zinc monocrystal to temperature and deformation rate," Kristallografiya <u>3</u>(1): 64-70 (1958).

Regel', V.R., Govorkov, V.G., and Dobrzhanskii, G.F. "Relation of the parameters of the extension curve for AgCl monocrystals to temperature and rate of deformation," Optiko-Mekh. Prom. No. 6: 28-32 (1958).

Regel', V.R. and Zemtsov, A.B. "Effects of crystallographic orientation of TlBr-TlI monocrystals on the yield point in tension," Tr. In-ta kristallografii Akad. Nauk SSSR No. 11: 166-171 (1955).

Reusch, E. "Über eine besondere Gattung von Durchgängen im Steinsalz und Kalkspat," Pogg. Ann. <u>132</u>: 441 (1867).

Rhodes, R.G. "Twinning structure in crystals of tungsten trioxide," Nature <u>170</u>(4): 322, 369 (1952).

Roitburd, A.L. "Theory of nucleation in the martensite transformation," Fiz. Metal. i Metalloved. <u>10</u>(2) 161-168 (1960).

Roitburd, A.L. "Some features of the growth of crystals in condensed systems," Kristallografiya <u>7</u>(2): 291-299 (1962).

Röm, F. and Kochendörfer, A. "Neue Ergebnisse über die Verfestigung bei der plastischen Verformung von Kristallen," Z. Metallk. <u>41</u>: 265 (1950).

Rose, G. "Über die im Kalkspat vorkommenden hohlen Kanäle," Physik. Abhandl. kön. Akad. Wiss. Berlin <u>57</u> (1868).

Rosenthal, D. and Woolsey, C.C. Welding J. (N.Y.) <u>31</u>: 475 (1952).

Rosi, F.,Dube,C., and Alexander, B. "Mechanism of plastic flow in titanium. Determination of slip and twinning elements," J. Metals <u>5</u>(2): 257-264 (1953).

Rozhanskii, V.N. "Unevenness in plastic deformation of crystals," Uspekhi Fiz. Nauk <u>65</u>(3), 387-406 (1958)

Rybalko, F.P. and Yakutovich, M.V. "Deformation of flat aluminum crystals," Zhur. Tekh. Fiz. <u>18</u>(7): 915-921 (1948).

Rybalko, F.P. and Yakutovich, M.V. "Localization of deformation and determination of the plasticity of steel in torsion and extension," Zhur. Tekh. Fiz. <u>23</u>(10): 766-771 (1953a).

Rybalko, F.P. and Yakutovich, M.V. "Effects of surface finish on the plasticity of steel in torsion tests," Zhur. Tekh. Fiz. <u>23</u>(10): 765-766 (1953b).

Sachse, H. Ferroelektrika, Munich (1956), pp. 98-101, 115-129.

Salkovitz, E.I. "Energy absorption during twin formation in zinc single crystals," Phys. Rev. <u>85</u>: 1046 (1952).

Scheil, E. Z. anorg. u. allgem. Chem. <u>183</u>(1-2): 98-120 (1929).

Schmid, E. and Wassermann, G. "Zur Bewegung der Gitterpunkte sekundärer Translation im Zwilling," Z. Physik <u>48</u>: 370 (1928).

Seeger, A. "The mechanism of glide and work hardening in face-centered cubic and hexagonal close-packed metals," International Conference on Dislocations and Mechanical Properties of Crystals, held at Lake Placid, September 6-8, 1956, pp. 243-349 (1957).

Seeger, A., Berner, R., and Wolf, H. "Die experimentelle Bestimmung von Stapelfehlerenergien kubisch flachenzentrierter Metalle," Z. Physik <u>155</u>(2): 247-262 (1959).

Seeger, A. and Schoeck, G. "Die Aufspaltung von Versetzungen in Metallen dichtester Kugelpackung," Acta Met. <u>1</u>: 519-530 (1953).

Seidle, L.L., Rosi, F.D., Parkins, F.C., Sama, L., and Opinsky, I. "Plastic flow and recrystallization of titanium: progress report," Silvania Elect. Products. Inc., Met. Lab. Rep. 1953.

Seifert, H. "Über Schiebungen an Bleiglanz," Neues Jahrb. Mineral, <u>57A</u>: 665-742 (1928).

Seifert, H. "Verformung von Salzkristallen," Z. Electrochem. <u>50</u>(5/4) (1944).

Sherrill, F.A., Wittels, M.C., and Blewitt, T.H. "X-ray determination of conjugate deformation twins in copper," J. Appl. Phys. <u>28</u>(5): 526-529 (1957).

Shevandin, E.M. "Growth of brittle-fracture cracks," Zhur. Tekh. Fiz. <u>8</u>(5): 441-446 (1938); <u>9</u>(8): 745-747 (1939a).

Shevandin, E.M. "On twinning and brittleness," Zhur. Tekh. Fiz. <u>9</u>(8): 745 (1939b).

Shevandin, E.M. "Twinning of iron at low temperatures and high loading rates," Zhur. Tekh. Fiz. No. 5: 402-405 (1940).

Shirane, G., Jona, F., and Pepinsky, R. "Some aspects of ferroelectricity," Proc. IRE 43(12): 1738-1793 (1955).

Shockley, W., Hollomon, J.H., Maurer, R., and Seitz, F. (editors). Imperfections in Nearly Perfect Crystals, New York and London, 1952.

Shternberg, A.A. "Removal of mechanical twins from crystals of optical calcite," Uchenye Zapiski Leningrad Gosudarst un-ta, ser. geol.-pochv. nauk 13(65): 51-55 (1944).

Shubnikov, A.V. Quartz and its Uses, Moscow-Leningrad, Izd-vo Akad. Nauk SSSR, 1940.

Shubnikov, A.V., Flint, E.E., and Bokii, G.B. Principles of Crystallography, Moscow-Leningrad, Izd-vo Akad. Nauk SSSR, 1940.

Shubnikov, A.V. and Tsinzerling, E.V. "Impact and pressure figures, and mechanical twins in quartz," Tr. Lomonosov. Inst. Akad. Nauk SSSR 3: 67-74 (1933); Z. Krist. 83: 243-264 (1932).

Shur, Ya. S., Shtol'ts, E.V., and Glazer, A.A. "Reconstruction of the domain structure in a magnetically uniaxial ferromagnetic in a magnetic field," Fiz. Metal. i Metalloved. 8(5): 685-688 (1959).

Shur, Ya. S. and Zaikova, V.A. "Effects of elastic stresses on the magnetic structure of crystals of silicon iron," Fiz. Metal. i Metalloved. 6(3): 545-555 (1958).

Shur, Ya. S. and Zaikova, V.A. "Causes of the increase in coercive force accompanying reduction in the thickness of ferromagnetic sheets," Fiz. Metal. i Metalloved. 10(3): 350-358 (1960).

Siems, R., Delavignette, P., and Amelinckx, S. "Die direkte Messung von Stapelfehlerenergien," Z. Physik 165(5): 502-532 (1961).

Siems, R. and Haasen, P. "Die Geschwindigkeit der Zwillingsbildung in Zink-Einkristallen," Z. Metallk. 49(5): 213-220 (1958).

Simonsen, E.B. "Twinning in iron," Acta Met. 8(11): 809-810 (1960).

Simonsen, E.B. "Annihilation of a coherent bounding plane by twinning in twins," Acta Met. 10(2): 172-174 (1962).

Sleeswyk, A.W. "Twinning and the origin of cleavage nuclei in α-iron," Acta Met. 10(9): 803-812 (1962).

Sleeswyk, A.W., and Helle, I.N. "Zigzag configurations of twins in α-iron," Acta Met. 9(4): 344 (1961).

Smith, C.S. "Grains, phases and interfaces: an interpretation of microstructure," Trans. AIMME 175: 15 (1948).

Smith, C.S. Trans. Met. Soc. AIME 212: 574 (1959).

Smith, R.L. and Rutherford, J.L. "Tensile properties of zone-refined iron in the temperature range from 298 to 4.2°K," Trans. AIMME 209: 857-864 (1957).

Smith, S.W.I., Dee, A.A., and Young, I. Proc. Roy. Soc. (London) A 121: 477 (1928).

Soifer, L.M. and Startsev, V.I. "Motion of dislocations in crystals of antimony," Doklady Akad. Nauk SSSR 138(5): 1084-1087 (1961).

Sosman, R.B. The Properties of Silica, 1927.

Sossinka, H.G., Schmidt, B., and Sauerwald, F. "Zur gittermässigen Auswahl von Translations- und Schiebungselementen," Z. Physik 85: 761 (1933).

Stanley, R.C. "Etch pits on calcite cleavage faces," Nature 183(4674): 1548 (1959).

Startsev, V.I. "A study of lattice rotation in plastic deformation," Zhur. Eksptl. i Tekh. Fiz. 10(6): 703 (1940).

Startsev, V.I. "The intermediate regions in plastically deformed crystals of rocksalt," Doklady Akad. Nauk SSSR 30(2): 124-125 (1941).

Startsev, V.I., Bengus, V.Z., Lavrent'ev, F.F., and Soifer, L.M. "Detection of dislocations in calcite crystals," Kristallografiya 5(3): (1960a).

Startsev, V.I., Bengus, V.Z., Lavrent'ev, F.F., and Soifer, L.M. "Formation of dislocations in the twinning of calcite," Kristallografiya (1960b).

Startsev, V.I. and Kosevich, V.M. "Elastic twinning of crystals," Doklady Akad. Nauk SSSR 101(5): 861-864 (1955).

Startsev, V.I., Kosevich, V.M., and Tomlenko, Yu. S. "A study of the intersection of twin bands in calcite monocrystals," Kristallografiya 1(4): 425-435 (1956).

Startsev, V.I. and Lavrent'ev, F.F. "An x-ray study of the accommodation regions in the twinning of zinc," Kristallografiya 3(3): 329-333 (1958).

Startsev, V.I., Lavrent'ev, F.F., and Soifer, L.M. "Thermal etching and annealing of twin bands in antimony crystals," Kristallografiya (1960).

Startsev, V.I. and Soifer, L.M. "Some effects observed in the deformation of antimony monocrystals," Doklady Akad. Nauk SSSR (1960).

Stepanov, A.V. Phys. Z. Sowjetunion 2: 537 (1932).

Stepanov, A.V. "Causes of early rupture," Izvest. Akad. Nauk SSSR, Ser. Fiz. Nos. 4-5: 797 (1937).

Stepanov, A.V. "The phenomenon of mechanical twinning," Zhur. Eksptl. i Teoret. Fiz. 17(8): 713-723 (1947).

Stepanov, A.V. "Mechanical twinning of quartz," Zhur. Eksptl. i Teoret. Fiz. 20: 438-441 (1950).

Stepanov, A.V. and Donskoi, A.V. "A new mechanism of plastic deformation of crystals," Zhur. Tekh. Fiz. 24(2): 161 (1954).

Stroh, A.N. "The strength of Lomer-Cottrell sessile dislocations," Phil. Mag. 1(6): 489-502 (1956).

Swenson, C.A. "Phase transition in solid Hg," Phys. Rev. 111(1): 72-91 (1958).

Tamsaki, S. "Effect of external stress on the domain structure of WO_3," J. Phys. Soc. Japan. 13(14): 363-366 (1958).

Tavonen, P.E. Ann. Acad. Sci. Fennicae, Ser. A 1: 42 (1947).

Taylor, G.J. "Resistance to shear in metal crystals," Trans. Faraday Soc. No. 4: 121 (1928).

Tertsch, H. Die Festigkeitserscheinungen der Kristalle, Vienna, Springer, 1949.

Thomas, L.A. and Wooster, W.A. "Piezocrescence: the growth of Dauphiné twinning in quartz under stress," Proc. Roy. Soc. (London) A 208(1092): 43-62 (1951).

Thompson, N. and Hingley, M. "The formation of mechanical twins," Acta Met. 3(3): 289-291 (1955).

Thompson, N. and Millard, D.J. "Twin formation in cadmium," Phil. Mag. 43: 421-440 (1952).

Thompson, N. and Millard, W.R. J. Metals 4: 295 (1952).

Thornton, P.R. and Hirsch, P.B. "The effect of stacking-fault energy on low temperature creep in pure metals," Phil. Mag. 3: 738 (1958).

Tiedema, T.J. "X-ray investigation of nuclear spot of crystals obtained by recrystallization," Proc. Koninkl. Ned. Akad. Wetenschap. 53: 1422 (1950).

Tipper, C.F. and Sullivan, A.M. Trans. Am. Soc. Metals 43. 906 (1951).

Toman, K. and Simerska, M. "The deformation texture of β-Sn, III. Derivation of texture from elements of plastic deformation," Czechoslov. J. Phys. 8: 233-245 (1958).

Tsinzerling, E.V. "The plasticity of quartz," Trudy Lomonosov. Inst. Akad. Nauk No. 3: 67 (1933); Z. Krist. (A) 85: 454-461 (1933).

Tsinzerling, E.V. "Mechanical twinning of quartz," Trudy Inst. Krist. Akad. Nauk SSSR No. 2: 149-166 (1940).

Tsinzerling, E.V. "Control of the twinning of quartz in β→α→β transitions," Doklady Akad. Nauk SSSR 33(5): 365-367 (1941a).

Tsinzerling, E.V. "Relation of the coloration of quartz to the tendency to twin in the α→β transition," Doklady Akad. Nauk SSSR 33(5): 368 (1941b).

Tsinzerling, E.V. "Twinning of quartz in an electric field," Doklady Akad. Nauk SSSR 33(6): 421 (1941c).

Tsinzerling, E.V. Brittleness and Twinning of Quartz in Response to Impact and Pressure, Dissertation, Moscow, 1943.

Tsinzerling, E.V. "Methods of determining the suitability of quartz for use in the artificial enlargement of monocrystalline regions," Trudy Inst. Krist. Akad. Nauk SSSR No. 4: 199 (1948).

Tsinzerling, E.V. "Twinning of curved quartz in β→α→β transitions," Trudy Inst. Krist. No. 4: 205-207 (1948b).

Tsinzerling, E.V. "Artificial stabilization of the lattice of quartz," Doklady Akad. Nauk SSSR 95(3): 529 (1954a).

Tsinzerling, E.V. "Stabilization of the lattice in zone quartz," Doklady Akad. Nauk SSSR 95(4): 801 (1954b).

Tsinzerling, E.V. "Amorphologic study of artificial twinning in quartz produced by various factors," Author's abstract of dissertation, Moscow, 1958.

Tsinzerling, E.V. "Artificial twinning of quartz," Izvest. Akad. Nauk SSSR (1961).

Tsinzerling, E.V. and Mironova, Z.A. "Detection of dislocations in quartz by selective etching," Kristallografiya 8(1): 117 (1963).

Tsinzerling, E.V. and Perekalina, Z.B. "Strength of quartz in torsion," Trudy, Inst. Krist. Akad. Nauk SSSR No. 11: 172 (1955).

Tsinzerling, E.V. and Shubnikov, A.V. "Über die Plastizität des Quarzes," Z. Krist. (A) 85: 456-461 (1933).

Tsinzerling, E.V. and Shubnikov, A.V. "Impact and pressure figures in mechanical twins of quartz," Trudy Lomonosov. Inst. Akad. Nauk No. 3: 1 (1933); Z. Krist. (A) 83(3-4): 243-264 (1932).

Tsinzerling, E.V., Urusovskaya, A.A., and Govorkov, V.G. "Are Japanese-law mechanical twins in quartz possible?" Zapiski Vsesoyuz. Mineral. Obshchestva, 1960 (in press).

Turner, F.J. and Ch'ih, C.S. Bull. Geol. Soc. Am. 62: 887 (1951).

Turner, K.G., Griggs, D.T., and Heard, H. Bull. Geol. Soc. Am. 1954.

Tyul'panov, A.A. Technology of the Making of Quartz Plates, Moscow-Leningrad, Gosenergoizdat, 1955, pp. 34-40.

Ubbelohde, A.R. and Woodward, I. "Structure and thermal properties of crystals, VI. The role of hydrogen bonds in Rochelle salt," Proc. Roy. Soc. (London) A 185: 448-465 (1946).

Urusovskaya, A.A. "Mechanism of formation of transverse impact figures in zinc," Trudy Inst. Krist. Akad. Nauk SSSR No. 12: 180-185 (1956a).

Urusovskaya, A.A. "Plastic-deformation figures observed in crystals of TlBr-TlI, CsI and CsBr," Trudy Inst. Krist. Akad. Nauk SSSR No. 12: 172-179 (1956b).

Urusovskaya, A.A. "Formation of regions of altered lattice orientation in the plastic deformation of monocrystals and polycrystals," Coll.: Some Aspects of the Physics of Plasticity, Moscow, Izd-vo Akad. Nauk SSSR 1960.

Vassemillet, L.E. "Stacking-fault probability of noble metal-zinc alloys," J. Appl. Phys. 32(5): 778 (1961).

Veit, R. "Künstliche Schiebungen und Translationen in Mineralen," Neues Jahrb. Mineral. Geol. 45: 121 (1922).

Venables, I.D. and Broady, R.M. "Dislocations and selective etch pits in InSb," J. Appl. Phys. 29(7): 1025 (1958).

Vernadskii, V.I. Slip Phenomena in Crystalline Matter, Moscow, Universitet, 1897.

Verneuil, M.A. "Mémoire sur la reproduction artificielle du rubis par fusion," Ann. chim. et phys. 3: 20-48 (1904).

Vladimirskii, K.V. "Twinning in calcite," Zhur. Eksptl. i Teoret. Fiz. 17(6): 530-536 (1947).

Vogel, F.L. and Brick, R.M. Trans. AIMME 197: 700 (1953).

Votava, E. and Berghezan, A. Appearance of Dislocations in Metal Crystals on Evaporation: "Twinning, dislocations and their relation to annealing twins in copper," Acta Met. 7(6): 392 (1959).

Vyrubov, M. Bull. soc. franc. minéral. 9: 262 (1891), 14: 233 (1891).

Wachtman, J.B. and Maxwell, L.H. "Plastic deformation of ceramic oxide single crystals, II," J. Am. Ceram. Soc. 40(11): 377-385 (1957).

Wallerant, F. "Bedingungen für Schiebungsfähigkeit," Bull. soc. franç. minéral. 27: 169 (1904).

Washburn, J. and Parker, E. "Kinking in zinc single-crystal tension specimens," J. Metals 4(10): 1076-1078 (1952).

Watts, H. "Etch pits on calcite cleavage faces," Nature 183(4657): 314 (1959).

Weaver, C.W. "Twinning and fracture in chromium," Nature 193(4812): 265 (1962).

Whelan, M.J. Proc. Roy. Soc. (London) A 249: 114 (1959).

Whelan, M.J., Hirsch, P.B., Horne, R.W., and Bollman, W. "Dislocations and stacking faults in stainless steel," Proc. Roy. Soc. (London) A 240: 524-538 (1957).

Whitwham, D. and Lacombe, P. "Les joints des grains "coherents" et "incoherents" dans les macles des metaux cubiques," 4 Colloque de Metallurgie, Propriétes des joints des grains, France, 1960.

Wood, E.A. and Holden, A.N. "Monoclinc glycine sulfate: crystallographic data," Acta Cryst. 10: 145-148 (1957).

Wood, W.A. and Scrutton, R.F. "Mechanism of primary creep in metals," J. Inst. Metals July, pp. 423-435 (1950).

Woolley, W.L. "Twinning and untwinning in polycrystalline magnesium," J. Inst. Metals 83(2): 57-58 (1954).

Wooster, W.A. and Wooster, Nora. "Control of electrical twinning in quartz," Nature 157(3987): 405-406 (1946a).

Wooster, W.A. and Wooster, Nora. "Control of electrical twinning in quartz," Nature 157: 406 (1946b).

Wooster, W.A., Wooster, Nora, Ryecroft, I.L., and Thomas, L.A. "The control and elimination of electrical (Dauphiné) twinning in quartz," J. Inst. Elec. Engs. 94(Pt. III), (16): 926 (1947).

Worrell, F.T. "Twinning in tetragonal alloys of copper and manganese," J. Appl. Phys. 19: 929 (1948).

Worrell, F.T. and Sievert, A.V. "The role of tetragonal twins in internal friction," J. Appl. Phys. 22(10): 1257-1259 (1951).

Yakovleva, É.S. and Yakutovich, M.V. "Relation to diameter of critical shearing stress for twinning and slip for cadmium crystals," Zhur. Eksptl. i Teoret. Fiz. 10: 1146-1150 (1940).

Yakovleva, É. S. and Yakutovich, M. V. "Effects of twinning on the brittle fracture of zinc crystals," Zhur. Tekh. Fiz. <u>20</u>(4): 420-423 (1950).

Yakutovich, M. V. and Yakovleva, É. S. "Kinetics of mechanical twinning of crystals," Zhur. Tekh. Fiz. <u>5</u>: 1171-1177 (1935).

Yakutovich, M. V. and Yakovleva, É. S. "The shape of a mechanical twin and the causes giving rise to it," Zhur. Eksptl. i Teoret. Fiz. <u>9</u>(7): 884-888 (1939).

Yamaguchi, K. "Internal strain of uniformly distorted aluminum crystals," Sci. Papers, Inst. Phys. Chem. Research (Tokyo) <u>11</u>: 151 (1929).

Zankelies, D. A. "Observation of slip in nylon 66 and 610 and its interpretation in terms of new models," J. Appl. Phys. <u>33</u>(9): 2797-2803 (1962).

Zapffe, C. A. Z. Metallk. <u>44</u>: 397 (1953).

Zemtsov, A. B., Klassen-Neklyudova, M. V., and Urusovskaya, A. A. "Complex presentation of plastic deformation in monocrystals," Doklady Akad. Nauk SSSR <u>91</u>(4): 813-816 (1953).

Zener, C. "Conference on cold working of metals," Am. Soc. Metals 1949.

Zhdanov, G. S. and Umanskii, Ya. S. X-Radiography of Metals, Metallurgizdat; Vol. 1, 1937; Vol. 2, 1938.

Zhirnov, V. A. "Theory of domain boundaries in ferroelectrics," Zhur. Eksptl. i Teoret. Fiz. <u>35</u>(5) (11): 1175 (1958).

Zolotov, V. A. Compt. rend. acad. sci. URSS <u>38</u>: 140 (1943).

AUTHOR INDEX

SUBJECT INDEX

212